Botanical Microscopy 1985

Botanical Microscopy
1985

Edited by

ANTHONY W. ROBARDS

Reader in Biology,
University of York

OXFORD   NEW YORK   TOKYO
OXFORD UNIVERSITY PRESS
1985

Oxford University Press, Walton Street, Oxford OX2 6DP
Oxford New York Toronto
Delhi Bombay Calcutta Madras Karachi
Kuala Lumpur Singapore Hong Kong Tokyo
Nairobi Dar es Salaam Cape Town
Melbourne Auckland
and associated companies in
Beirut Berlin Ibadan Nicosia

Oxford is a trade mark of Oxford University Press

Published in the United States
by Oxford University Press, New York

British Library Cataloguing in Publication Data

Botanical microscopy 1985.
1. Plant cells and tissues   2. Microscope and microscopy—Technique
I. Robards, Anthony W.
578   QK673

ISBN 0-19-854587-8

Printed in Great Britain by
St Edmundsbury Press
Bury St Edmunds, Suffolk

Preface

At the end of March 1967, a meeting on 'Plant Fine Structure' was organized at Bodington Hall, University of Leeds, by Professor I. Manton, F.R.S. The cordial atmosphere at that meeting, and the high quality of the science discussed, was such that many of those participating expressed the view that there should be another similar meeting in due course. In fact, it took twelve years before the 'Second International Botanical Microscopy Meeting' was organized by the Royal Microscopical Society in York during July 1979. Once again, rather more than 100 botanical microscopists came together in a relaxed and welcoming atmosphere to communicate their scientific results to each other. On that occasion, one of the subsequent criticisms was that none of the papers presented had been published as part of a conference proceedings. In particular, many of the participants felt that it was something of a loss that the invited papers had not been published together so that botanists other than just those who had been lucky enough to attend the meeting could benefit. In consequence, while organizing the 'Third International Botanical Microscopy Meeting', which took place in York during July 1985 (again under the auspices of the Royal Microscopical Society), I had it firmly in mind that it would be a useful service to science if I could prevail upon the invited speakers to write up their papers for prompt publication. The results of their efforts are represented in this book.

It might well be asked why there should need to be a meeting on botanical microscopy at all and, therefore, why produce a book such as this? Many botanists use microscopes and many microscopists look at plant specimens. Nevertheless, relatively little botany is discussed at electron microscopy congresses, and botanical microscopy plays a small, and often disseminated, part in cell biology meetings. In consequence, it has been useful to get together this relatively small 'club' of people to discuss matters of joint interest. The success was most evident from the vigorous discussion sessions at the meeting where a sufficient 'critical mass' of knowledgeable botanists had come together to make positive contributions to many different topics. As an aside, Professor Manton's own presence and contribution, 18 years on from her earlier meeting in Leeds, set the conference off to a noteworthy start and gave much pleasure to many of the delegates.

In drawing up a list of the main speakers for the 1985 botanical microscopy meeting, my main aim was to accentuate those aspects that were either of fundamental importance and/or represented innovative science. The reader must judge for him- or herself whether I have been successful. Inevitably, in the year or so between the invitation and the conference, changed circumstances have caused some alterations. Indeed, there have been one or two last minute 'nightmares' for the editor! Nevertheless, the papers presented here represent a cross-section of the important and exciting growth areas of botanical microscopy. The authors are to be congratulated on producing their camera-ready copy to a specific time-shedule. For their part, the publishers have undertaken to make the book available before the end of the current calendar year: an aspect that adds enormously to the value of contemporary work such as this.

Not only are the major papers presented in this volume, but some of the contributed papers will appear in a special issue of the *Nordic Journal of Botany*. Thus, the presentations for *Botanical microscopy 1985* will be well-served in the literature. World-wide pressure on research budgets is having a clear effect on limiting the ability of scientists to get together to discuss their work. It is, therefore, even more important that steps should be taken to try and disseminate scientific information more effectively. The contents of this book should by no means be of interest only to the 'state-of-the-art' research worker. Many of the different contributions will be directly relevant to cell biologists more generally, to university and college lecturers and to schoolteachers at senior level. Novel techniques are constantly being introduced to provide new information about cells and how they work. It is important that details of both techniques and results are communicated from top to bottom of the scientific 'family tree'. The purpose of this book is, in its own small way, to help in that process.

York   A.W.R.
August 1985

Contents

Contributors

Ms. R. M. Abeysekera, Department of Biology, Dalhousie University, Halifax, Nova Scotia, Canada.

Dr L. G. Briarty, Botany Department, University of Nottingham, UK.

Dr J. Burgess, John Innes Institute, Norwich, U.K.

Dr M. N. Christiansen, Plant Virology Laboratory and Plant Stress Laboratory, Agricultural Research Service, Beltsville, Maryland, USA.

Professor J. Coetzee, Electron Microscopy Unit, University of Pretoria, South Africa.

Dr Graeme R. A. Dunbar, Department of Botany, University of Aberdeen, UK.

Dr E. F. Erbe, Plant Virology Laboratory and Plant Stress Laboratory, Agricultural Research Service, Beltsville, Maryland, USA.

Dr M. G. Erwee, Department of Agronomy and Agricultural Science, University of Sydney, Australia.

Dr P. B. Goodwin, Department of Agronomy and Agricultural Science, University of Sydney, Australia.

Dr C. Hawes, School of Botany, University of Oxford, UK.

Professor P. K. Hepler, Department of Botany, University of Massachusetts, Amherst, Massachusetts, USA.

Professor W. Herth, Zellenlehre, Faculty of Biology, University of Heidelberg, Federal Republic of Germany.

Professor J. Heslop–Harrison, Welsh Plant Breeding Station, Plas Gogerddan, Nr. Aberystwyth, UK.

Dr R. P. C. Johnson, Department of Botany, University of Aberdeen, UK.

Professor R. B. Knox, Plant Cell Biology Centre, Botany School, University of Melbourne, Australia.

Dr P. Linstead, John Innes Institute, Norwich, UK.

Dr A. W. Robards, Department of Biology, University of York, UK.

Dr I. Roberts, John Innes Institute, Norwich, UK.

Dr K. Roberts, John Innes Institute, Norwich, UK.

Dr M. B. Singh, Plant Cell Biology Centre, Botany School, University of Melbourne, Australia.

Professor M. W. Steer, Department of Botany, University College, Belfield, Dublin, Eire.

Dr R. L. Steere, Plant Virology Laboratory and Plant Stress Laboratory, Agricultural Research Service, Beltsville, Maryland, USA.

Dr E. L. Vigil, Department of Horticulture, University of Maryland, USA.

Dr J. H. M. Willison, Department of Biology, Dalhousie University, Halifax, Nova Scotia, Canada.

1

Botanical microscopy — the current state of the art

J. Heslop-Harrison

My charge in opening this symposium is to attempt a
preliminary overview of the art of microscopy in its diverse
applications to the plant sciences. I begin by denying
immediately any claim to comprehensiveness: the subject is
obviously overwhelmingly too vast for the time available.
This commentary will therefore be highly selective, and
necessarily it will also reflect personal experience and bias.

Over the years there has been no lack of published
information dealing with every aspect of microscope
technology, both in the specialised press and in the more
general journals such as those of the Royal Microscopical
Society itself, where the requirements of the instrument-user
aside from those of the instrument-designer have always been
generously served. That element of the biological community
whose bread-and-butter has depended to a greater or lesser
degree on microscopy in teaching or research has varied in its
attitude to this literature. Some have made whatever they
could of it; the bulk, I suspect, have either been essentially
unaware of its existence, or have regarded it as not
especially pertinent to their professional interest in the
microscope as a working tool - any more, say, than
glass-making technology might be simply because their
laboratories happen to be well-endowed with Pyrex ware. And
there has been some justification for this attitude; after
all, throughout the inter-war years and for some time after
the second war, the light microscope (LM) existed as a fairly
stable element of the hardware of the botanical laboratory -
heavily used maybe, but scarcely in itself a pivotal factor in
the development of the subject through its own continued
evolution. I do not have to emphasise that the scene has
changed dramatically in the last three decades. It is
arguable that the revolution can be dated from the advent of

the transmission electron microscope - the "conventional" TEM,
CTEM, as we should now call it - as a machine widely available
to biologists. The impact of this instrument coincided with
the beginnings of a radically new cell biology; necessarily
so, because in itself it played no small part in creating the
young discipline, which rapidly burgeoned as a synthesis of
biochemistry, cytology, genetics and developmental physiology.
In the atmosphere generated by the new cell biology and its
outward ramifications in the fields of anatomy, morphology and
morphogenesis, microscopy has today attained a new stature -
and microscopy in the widest sense, because the venerable LM,
even in its classical form, has itself entered into fresh
fields of technical innovation and application in the wake of
the EM revolution.

There is no doubt that these developments are producing a
new generation of highly informed, well-equipped microscopists
in biological laboratories. Their influence is now feeding
through into publication, where technical commentaries find a
place far more frequently than hitherto. Yet, at the same
time, one suspects that in the botanical sciences, where
microscopy has had such a long and distinguished tradition,
much of what is happening in the field is still passing by
many of those who could benefit most from the new
technologies. Meetings such as this one provide an
opportunity for those who acknowledge themselves as
professional microscopists to take stock; but the publications
they generate may also have some value in high-lighting
neglected areas of application for colleagues whose vision may
not yet have encompassed the full range of current potentials.

In the plant sciences, as in biology in general, microscopy
is an adjunct to research - a means, rather than an end in
itself. This is not to say that new knowledge may not come
serendipitously simply as a reward for looking and seeing;
after all in the early days of electron microscopy this is how
most of the major discoveries were made, from the ribosome,
endoplasmic reticulum and microtubule to the fine-structural
organisation of the plant cell wall. But increasingly
exploitation of the currently available hardware in light,
electron and other forms of microscopy requires the
formulation of some kind of research plan, however loose this
may be. Were one to set oneself the task of laying out an
ideal scheme - such for example as that that might be required
in a grant application - a certain sequence would fall out.
Ideally this might be: (a) formulation of the problem; (b)
identification of the material of which to ask the questions;
(c) preliminary review of the methods available; (d)
development of the handling techniques, including specimen
preparation; (e) use of the appropriate instrument or

instruments to acquire data; (f) processing and evaluation of
the data; (g) interpretation, and (h) the preparation of the
results in publishable form for communication to others in the
field.

Of course, we all know that no-one really follows the ideal
scheme, whatever may be written in grant applications. We
come in at different points; we skip or fuse steps; and some
of us, despite mighty efforts up to (g), never get around to
(h), so that whatever we may have found out remains fossilised
in laboratory notebooks.

Most significantly, however, in practice one usually
departs from schemes such as this simply because a programme
is never treated as a linear continuum, since the research
process is always replete with dead-ends, new starts and
feedback loops. But what we must always accept - to repeat my
earlier point with slightly different emphasis - is that the
product of research should be new knowledge, not endless happy
tinkering, however entertaining the equipment put at our
disposal by the tax-payers might be.

What, then, is the state of microscopy so far as it
impinges, or could impinge, on the work of the plant scient-
ist? Leaving aside the crucial matter of handling and
processing the material, we can recognise four principal ways
in which notable progress has been seen in recent years
through developments in the hardware. They are, (a) the
improvement or extension of the basic function of image-prod-
uction through advances in light and electron optics; (b) the
appearance of new technologies for recording, processing, man-
ipulating, quantifying and communicating microscope image data
of all kinds, and (c) the emergence of microscopes - light,
electron and others - as analytic tools of great power in
their own right, and (d) the introduction of altogether new
means of imaging the very small. In offering this list, I do
not overlook the advances that have also been made in refining
traditional systems of microscopy to improve handling convenience. The efforts of manufacturers in this field have made it
possible for the non-specialist to attain higher and higher
levels of performance from his instrumentation while not burd-
ening him with an unreasonable load of technical detail. For
the botanical teacher or research worker who has no desire or
time to become an instrument engineer this is a most important
contribution, by no means to be depreciated. The process has
by no means come to an end. The first computerised and virtually fully automated research LM stand has already appeared
on the market, and it is surely the precursor of many such to
come.

A list of fundamental as apart from ergonomic developments in the LM in the last three decades would certainly include the progressive optimisation of the optical train, following the trend initiated in post-war years by the introduction of new glasses and computerised lens design. How sensitive the botanical user might be to this is very largely dependent on the nature of his work. For example, the enhanced image quality offered by CF lenses and their equivalents may be insignificant at one level of application, crucial at another. The same might be said of flat-field objectives and wide-field viewing systems. But there is no doubt that advances in the design of the basic components of the "standard" microscope have opened up new possibilities in research. One that comes to mind is the introduction of very high aperture objectives in the 40-60 X range with great promise in fluorescence microscopy.

In emphasising the power of the modern LM even in its standard bright-field mode, I am to some extent reacting to what I perceive as an unfortunate current tendency to leapfrog it by taking material directly to the EM. This not only extravagant but often downright foolish, since it sacrifices a great field of potential techniques and so throws away immediately a whole dimension of understanding. The point is readily enough illustrated by some of the papers appearing in current botanical journals, where one can see electron microscopy applied as a first resort in contexts where the LM could have provided at least a better initial orientation, and in some instances even a better yield of information over all.

It needs no emphasis that no matter how ergonomically excellent the design may be, extracting the optimum performance from the optics of a modern LM does require the most careful attention to setting up, and to all of the ancillary factors. I have myself worked in botanical laboratories where instruments and associated accessories with total values amounting to £50,000 or more have been routinely used in so poor a state of adjustment that their performance was no better than might have been obtained from a simple student microscope. Where this is the state of affairs, one wonders why all the money was spent in the first place.

The LM has of course several other modes of operation than bright-field. Polarising, dark-field and phase-contrast systems can scarcely be viewed now as innovations, but differential interference contrast (DIC) on the other hand, still seems to be relatively novel in the hands of botanists, judging from the infrequency of its use in published work. The basic Smith system of forty years ago as modified by Nomarski is now the standard, and is widely available from

different manufacturers, albeit always at a price. Although
DIC systems have considerable value as teaching tools, their
greatest significance is undoubtedly in research, and especially hitherto in imaging resin-embedded material sectioned in
the 1.5-2 μm range. DIC is also remarkably effective for
observing surface features, given appropriate illumination
systems. I know of no other microscope mode - light or electron - in general use that offers comparable facility for
viewing and quantifying the features of living plant surfaces,
and giving at the same time with high aperture lenses the
potential for imaging features in different focal planes
through sub-adjacent walls. Much valuable information can be
gained from this kind of application, and the new data could
soon begin to revolutionise plant micro-anatomy.

In commenting on the development of the LM one should
certainly not overlook true ultraviolet microscopy, by no
means a new art. Although it has languished in botanical
applications in recent times, UV-microscopy may well now be
due for revival. Some twenty years ago I had the privilege of
working with a splendid Cooke, Troughton & Simms stand, not so
much for the advantage it gave in resolution, but because of
the capacity it offered for viewing UV-absorbing constituents in plant cells and in microelectrophoresis fibres. The
quartz optics were first class, but operation was difficult
because of illumination problems and the need to focus on a
barely-visible image on a fluorescent screen. Both limitations have now gone with the availability of high-intensity
xenon-arc light sources, monochromators covering the appropriate wavelength range, modern UV detectors and electronic
image intensification systems. Perhaps the way is now open
for a new cytochemistry, based upon direct localisation and
quantitation of UV-absorbing cell components or stained
derivatives, taking up where Caspersson finished.

Operation of the LM in the scanning mode is another technique with a longish history but little current application in
the plant sciences. Its future is, however, potentially
bright. Many will recall the Cintel "Flying Spot" microscope
of the 1950s, something of a revolutionary concept then, well
ahead of its day when it first appeared in exhibitions in
London, but without a serious market at that time. Scanning
laser microscopy really takes over where it left off. Many
instruments of this latter type have been built in recent
years, but I have not come across any botanical applications,
and it is probably early yet to evaluate the usefulness of
such systems in the plant sciences. Still other systems are
on the way. Readers of the RMS Proceedings will have been
intrigued by the accounts in the May, 1985 issue of the Tandem
Scanning Reflected Light Microscope due to Petran, regarded by

the Editor "..as one of the most significant developments of
recent years." The principle seems remarkably simple, depending upon scanning by a point source, with the detector, the
field of which is ingeniously coupled to that of the scanner,
collecting only light from the conjugate image point. Since
scattered light from other focal planes is attenuated, it
contributes little to the image. A method such as this that
makes it possible to select successive focal planes and
view them in sharp focus and high contrast throughout the
depth of translucent sections or specimens of thicknesses up
to 200 μm must have important applications in botany,
especially in conjunction with electronic image storage and
processing systems discussed later.

   The history of "conventional" electron microscopy in botanical work during the last fifteen years illustrates an
interesting point. During this period the instruments themselves have evolved; dramatically in convenience of use, but
also in performance through advances in the design of vacuum
systems, power supplies and the electron-optics, and through
the better control of specimen contamination. Resolutions of
0.4 nm are now routinely available on suitable specimens with
a number of commercial machines. But a comparison of the
CTEM-micrographs of sectioned plant material in the journals
of today with those of 1970 shows little in the way of advance
that can be attributed to the instrumentation. In fact, something of a plateau was reached in the late 1960s, when after a
rapid period of evolution a series of standards emerged in
specimen preparation that have not really been superseded -
standards covering fixation, embedment, sectioning and post-
staining. There must be many laboratories where essentially
the same schedules have been in use for the better part of
twenty years. The adjustments have been minor - perhaps in
the use of better quality glutaraldehyde and other reagents,
the substitution of low viscosity resins, and modifications in
the way of making up the post-stains. One may deduce from
this that the standards must have been quite good in the first
instance still to sustain so important a role in the
continuing exploration of the fine-structure of plants. But
there is another conclusion to be drawn, namely that it is
specimen preparation procedures that are now setting the
limits to what is seen, not the hardware. I shall revert to
this matter.

   Undoubtedly the most spectacular advance in TEM instrumentation in the last twenty years has been the introduction of
the high-voltage machine (HVEM), pioneered in France and
developed commercially in the UK and Japan. Botanical applications have been investigated in this country mainly at
Oxford, and some of the remarkable results will be described

and discussed later in this symposium. The HVEM is no doubt
an instrument that will only ever be accessible to a few, but
this is far from meaning that it does not have substantial
potential as a botanical research tool, and probably in
several contexts. Some years ago we put a modest amount of
effort into developing an environmental stage for the NPL
instrument for the purpose of viewing living fungi. Success
was limited at that time, but one can see that this could well
be one line of development, not only for fungi but for various
other filamentous or paucicellular plants.

Since its first appearance twenty years ago, the scanning
electron microscope (SEM) has found itself a place near to the
heart of many botanical users who probably would never have
contemplated getting into CTEM work. The interest has been
principally in secondary-electron imaging, the instrument in
this mode fulfilling essentially the role of an immensely
powerful pocket-lens with splendid depth of field. I have no
doubt that this will continue to be the main use for the SEM
in biology in general, and there is every reason to applaud
such application, for the contribution it can make to micro-
morphology, anatomy and taxonomy is substantial indeed.
Again, handling and image recording have been simplified in
recent years, increasingly through automation. Yet this trend
has not gone far enough, for the design of even modern
instruments leaves much to be desired from an ergonomic point
of view. The eccentric column and awkward micrometer stage
controls have been with us from the time of the first
Cambridge instruments. The column position is no doubt an
inherent character of the breed, but centralised motorised
stage movements under microprocessor control, surely within
the technical competence of manufacturers, would considerably
improve the user's lot.

The SEM has seen notable evolution in image resolution,
mainly through the development of higher and cleaner vacuum
systems, the introduction of new types of filament, and the
optimisation of gun design and associated electronics. Whether improvement in ultimate performance parameters beyond a
certain point affects, or is even noticed by, the botanical
SEM snapshotter is a matter of speculation. The maximum resolution of the instrument obtainable on a physical specimen is
often taken as a criterion of its quality and utility in biological applications; yet, precisely as with the CTEM, with
modern instruments it is rarely the properties of the elec-
tron-optical system which limit the resolution obtainable, but
rather the method of specimen preparation.

The interaction of an electron beam with a specimen generates a number of emissions other than secondary electrons, and materials scientists and others are increasingly concerned with the use of these for imaging. We still await exploration of the various possibilities in botanical applications. Cathodoluminescence, where the image is built up from the light emitted from luminescent parts of the specimen under the incident electron beam, would seem to have distinct possibilities should it be feasible to develop suitable luminescent stains. Imaging using back-scattered electrons (BSE), reflected from the incident beam by the specimen, has already seen some use in clinical work. The advantage to be gained from this mode arises from the fact that the image is created from electrons reflected even within the specimen, making it feasible to "look into" embedded sections of quite considerable thickness. The reflection is greatest from elements with higher atomic number, suggesting that the method might be valuable in localising in plant tissue iron and gold labels on antibodies or lectins.

I now turn to image processing and quantitation. By definition the product of microscopy, light or electron, is an image, and the job of the research microscopist is to acquire images and use them - directly or in some derivative state - to convey biologically significant information in an appropriate way to the scientific community at large. For the LM, image capture hitherto has been mainly a matter of drawing or photography; image processing, the application of judicious artistry either in the darkroom or on the drawing bench; and quantitation, the tedious exercise of manual mensuration either through the eyepiece or on the photograph. The procedures have been much the same for the EM, although presumably no-one actually draws EM images. The time-hallowed methods persist, but electronic technology is changing the scene with increasing vigour.

I distinguish between image capture, processing and quantitation not because the technologies are unrelated but in acknowledgement of the fact that each has some unique significance for the botanical microscopist in the currently rapidly evolving scene. The venerable art of photomicrography has of course now achieved an advanced level of sophistication through the refinement and automation of camera systems and the advance of emulsion chemistry and processing procedures. It is in no sense superseded. The case is different for LM microcinematography. This has been with us for many years, but as the province of a few, mainly because of the cost of materials and equipment, and the tediousness of putting satisfactory systems together. In some applications - for example in high-speed filming, some forms of time-lapse work, and where the resolution of 35 mm film is required - it remains unsurpassed. However, for the average laboratory things have changed radically in the last few years. For the cost of some forty spools of 16 mm film a microscopist can acquire a CCTV camera and VCR which will enable him to record two hours of events in the microscope for about £5, and to do so with reasonably good colour fidelity and resolution. As many of us already know, such equipment provides a valuable teaching tool. But beyond this it offers a wide range of possibilities in research. Analysis of events in living systems in real time becomes almost embarrassingly easy compared with the hassle involved in time-lapse still photo-micrography, or even automated microscope cinematography. Cameras with 1 lux sensivity or greater are now available at relatively low cost, so that quite weak fluorescers and emitters, including those with application in vital staining, can be picked up. This greatly simplifies hitherto demanding procedures such as tracking the movement of fluorescent mark-ers within and between cells, scarcely feasible with 16 mm cinematography. But the special value surely comes with the facility for continuous recording over periods of several hours. Dynamic phenomena such as hyphal and pollen tube extension, spore discharge, gamete movement and fusion, cycl-osis, organelle movement, nuclear division, cytokinesis, wall growth and tropic and tactic behaviour are all amenable to video tape recording, and then become available for subsequent leisurely analysis in a manner hardly feasible at all hither-to. With environmental microscope stages the range is extend-ed to embrace a whole range of experimental treatments, from the investigation of light effects on chloroplast movement to tracing the long-term effects of laser surgery on unicells.

Analogue recordings of microscope images are themselves open to some degree of processing, for example in the control of black level and contrast range as in the AVEC system, the latter capable of generating images from the LM that can be better for visual observation and interpretation. But the real advance in image processing has come with the development of digital systems, which can be run directly on line from the microscope, LM or EM, or from secondary sources such as photographs or video recordings. The essential steps are now well enough known; they involve the translation of analogue signals into a digital data stream, the storage in memory of the information, and subsequent manipulation in numerous possible ways. Some of these are concerned with quantitation, to which I will shortly revert. But for the moment let us consider image enhancement and its implications.

The resolution achieved in the stored image and its deriv-atives is determined not by the properties of the microscope,

whether LM or EM, but by the memory available in the processing system itself. Practical systems for the average laboratory offer from 128 X 128 picture elements (pixels) with the
BBC micro up to about 1024 X 1024 for commonly available
dedicated devices using minicomputers. However, no digitised
image offers any advantage over a continuous tone image except
in two contexts - for the purpose of electronic storage and
transmission, and as the raw material for manipulation. The
significance for the research microscopist is greatest in the
second context. As with image resolution, the kinds of manipulation possible are also memory-limited, but even some of the
currently marketed low-cost systems offer surprising
facilities, including programmes for grey-scaling and gating to
enable isolation of elements of the field for independent
analysis, and the generation of false-colour images to enhance
selected features. The more sophisticated equipments allow a
remarkable degree of control, and indeed can extend the
working range of the microscope itself. A striking example is
seen in frame-store systems for the LM which enable the
integration of information from the specimen over a period of
time with a consequent smoothing out of the random element in
the signal. These are particularly effective in application
to noisy images such as those obtained in the microscopy of
weak fluorochromes. There can be few more satisfying experiences for a microscopist interested in localising weakly
fluorescent elements in the cell than witnessing the gradual
emergence of a consistent image of what he wants to see from a
screen initially full of meaningless hash.

    This is but one example of electronic image enhancement,
but it happens to be one where the involvement of the operator
is minimal and the result, accordingly, reasonably objective.
Others include the facility for the deliberate, selective
enhancement of features of the image, LM or EM derived, such
as boundaries and transitions, and the ability to suppress
"unwanted" elements. I may say that I scent hazard here,
since there is little to choose in an ethical sense between he
who manipulates an electronic image whether with conscious or
subconscious intent to show what he thinks might be there, and
he who attains the same goal by pencilling in the odd line or
two on a photomicrograph.

    Image enhancement is by no means the only form of electron-
ic manipulation currently available. An important application
is in enabling three-dimensional reconstruction of LM or EM
data and, if required, the subsequent manipulation of the
product so as to reveal new aspects, for example with rotation
programmes. "Manual" means for achieving these ends have long
been part of the microscopist's armoury, mostly based upon
serial sectioning and the assembly as 3-D models of the images

from each section plane. Such means have been used in the
past to establish the true architecture of complex anatomical
features such as nodal anatomy. These methods are now joined
by procedures where the reconstruction from the microscope
data is performed by the computer, as in recent demonstrations
of the organisation and disposition of the male gametes of
angiosperms from serial EM sections.

An interesting and again potentially important extension of
the method for the LM involves the acquisition of Z-dimension
data not from a series of physical sections, but from a sequence of focal planes through the same thick section. The AVEC
and TSRLM systems would presumably lend themselves to this,
and others have been developed.

Image quantitation has always been one of the aims of botanical microscopists. Cumbersome manual procedures have
been adopted hitherto, depending on eyepiece micrometer
measurement, or on mensuration of photographs or projected
images. Areas have been determined by squared-paper methods or
by outline-clipping and weighing. Such methods still have
application, but they are now progressively being superseded by
partially or completely computerised systems based upon image
digitising. Several types of instrument are currently available, varying in level of sophistication and cost. The simplest
are based upon microcomputers. In these systems, photographs,
camera lucida or other images are traced manually on a digitising pad, the numerical data being stored in memory for subsequent manipulation to yield lengths, perimeters and areas. At a
somewhat higher level of sophistication, an image displayed on a
monitor from the microscope or from a secondary source is digitised by tracing it with a cursor by means of a "mouse" or similar device. Automated systems acquire data directly on line from
the microscope, or again from a secondary source via a CCTV
camera. Subsequent processing gives object-counts and such
other metrical data as may be required. The automated systems
are especially valuable for repetitive quantitation, such as
might be needed in karyotype analysis and comparison, for which
various powerful software packages are available. However,
automated instruments by no means solve all image quantitation
problems, since not all LM or EM images are amenable to machine
interpretation. Human judgment is invariably involved in
quantifying continuous-tone images, for example in the setting
of black levels and grey-scaling. Manual systems using digit-
ising-pads may be just as effective in such instances.

Special difficulties arise in the mensuration of complex
entities in plant cells and tissues, as in determining the
volumes of organelles and apoplastic spaces, or the areas of
membranes or cell surfaces. Sometimes it is feasible to produce

rough estimates by various geometrical simplifications, but more
precise information can often be obtained by stereological analysis. Stereology employs a suite of mathematical formulae to
relate the profiles of structures as observed in the samples
represented by two-dimensional sections to the form of the
actual object. Increasingly the method is finding application
with plant material.

The third field in which microscopy, light and electron, has
advanced latterly is in qualitative and quantitative chemical
analysis. The use of the LM in interpreting the chemistry of
plant constituents has a long history, dating from Raspail's
work of more than a century-and-a-half ago. Today, qualitative
histochemistry (or cytochemistry, where the cell is the target)
deploys a wide range of techniques, enlarged recently by the
introduction of fluorescence methods, some of high specificity.
There is still great potential for further advance here in the
development of new fluorochromes.

Where the stoichiometrical relations of histochemical
reactions are known, quantitation becomes possible through
microspectrophotometry and microdensitometry. Spot- and
scanning-microdensitometers - dedicated, or based on
general-purpose stands - have been around for some years now.
Newly available computerised versions of the scanning instrument provide greater convenience in use, if not necessarily
higher accuracy. Microfluorimetry offers another means for
chemical quantitation of cell constituents. This technique has
been facilitated latterly by the availability of high-sensit-
ivity photomultipliers, and again we may expect to see its
power increasingly exploited with plant material.

While analytical methods using the LM are familiar enough to
most plant scientists, X-ray microanalysis with the various
types of electron microscope still wears an air of novelty.
The method, which depends on the spectrometry of the X-rays
emitted when an electron beam interacts with a specimen, has
nevertheless a considerable history. I imagine most RMS
Fellows will be well aware of the story of the early research
by Coslett and his colleagues at Cambridge which led to the
first Cambridge Instruments machines, and also of the essentially independent work by AEI that produced the famous EMMA
some quarter-century ago. Exploiting these pioneer instruments
was mainly the province of specialists, but with the introduction in more recent years of new types of solid-state X-ray
detectors and high-resolution spectrometers simultaneously with
the development of multi-function SEMs and the scanning-trans-
mission electron microscope (STEM), the technique has been
greatly refined. In consequence of all this, application has
been substantially simplified for the biological user. The

dedicated minicomputer, also, has had a vital part in making
complex systems accessible for biologists. The computer in
itself would have no value without the appropriate software,
but this has now achieved a high level of sophistication,
offering the ability to process, record and present qualitative and quantitive data on element distribution in a specimen
quite as readily as producing a photographic image of it.

Moreover, notwithstanding the cost of comprehensive install-
ations, instruments are today becoming more widely accessible.
In botanical contexts it is no doubt true that serious application is still only its infancy, but already the technique has
yielded notable contributions, including new evidence bearing
on the potassium and chloride ion shuttle associated with
stomatal function, the distribution and movement of calcium
associated with the growth of certain cell types, the accretion
of phosphorus in the maturation of bryophyte sperms, and the
process of silicon deposition in the surface trichomes of grass
leaves.

Predictably, at this stage of the development of the art in
application to plants the limitations arise not primarily from
the hardware, but more in identifying suitable problems, in
adopting, developing or adapting specimen-preparation methods
appropriate for each case, and in interpreting the results
obtained. I am sure that few dedicated microscopists will be
unaware of what can now be done with X-ray microanalysis, but I
suspect that its value is as yet inadequately appreciated by a
large sector of the plant-physiological and biochemical cons-
tituency. Yet its potential as a powerful tool for those whose
concern is with nutrition, transport, secretion, growth and
movement - all processes which involve ion accumulation or
redistribution - is clear enough, and beyond this the technique has obvious possibilities in tracer work. We may surely
anticipate a flood of papers on these topics in the years
immediately ahead as the possibilities are exploited.

With regard to specimen preparation, experience has already
shown that before rational procedures can be worked out for the
use of X-ray microanalysis, the research problem itself must be
well defined, emphasising a point I have made earlier. The
conventional preparation methods for the CTEM and SEM have some
value; but they are limited, for the obvious reason that many
of the elements of biological signicance in plant cells which
are amenable to localisation and quantitation by X-ray microanalysis are extracted or displaced by the processing involved.
Yet there are "fixation" procedures that can be used to precipitate or bind elements of interest, and these can be used in
association with conventional processing. Interestingly,
certain quite hoary LM histochemical procedures lend themselves

immediately for adaption. For example, we have found it
possible to track the movement of potassium with reasonable
resolution in massive tissues by the sodium cobaltinitrite
technique, and similarly to localise chloride by silver precipitation. In other instances potassium pyroantimonate has been
used for calcium localisation. A search for other precipitation techniques beyond those already in service in LM cytochemistry would obviously now be well worth while.

As a first approach to many problems concerning icn localisation in plant tissues these essentially qualitative methods
are undoubtedly valuable. They necessarily fail where the
spatial resolution they provide is inadequate, and it will be a
rare case where they can be used in quantitation with any great
degree of confidence. When the structural integrity of individual cells is not a first consideration, rapid heat-drying or
even ashing provide effective means for preparing tissues for
quantitation. Present indications are, however, that freezing
methods have the greatest potential. The familiar freeze-
drying procedure is effective for isolated cells and small,
well contained tissue masses, notably for X-ray microanalysis
in the SEM. The technique appears to be less promising for
larger tissue fragments and sections. In some instances it
has been found that even with rapid freeze-drying of cryostat
sections actually in the microscope, ion migration may occur
sufficiently to invalidate, or at least confuse, the results.
The nearest approach to an ideal at present, especially for
detailed quantitative studies, seems to be the use of frozen-
hydrated specimens or sections, and commercial instruments and
add-ons offering facilities for this are becoming more widely
available.

Many of the problems of interpretation and quantitation
arising in biological X-ray microanalysis haunt physical pract-
itioners also, and most are well documented in the technical
literature. In general the biological user tends to be at the
mercy of the instrument designer and manufacturer, and is
increasingly beholden to the software writer. If the user is
himself a novice in the game, he is likely also to be at the
mercy of the operator. Yet at the same time some of the major
hazards are inherent in biological material itself. Cell
architecture, and especially the topography of dried or cut
surfaces, has been shown to be capable of affecting derived
spectra, a clear warning of lurking peril when quantitative
comparisons are in question. During the present period of
exploration and improvisation, the need clearly is for the
testing and comparison of as many as feasible of the available
preparation techniques, and of course for liberal replication.

Finally, we may note in this context also the potential
dangers that lie in the processing of results. The writing of
software for analytical machines of all types has become something of an art form, and the virtuosity of some software
writers is astonishing. There is an irresistible fascination
in exploiting the powers they have handed to users for manipul-
ating, smoothing, transposing and converting data; and there is
certainly no denying that multicoloured spectra and element
maps spectularly displayed on colour monitors have a strong
aesthetic appeal. But it is the relevance, reliability and
communicability of the product in relation to the biological
problem that is in the end important. Commenting on spectrum
treatment, a physical writer has recently observed that a point
is soon reached where further sophisticated manipulation of
data is profitless - and even misleading - when the sole consequence is to conceal the true variation, or to smooth away the
errors in the raw data. The stricture certainly applies in
application to plant material in the present state of the art.
   The last of the four trends in microscopy I mentioned at the
outset is the development of wholly new methods for imaging.
None has yet reached a point where early application in botanical investigation can be anticipated, but each has promise.
The list includes X-ray microscopy, photoemission electron
microscopy, scanning acoustic microscopy and scanning harmonic
optical microscopy. Although at present there is not a great
deal of practical evidence to judge from, the biological
potential often comes through very clearly from the published
commentaries on the various systems. Thus X-ray microscopy is
likely to be far out of reach of the generality of biologists
for some considerable time because of the sparsity of suitable
X-ray sources; but it has been pointed out that were it ever
practicable to develop a suitable machine for general use, the
fact that protein molecules are effectively "coloured" in soft
X-radiation could give a new range of powers for viewing unfixed, unstained biological specimens at high resolution.
Photoemission electron microscopy, where imaging uses the electrons released from a specimen exposed to high intensity light,
has already been employed to photograph individual DNA
molecules, exploiting the fact that nucleic acids are relatively efficient photo-emitters compared, for example, with
proteins. Scanning acoustic microscopy has now attained new
levels of resolution with the development of generators capable
of producing sound of extremely short wavelength; its special
value lies in the potential it offers for looking "into" tissues, much in the manner of the acoustic body-scanners already
in general clinical use. Recent experience suggests that the
lead-time between laboratory prototypes of scientific instruments and their becoming generally available is continuously
diminishing, so that today's biologists may not have so long to
wait to get their hands on some of the new kinds of instruments

as did their predecessors through the long period following
Ruska's pioneering EM research.

Repeatedly in the foregoing I have commented on the limitations set in botanical microscopy by the specimen itself. The
topic of specimen preparation is one that will recur throughout
this symposium, and I do not propose to review its many aspects
here. This does not mean, however, that I in any way underestimate its significance for botanical microscopists. Quite the
reverse: I regard the evolution of this part of microscope
technology as the very key to successful progress in the future
in exploiting the manifold opportunities provided by the instrumentation now at our disposal. For botanical microscopists,
the real state-of-the-art is defined by what they do, or can
do, with their material; and it is in the handling of the
material that they are much more in control of the situation.
However, in contemplating where things are likely to move next,
one fact has to be acknowledged - namely that for the bulk of
the currently most widely used preparation techniques plant
scientists are indebted to workers on the animal side, especially those in clinical laboratories. From them most of
current histochemical and cytochemical technique has been
borrowed, and much else as well. This has been inevitable,
given the preponderance of workers on the animal side and the
greater resources at their command. Also, it is they that
tend to create the market towards which equipment and chemical
manufacturers are oriented; all other demands are essentially
incidental. Since there is no real expectation that these
circumstances are likely to change, it is safe to predict that
botanical microtechnique will continue to be largely parasitic.
But the translation and adaptation of the borrowed technology
to suit the different characteristics of plant material will
always provide a challenge, and there will continue to be
plenty of unique problems that will need to be tackled independently. I will end with one intriguing example where rapid
evolution of plant microtechnique is required. Existing
methods are proving far from adequate for handling the processing of plant tissues in spacecraft. There is an embargo on
crews handling toxic fixatives in the open atmosphere of the
craft, and in any event they mostly do not have the time for
much manipulation. Yet, if plants are to play a major part in
life-support systems for long-term space missions, much more
must be learned about how they react to weightlessness and the
spacecraft environment. For the now urgent cell biological
studies, new methods of fixation are going to be needed, and
new kinds of sealed, miniaturised, automated systems for processing tissues. An interesting set of problems here for
anyone, or any group, that can deploy the necessary combination
of botanical, chemical, mechanical and electronic skills!

2

Fixation of plant cells for electron microscopy

J. Coetzee

2.1 INTRODUCTION

When plant material is to be examined with the optical microscope, it is often possible and indeed, nearly always desirable, to study tissue which has been exposed to the absolute minimum of preparative procedures. In this way the formation of artefacts can hopefully be minimized, even if not completely eliminated.

In many cases however, the nature of the tissue or the type of information desired makes the use of thin sections necessary, and in transmission electron microscopy, the use of ultrathin sections is usually obligatory. Preparation techniques for thin sections are essentially the same for both light and electron microscopy. Cryo-sectioning is the only preparation method with which thin sections can be prepared which does not necessarily involve the application of artefact-inducing chemicals. This is therefore the only sectioning method with which plant cells can be prepared which would allow us to assume with any degree of certainty that the structure we see is really that which is present *in vivo*.

The tremendous technical difficulties associated with the successful preparation of usable ultra-thin cryo-sections have, however, had a nearly totally inhibitory effect on their use. Even in the case of animal tissues, where these difficulties are very much less, cryo-sections presently cannot represent more than a very modest proportion of those sections which are cut with the primary object of obtaining structural information. Freeze-fracturing methods have certain distinct advantages, but also some specific disadvantages, such as the inability of the method to guarantee specific face orientation, as well as the impossibility of obtaining true serial views.

At the present time the most practical thin-sectioning preparation protocol is therefore still the traditional method involving, in the first case, chemical fixation whereby proteins are initially stabilized by crosslinking with the aid of an aldehyde, and lipids are subsequently immobilised by treatment with osmium tetroxide. This two-stage fixation procedure can then be followed by any one of a variety of embedding methods.

To obtain the maximum fidelity in the preservation of structural detail during the fixation processes, a number of parameters have to be optimised. The parameters which are associated with the aldehyde fix include the following which seem to have greater or lesser influences on the structure of the material: the choice of aldehyde, the concentration of the aldehyde, the duration of the primary aldehyde fixation, the temperature during this fixation, the choice of buffer, the pH of the buffer, the molarity of the buffer, and finally, the effective osmotic pressure of the fixative. Known parameters associated with the osmium fix are: the concentration of the osmium and the fixation time. Daunting though this list may appear, there is no guarantee that it is exhaustive.

Each of the above factors influences the architecture of the tissue, and will necessarily have some bearing on the deductions which are based on the observed structure. It is therefore of the utmost importance that researchers have some knowledge of the influence of the most important structure-modifying parameters. The influence of some of these variables are very well known, due to the important contributions of a number of investigators over the years. The contribution of other factors are not so well known, so much so that critical evaluation of the literature on the crucially important fixation procedure reveals large gaps in our knowledge. Much more work has to be done before we can become complacent about our preparatory procedures.

Botanical microscopists have, with some notable exceptions, been very ready to uncritically accept the microtechnical methods of non-botanists. Because of the fundamental structural differences between eukaryotic plant cells and the majority of all other cell types, knowledge of fixation techniques is not always directly transferable between these disciplines. Two complicating structures are responsible for this state of affairs, namely the cell wall and the vacuole. These structures have a profound influence on the differing response of plant versus other cells to, for instance, variations in external osmotic potential. The vacuole, due to its composition and pH, may have additional influences on the course of the fixation process.

2.2 FACTORS INFLUENCING FIXATION QUALITY

2.2.1 The choice of aldehyde

Three aldehydes are in reasonably widespread use. They are, in order of increasing molecular mass, formaldehyde (molecular mass 30.03), acrolein (56.06), and glutaraldehyde (100.12). The rate at which each of these three chemicals diffuse into the tissue is governed not only by the molecular mass and molecular dimensions, but also by additional factors such as the formation of a "fixation barrier" which develops during fixation. Such a barrier is formed by the presence of fixation products within the tissue. The aldehyde which is more efficient at crosslinking proteins (and possibly a few other chemical species) exhibits this phenomenon much more clearly than does less efficient crosslinkers. Due to the dialdehydic nature of glutaraldehyde, the protein crosslinks formed by this fixative are much more stable than those associated with the use of either formaldehyde or acrolein. The fixation barrier retards the penetration of fresh fixative into the tissue, so that the interior of a tissue block may become "starved" of sufficient fixative. Whether the presence of this barrier, which has been demonstrated in model systems after relatively long fixation periods in glutaraldehyde (Coetzee & Van der Merwe, 1984b), is of great practical signi-ficance during the fixation of small blocks of tissue for periods not exceeding a few hours, is not clear. This effect may, however, be important if large blocks of tissue are to be fixed.

The differential penetration rates of different aldehydes into hair cells have been elegantly demonstrated in the pioneering work of Mersey & McCully (1978), who came to the conclusion that acrolein penetrates more rapidly than glutaraldehyde. Whether the effective penetration of fixative into tissue blocks is exactly analogous, is very difficult to decide. It must always be kept in mind that the penetration of the fixative into the tissue is only one side of the coin. The other side of this coin is the rate at which this aldehyde "fixes" the tissue. Penetration *per se*, is of little impor-tance if it is not accompanied by nearly immediate crosslinking. If both these factors are taken into account, it appears as if glutaraldehyde is the "fastest" fixative of the three, and acrolein the "slowest" (Coetzee & Van der Merwe, 1984b).

The behaviour of the aldehydes are, however, influenced by the choice of buffer (Coetzee & Van der Merwe, 1985c). The choice of aldehyde, even if purely structural (as opposed to functional) preservation of the tissue is the object of the

fixation, is therefore not nearly as straightforward as one
would like.

2.2.2 Concentration of the aldehyde

The three aldehydes in question are normally applied at
concentrations somewhere between 1% and 10%, usually in a
buffer. Depending on the molecular mass of each aldehyde,
this represents molar concentrations between 0.1M (1% glutaraldehyde) and 3.3M (10% formaldehyde). The choice of
concentration influences the penetration rate, the crosslinking rate, the osmotic value of the aldehyde, and the
osmotic value of the buffer. Higher concentrations of
aldehydes also need higher buffer concentrations to keep the
pH at the selected value.

        Protein crosslinking rates (Coetzee & Van der Merwe,
1984b) suggest that the lowest practical concentrations to
employ are (very approximately): 1% glutaraldehyde, 3%
acrolein and rather more than 5% formaldehyde. At these
concentrations the protein crosslinking rate of glutaraldehyde
in phosphate buffer is three times higher than that of acrolein
and nine times that of formaldehyde. The choice of buffer may
influence these rates unpredictably.

        A survey of the literature suggests that concentrations
of glutaraldehyde, lower than those which were in general use
only a few years ago, are becoming generally accepted.
Experience as well as the very limited amount of experimental
results available both suggest that, in the past, over-
fixation has been the norm rather than the exception, so that
the use of these lower (1%-2.5%) concentrations is a step in
the right direction.

2.2.3 Duration of the aldehyde fixation

The effect of longer than optimal fixation in glutaraldehyde is
firstly extraction of substances from the tissue (Table 2.1)
and secondly, if the fixation is even more protracted, the
destruction of membranes. The result is that electron
micrographs of material so treated present a low-contrast
image with damaged membranes (Figs. 2.1, 2.2, 2.3). Granular
deposits of unknown origin are also often present. There is
apparently no specific critical time which must not be
exceeded before these unacceptable artefacts become obvious.
The onset as well as the appearance of such artefacts are
influenced by the choice of fixative buffer vehicle (Coetzee
& Van der Merwe, 1985a).

Table 2.1

Percentage loss of ^{14}C label from bean leaf tissue (exposed to $^{14}CO_2$ for 2h) in 2.5% glutaraldehyde in 0.1M Na-cacodylate buffer

Fixing time	% loss of label
1h	2.0 ± 0.35
4h	6.2 ± 0.85
24h	16.8 ± 0.64

means $\pm$ standard error

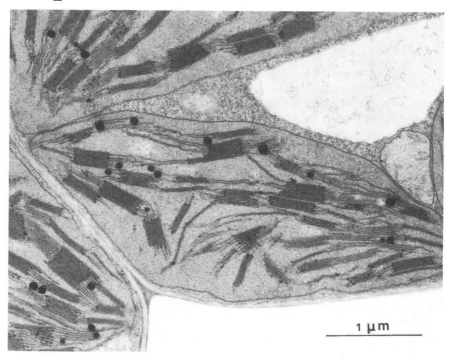

1 μm

Fig. 2.1   Micrograph of bean leaf tissue which was fixed for 1 h in 2.5% glutaraldehyde in 0.1M Na-cacodylate buffer at pH 7.2, washed in three 5 min rinses in the same buffer, post-fixed for 1h in 1% OsO_4 in the same buffer, washed in three 5 min buffer rinses, dehydrated in acetone, and embedded in Spurr's (1969) resin.

It seems as if superior structural preservation is attainable if the fixation is allowed to proceed for the minimum time which is compatible with adequate penetration of the aldehyde into the tissue. The average penetration rate of glutaraldehyde into carrot tissue (after 18h fixation) is approximately 2μm per minute at room temperature, but this rate is profoundly influenced by the concentration of the aldehyde as well as the type of buffer employed (Coetzee & Van der Merwe, 1985c). Because the diffusion constant of the fixative (Medawar, 1941) decreases with an amount proportional to the square of the fixation time the penetration rate may be very much higher if the average is taken over shorter fixation times. In practice, the result is that fixation times (of small blocks of "typical" plant tissue in 2% glutaraldehyde in phosphate buffer) of one hour or less usually gives entirely satisfactory results. It has to be kept in mind that the 18h penetration rate of glutaraldehyde implies that the penetration depth of the fixative limits the maximum size of any piece of plant tissue to a block with no dimension larger than about 2mm. Even in a block of tissue of this size, the inner part of the tissue block would have had a long time to develop autolytic symptoms, whilst the outer layers would tend to be overfixed.

2.2.4 Temperature during fixation

The use of low (0-4°C) temperatures during the fixation cycle may be advantageous as far as the "stiffening" of the cytoplasmic gel is concerned, but the effect of the low temperature on the rate of the chemical reaction between aldehyde and protein must be allowed for. Even after 4h in glutaraldehyde at 4°C, the fixation of bean leaf material is not as complete as when the fixation is done at 20°C. This can be deduced from the loss of ^{14}C-labelled substances from the tissue during subsequent processing steps (Fig. 2.4), which shows that slightly more label is extracted during the buffer rinses preceding the osmication step from material which was fixed at low temperature, than from the material which was fixed at 20°C. After osmication there are no significant differences in extraction between the treatments.

The rate of penetration of the aldehyde into the tissue is reduced at lower temperatures. Mersey & McCully (1978) found that lowering the temperature approximately 20°C to near freezing, reduced the penetration rate of glutaraldehyde by nearly 60%. No figures are available, but it seems logical to assume that the slower penetration and the less active crosslinking at low temperatures have a cumulative effect.

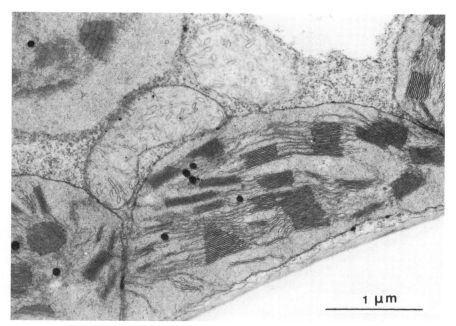

Fig. 2.2.   Micrograph of tissue similar to that
shown in Fig. 2.1, prepared with the same protocol,
but fixed for 16h in glutaraldehyde.

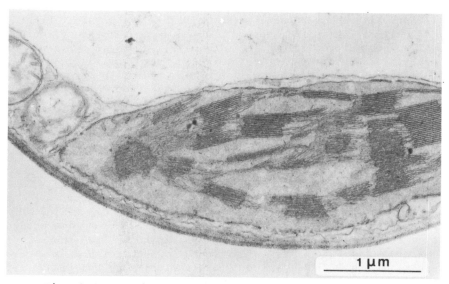

Fig. 2.3.   Micrograph of part of a bean leaf cell,
prepared as for Fig. 2.1, but fixed for 28d in
glutaraldehyde.

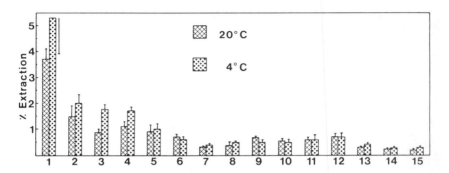

Fig. 2.4.   Extraction of ^{14}C-labelled substances
from bean leaf tissue processed at 4°C and at 20°C.
Bean plants were exposed to [14]CO_2 for 2h. Processing
steps: 1=2h fixation in 2.5% glutaraldehyde in 0.1M
Na-cacodylate; 2,3,4=3x15min buffer rinses; 5=1h
postfix in 1% OsO_4 in buffer; 6,7,8=3x15min buffer
rinses; 9-15=15min dehydration in each of 30, 50, 70,
90, 100, 100, 100% acetone. Bars= 2s.

The destructive effect of low temperature fixation on the
preservation of microtubules (Hayat, 1981) and possibly other
organelles, have to be kept in mind. High temperature fixation
(as suggested by O'Brien & McCully, 1981) may have certain
advantages. These high temperatures may very well increase the
rate of autolysis of the cells and make the cells more
susceptible to mechanical damage, but the enhancement of the
effectivity of the aldehyde (Bowes & Cater, 1968) may more than
compensate for this. Experimental results are much needed.

2.2.5 Choice of buffer

The buffer vehicle may have subtle effects on the general
preservation of the structure of plant tissue (Fineran, 1971,
Salema & Brandao, 1973). Some authors (Goodchild & Craig, 1982)
are of the opinion that the structural modifications introduced
by different buffers may be even more important than the
modifications which can be attributed to the choice of fixing
agent. Although the effect of these variables on the tissue may
be prominent when identical pieces of tissue, processed in
different buffers, are compared, they can be very difficult to
describe and to quantify because attributes such as local
contrast, extraction of certain cell components and the

"mordanting" of specific structures towards heavy metal contrasting, are influenced.

That the buffer can be an important modifier of the extractive behaviour of the fixative, and therefore of the resultant structural preservation (compare Figs. 2.5, 2.6 and 2.7), has been known since the work of Holt & Hicks (1961) and others. Luft & Wood (1963) determined the loss of certain components quantitatively, and found that the amount of different classes of chemical components, extracted during fixation, is governed primarily by the buffer involved. The work of Salema & Brandao (1973) and Coetzee & Van der Merwe (1984a) confirmed that this was also the case for plant tissues. One important fact, which can be deduced from our results as well as from the work of others on animal systems, is that the choice of vehicle influences not only the extraction of substances during fixation, but also the extraction during subsequent processing steps. We found that the lowest average extraction of components from bean leaf tissue during the fixation process resulted from the use of a sodium phosphate buffer. Other buffers (including PIPES) led to higher average extraction values. Because some classes of compounds are extracted differentially by different buffers, a buffer which may show a high extraction of one compound, may be the buffer of choice when another compound is to be retained (Table 2.2).

The abovementioned variables, however, are not the only factors influencing the choice of buffer vehicle. Practical issues, such as ease of use, price and toxicity are also important. Other factors such as the influence of the buffer on the osmotic value of the fixative should also be taken into account. The osmotic behaviour of the buffer is discussed under the appropriate heading.

2.2.6 Buffer pH

Traditionally the primary aldehyde fixative is buffered at approximately physiological pH, which for animal cells is usually taken as being near to neutrality. Plant cells, with their large vacuoles in which the pH may reach very low values, present a different situation.

Glutaraldehyde is chemically most reactive towards proteins at rather strongly alkaline pH values (Bowes & Cater, 1965). Essentially no crosslinking of proteins can take place if aqueous glutaraldehyde (pH approximately 3.4) is used as fixative, with the result that the tissue exhibits a great deal of structural damage, most of it produced during post-fixation

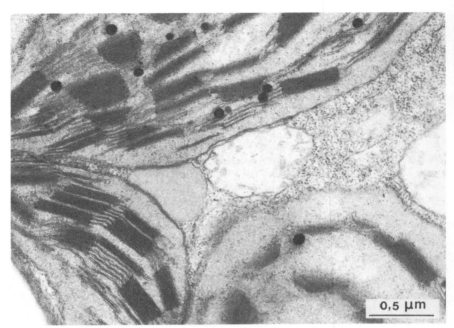

0,5 µm

Fig. 2.5.   Micrograph of bean leaf tissue, prepared
as for Fig. 2.1, with 4h fixation in 2.5% glutaraldehyde in 0.1M Na-cacodylate buffer.

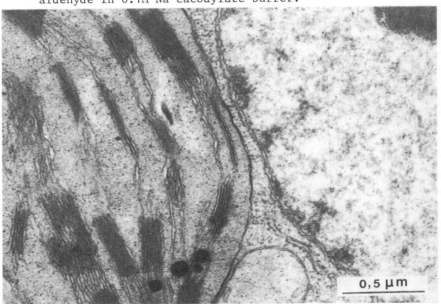

0,5 µm

Fig. 2.6.   Bean leaf cells prepared as for Fig. 2.5,
except for the use of 0.1M PIPES buffer.

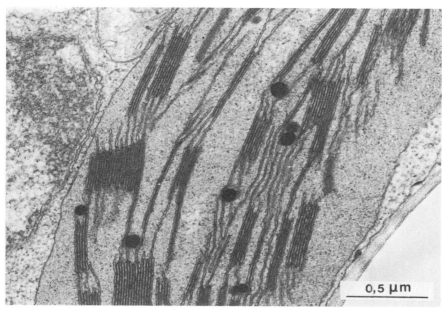

Fig. 2.7.   Bean leaf cells, prepared as for Fig. 2.5,
except for the use of Na-Na-phosphate buffer.

Table 2.2

Extraction of different substances from bean leaf tissue, fixed
for 4h in 2.5% glutaraldehyde in various 0.1M buffers at pH 7.2

Buffer type	Retention				
	Ca	Mg	Sugars	A. acids	Protein
Citrate	+		+++		
HEPES		+++	+		
MOPS			++		
Na-cacodylate					++
Na-K-phosphate	++			+++	+
Na-Na-phosphate	+++	+++		++	+
PIPES				+	

+++: best observed retention (least extraction)
+:   appreciably higher extraction

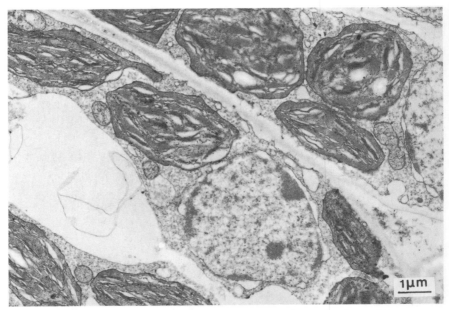

Fig. 2.8.   Micrograph of bean leaf palisade cells,
showing the typical damage sustained during post-
fixation processing when the tissue was "fixed" for
4h in 2.5% aqueous glutaraldehyde at pH 3.4.

processing (Fig. 2.8). If the aqueous glutaraldehyde is
neutralised by the addition of a very small amount of sodium
hydroxide, the osmolarity of the fixative remains practically
unchanged, but the resultant structural preservation is greatly
enhanced (Fig. 2.9), because, in this case crosslinking of
proteins, including those in the membranes, could take place
during fixation. The semi-permeable characteristics of the
membranes are therefore modified, and less osmotic damage is
induced during subsequent processing steps. Our own experience
suggests that excellent fixation is the exception rather than
the rule if the pH of the primary fixative is lower than 6.5.
At rather higher pH-values than the normal values of 7.2-7.6
the high activity of the aldehyde may make shorter fixation
schedules possible, which would minimize extraction from the
tissue.

2.2.7 Molarity of the buffer

A low buffer concentration will contribute less to the
osmolarity of the fixative than a higher concentration, but the
buffering capacity of generally used buffers, including that of
one of the zwitterionic buffers (Good *et al.*, 1966), is not

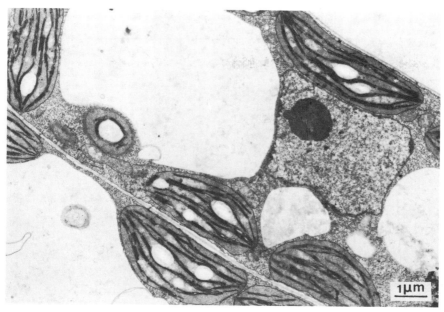

Fig. 2.9. Micrograph of tissue similar to that shown
in Fig. 2.8, similarly processed, but with the pH of
the fixative adjusted to 7.2 with NaOH.

adequate to stabilize the pH of fixatives containing 2.5%
glutaraldehyde (Fig. 2.10). This emphasizes the need to
measure the pH of the fixative immediately before use, even if
relatively high (0.1M) buffer concentrations are used.

2.2.8 Osmolarity of the buffer

The term "osmolarity" is not one that is traditionally used by
botanists, so that clarification of the term might be in order:
    A solution which exerts an osmotic effect, equal to
    that of a molar solution of a nonionic substance such
    as glucose (therefore with a dissociation constant of
    1.0) has an osmolarity of 1000 milliosmoles. A molar
    solution of a substance such as NaCl, which is
    completely dissociated in solution (dissociation
    constant of 2.0) would exhibit an osmolarity of
    2000 mOsm.
The osmolarity of a buffer therefore depends on the
concentration of the buffer, as well as on the value of the
dissociation constant, K. Maser et al. (1967) published
graphs showing the relationship between the osmolarity, the pH
and the molarity of a number of buffers and fixing agents.
These figures are very useful to determine the osmolarity

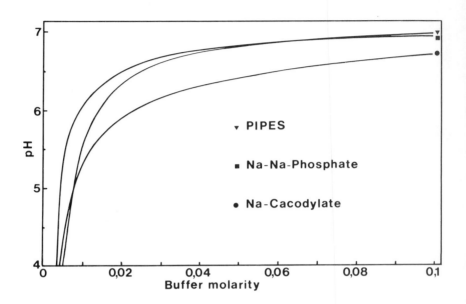

Fig. 2.10.   Graph illustrating the pH of different
buffers containing 2.5% glutaraldehyde. All the
buffers had an initial pH of 7.2.

of any of the components of a fixative, but if used to predict
the osmolarity of a mixture containing buffer as well as
glutaraldehyde, may give rise to misleading results because
these graphs do not allow for the changes in osmolarity which
take place when these components are mixed. The different
osmolarities which buffers exhibit when in aqueous solution
and when mixed with glutaraldehyde, are the result of the
influence of the aldehyde on the dissociation constant of
the buffer (Coetzee & Van der Merwe, 1985b). The result of this
phenomenon is that the osmolarity (and therefore also the
osmotic effect) of a buffer which is mixed with glutaraldehyde
is dependent on: the type of buffer, the molarity of the
buffer, the pH of the buffer as well as on the concentration of
the glutaraldehyde (Fig. 2.11). The K-values of this buffer may
be as low as 1.85 (0.1M, no glutaraldehyde), with a resultant
osmolarity of 185mOsm.100mM^{-1}, to a value as high as 4.0
(0.03M, 2.5% glutaraldehyde), with an osmolarity of
400mOsm.100mM^{-1}.

    Two pieces of the same tissue, fixed in two fixatives
differing only with respect to the buffer concentration, may
result in very different micrographs (Figs 2.12, 2.13). The

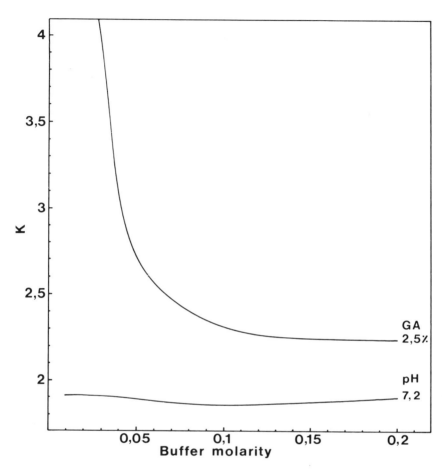

Fig. 2.11.   Smoothed curves illustrating the
relationship between the dissociation coefficient
(K) and the buffer molarity for Na-cacodylate buffer
at pH 7.2, with and without 2.5% glutaraldehyde.

differences in osmolarity, which are due to the contribution of
the buffer, in mixtures containing glutaraldehyde, can
therefore be of sufficient magnitude to negate careful
calculation of the relative water potentials involved. Because
the buffer usually contributes more than the glutaraldehyde
does to the total osmolarity of a fixing solution, it is of
worthwhile practical importance to determine what this
contribution is. Even if no further allowance is made for the
contribution of the glutaraldehyde, fixing conditions can be
improved by optimising the buffer molarity.

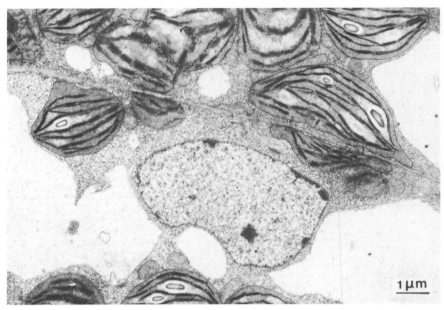

Fig. 2.12.   Micrograph of bean leaf tissue, processed as for Fig. 2.1, but with the use of 0.05M
buffer.

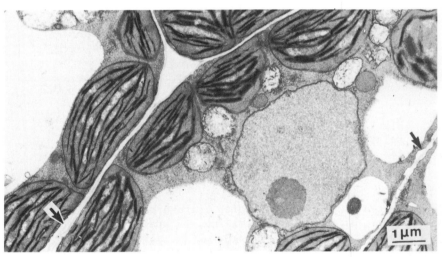

Fig. 2.13.   Material processed similarly to that in
the previous fig., but with the use of a very high
(0.3M) buffer concentration. General shrinkage of
organelles, as well as plasmolysed areas (arrowed)
where the plasmalemma has pulled away from the cell
wall, are visible.

2.2.9 Effective osmolarity of the fixative

Nett movement of water into, or out of the cell, will not take place only if the cell and the surrounding fixative have exactly the same water potential (or, are isotonic with respect to one another).

Attaining this ideal situation during fixation implies that the osmolarity of the two components of the system, ie. that of the cell as well as that of the fixative, are known. The water potential of the tissue can be measured, albeit not easily, with a method such as that of Chardakov (Salisbury & Ross, 1978) or with a vapour pressure psychrometer, whilst the total osmolarity of the fixative can be determined with an osmometer. If plant tissue (or, for that matter, animal tissue) of known water potential is fixed in a buffered glutaraldehyde fixative which has a similar total osmolarity, more often than not the result is osmotically damaged tissue. A number of authors, working on plant material, have mentioned this fact, (which is well known, and routinely allowed for during preparation of many animal tissues), but a survey of the literature suggests that this is hardly ever allowed for during the preparation of plant tissue.

This osmotic reaction of the tissue towards the fixative is due to the fact that the effective osmotic pressure (EOP) which the fixative exerts on the cell, is not equal to the osmotic effect which a non-reactive osmoticum of the same osmolarity would have. This is the result of the modifying effect which glutaraldehyde has on the proteins of the membranes. Young (1935) was the first to mention this, and suggested that the osmotic effect of a fixative can be adequately described if it is accepted that the aldehyde makes no contribution towards the EOP of the fixative. Barnard (1976) suggested that the measured osmotic value of glutaraldehyde should be halved, whilst Mathieu et al. (1978) suggested that there is a rather more complex relationship between the EOP of the fixative, the measured osmolarity of the total fixative and that of the glutaraldehyde.

If: MG=milliosmoles of glutaraldehyde
      MB=mOsm of buffer and MT= mOsm of total fixative,
the findings of these authors can be summarized as:

Young (1935):   EOP= MB

Barnard (1976):   EOP= MB/2 + MB

Mathieu et al. (1978): EOP= $1056 \cdot MT \dfrac{(1.3 \cdot 10^{-4} \, MG+0.81)}{MG+239}$

The increasing complexity of the perceived relationships between the parameters is obvious. These relationships have been suggested to clarify the situation in animal tissue, where, due to the absence of rigid cell walls and large vacuoles, the osmotic situation is less complicated than it is in plant tissues.

The presence of, firstly, the essentially inextensible cell wall, and secondly, the vacuole, causes still another parameter to be added to the formula. This parameter is the average osmotic value of the cell. Our own work suggests that these relationships are complex, nonlinear and are influenced additionally by the type of buffer vehicle present.

When a piece of plant tissue is immersed in a buffer, the addition of glutaraldehyde to the buffer causes an increase in the EOP of the solution, resulting in loss of water from the tissue (Fig. 2.14), which suggests that the glutaraldehyde is osmotically active. This is further borne out by the fact that high concentrations of glutaraldehyde cause a bigger increase in EOP than do low concentrations.

The behaviour of plant tissue in glutaraldehyde solutions may be anomalous, for instance, if a piece of plant tissue is immersed in an aqueous glutaraldehyde solution, the addition of a low concentration of buffer causes a decline in the EOP of the solution, with a resultant increase in the water content of the tissue (Fig. 2.15). This anomalous behaviour is reversed at higher buffer concentrations, where an increase in buffer molarity causes the expected increase in EOP, resulting in the observed loss of water from the specimen.

Balancing the water potentials of the tissue and the fixative so that very little or no osmotically-induced water movement takes place during fixation, can result in enhanced structural preservation. We have found that "normal" fixation procedures may easily involve a tissue in the gain or loss of more than half its total mobile water content. Gains or losses of this magnitude are clearly of structural importance.

2.3 CONCLUSIONS

Enough is known in a few selected areas of the field of botanical tissue fixation to allow a few general conclusions to be made. One of these conclusions is that the fidelity with which the ultrastructure of plant cells is preserved during preparation for electron microscopy, depends to a large extent on the primary aldehyde fixation parameters.

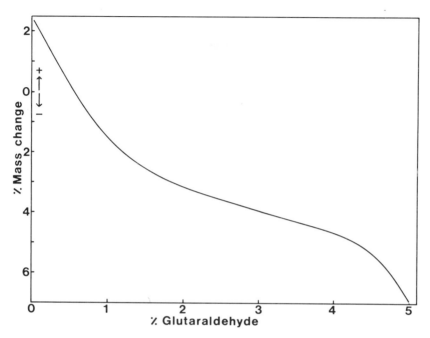

Fig. 2.14.   Percentage change in mass of carrot discs immersed in various concentrations of glutaraldehyde in 0.1M Na-Na-phosphate buffer.

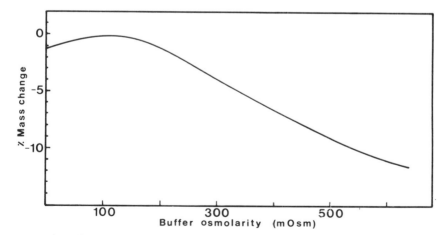

Fig. 2.15.   Percentage mass change of carrot discs immersed in different osmolar buffer solutions, all containing 2.5% glutaraldehyde.

The successful formulation of an aldehyde fixative for a specific purpose depends on knowledge of the water potential (osmolarity) of the fixative, as well as that of the cells in question. The effective osmotic pressure of the fixative is not necessarily (and usually is not) equal to the measured osmolarity of the complete fixative, or even to that of the buffer vehicle only. Plant cells differ tremendously from one another as far as water potential is concerned, which implies that there can be no single fixative formulation which will preserve all, or even most, plant cells without causing observable osmotically-induced artefacts.

Much additional work remains to be done on the interaction between fixative and plant tissue. At this stage it is, except for certain well-defined tissues, difficult to formulate an osmotically balanced aldehyde fixative for a specific application.

The type of buffer vehicle employed influences the contrast of the final image, the extraction of specific cell components and makes an important contribution to the osmolarity of the fixative. The buffer type and its final concentration is important in regulating the pH and therefore the crosslinking abilities of the aldehyde. The fixation temperature has an effect on the preservation of certain cell components, the rate of fixative penetration, the extraction of substances and the rate of protein denaturation.

2.4   REFERENCES

Barnard, T. (1976). An empirical relationship for the formulation of glutaraldehyde based fixatives based on measurements of cell volume changes. *J. Ultrastruct. Res.* <u>54</u>, 478.

Bowes, J.A. and Cater, C.W. (1968). The interaction of aldehydes with collagen. *Biochem. biophys. Acta* <u>168</u>, 341-352.

Bowes, J.A. and Cater, C.W. (1965). The reaction of glutaraldehyde with proteins and other biological materials. *J. R. Microsc Soc.* <u>85</u>, 193-200.

Coetzee, J. and Van der Merwe, C.F. (1985a). The influence of processing protocol on the structure of bean leaf cells. *S. Afr. J. Bot.* (in press).

Coetzee, J. and Van der Merwe, C.F. (1985b). Effect of glutaraldehyde on the osmolarity of the buffer vehicle. *J. Microsc.* 138, 99-106.

Coetzee, J. and Van der Merwe, C.F. (1985c). Penetration rate of glutaraldehyde in various buffers into plant tissue and gelatin gels. J. Microsc. 137, 129–136.

Coetzee, J. and Van der Merwe, C.F. (1984a). Extraction of substances during glutaraldehyde fixation of plant cells. J. Microsc. 135, 147–158.

Coetzee, J. and Van der Merwe, C.F. (1984b). Penetration of glutaraldehyde, formaldehyde and acrolein into gelatin gels and their crosslinking of albumin. Proc. Electron Microsc. Soc. South. Afr. 14, 11–12.

Fineran, B.A. (1971). Effect of various factors of fixation on the ultrastructural preservation of vacuoles in root tips. Cellule 68, 269–286.

Good, N.E., Winget, G.D., Winter, W., Connolly, T.N., Izawa, S. and Singh, R.M.M. (1966). Hydrogen ion buffers for biological research. Biochemistry 5, 467–477.

Goodchild, D.J. and Craig, S. (1982). Structural aspects of protein accumulation in developing pea cotyledons. IV. Effects of preparative procedures on ultrastructural integrity. Aust. J. Plant Physiol. 9, 689–704.

Hayat, M.A. (1981). Fixation for electron microscopy. Academic Press, New York.

Holt, S.I. and Hicks, R.M. (1961). Studies on formalin fixation for electron microscopy and cytochemical staining purposes. J. Biophys. Biochem. Cytol. 11, 31–45.

Luft, J.H. and Wood, R.L. (1963). The extraction of tissue protein during and after fixation with OsO_4 in various buffer systems. J. Cell Biol. 19, 46a.

Maser, M.D., Powell, T.E. and Philpott, C.W. (1967). Relationships among pH, osmolarity and concentration of fixative solutions. Stain Technol. 42, 175–182.

Mathieu, O., Claassen, H. and Weibel, E.R. (1978). Differential effect of glutaraldehyde and buffer osmolarity on cell dimensions: a study on lung tissue. J. Ultrastruct. Res. 63, 20–34.

Medawar, P.B. (1941). The rate of penetration of fixatives. J. R. Microsc. Soc. 61, 46–57.

Mersey, B. and McCully, M.E. (1978). Monitoring of the course of fixation of plant cells. *J. Microsc.* 114, 49-76.

O'Brien, T.P. and McCully, M.E. (1981). *The study of plant structure: principles and selected methods.* Termacarphy Pty. Ltd., Melbourne.

Salema, R. and Brandao, I. (1973). The use of PIPES buffer in the fixation of plant cells for electron microscopy. *J. Submicrosc. Cytol.* 5, 79-96.

Salisbury, F.B. and Ross, C.W. (1978). *Plant Physiology*. Wadsworth, Belmont, California.

Spurr, A.R. (1969). A low-viscosity epoxy resin embedding medium for electron microscopy. *J. Ultrastruct. Res.* 26, 31-43.

Young, J.Z. (1935). Osmotic pressure of fixing solutions. *Nature* 135, 823-824.

3

The use of low temperature methods for structural and analytical studies of plant transport processes

Anthony W. Robards

3.1 INTRODUCTION

The last few years have seen a dramatic surge of interest in the use of low temperature methods in biological electron microscopy. The reasons for this are now well-established and will not be repeated here; nor will undue attention be given to technical details of methods which have also been fully reviewed elsewhere (e.g. Robards and Sleytr 1985). The purpose of this paper is rather to point to some of the recent applications and contributions of low temperature techniques to botanical ultrastructural and analytical research and to highlight present advantages and limitations. This will be done against a background of work carried out in my own laboratory which relates particularly to ultrastructural aspects of transport processes in plants.

Plants can pose special problems when low temperature methods are used. For example, thick cell walls hinder rapid freezing and frozen cells can be extremely difficult to process through subsequent stages of cryosectioning, freeze-substitution or freeze-fracturing. Nevertheless, the goals are so attractive, and some results have already been so encouraging, that further work must surely provide more of the rewards that the techniques have to offer.

3.2 FREEZING

The freezing process is crucial to the success of any low temperature method in microscopy. Substantial progress has been made in both the theoretical understanding and practical achievement of rapid freezing in isolated cells, monolayers and very thin tissue slices. Unfortunately, however,

substantially large (>0.1 mm in all dimensions) plant specimens
still pose formidable problems.   This article concerns itself
mainly with the application of low temperature techniques to
tissues rather than to isolated cells and monolayers.   The
depth of 'good' freezing in uncryoprotected specimens is rarely
greater than 15-20 μm which means that only the outermost cell
layer or so is likely to be ideally preserved.   This is so
whatever method of cooling is used, whether rapid plunging,
propane jet cooling, or cold block freezing (Elder *et al.*
1982).   The situation is much more favourable for very small
specimens, especially if in suspension, when spray freezing or
'sandwich freezing' between propane jets can be used.   For
these reasons, it has commonly been the practice to cryoprotect
plant tissues, usually by infiltration with glycerol.   This is
discussed further below.

     The question may be asked as to which is the best cooling
method for tissue pieces.   Plunging into a liquid cryogen,
spraying with a jet of cold propane and freezing on the surface
of a cold metal block have all been used.   Unless there are
special reasons to the contrary, the simplicity and low cost of
plunging methods recommend them for use in most circumstances.
However, it is most important that the conditions of freezing
are optimised.   This subject has been reviewed fully elsewhere
(e.g. Elder *et al.* 1982; Costello 1980; Robards and Crosby
1983; Robards and Sleytr 1985) and will not be explored further
here except to note that propane or ethane are currently the
preferred cryogens; they should be used at their lowest
possible liquid temperature; and the specimen should be
inserted rapidly (>1.0 m s^{-1}).   Despite using such a method,
the bulk of the tissue will still contain large ice crystals.
This may not matter too much (e.g. freeze-fracture study of the
plasmalemma - see below); or it may totally preclude obtaining
the desired information (e.g. freeze-substitution studies of
cell ultrastructure).   It is probably always best at least to
attempt some studies on uncryoprotected material even if,
ultimately, cryoprotectants have to be used.

3.3 CRYOPROTECTION

It is important in any study, and no less in low temperature
techniques, to consider carefully what information is required
from the specimen prior to finalising the experimental methods.
For example, many freeze-fracture studies on plant cells are
concerned with the nature of the plasmalemma.   The effects of
glutaraldehyde and glycerol on intramembrane particle (IMP)
appearance and distribution are now very well documented (see
reviews by Sleytr and Robards 1982 and Willison and Brown 1979;
also Arancia *et al.* 1979; Fineran 1972, 1978; Hasty and Hay

1977; Hay and Hasty 1979; Niedermeyer and Moor 1976; Parish
1975) and,   therefore,   these chemicals are best avoided;
indeed,   even the rate of cooling has been shown to influence
IMP distribution (Lefcrt-Tran et al. 1978). We have noticed
that the cytoplasmic matrix immediately adjacent to the
plasmalemma sometimes has ice crystals that are smaller than
those deeper within the cell. This may be attributable to the
the presence of hydrophilic polymers, in the adjacent cell wall,
that behave in a similar fashion to non-penetrating
cryoprotectants such as hydroxyethyl starch or
polyvinylpyrrolidone. At all events, it means that it is
often perfectly possible to freeze uncryoprotected tissues for
freeze-fracture studies of the plasmalemma without undue
artefact production.

     It may be be decided that, for whatever reason, there is
no alternative but to cryoprotect tissues prior to freezing.
This, in turn, produces further problems because it may be
necessary to fix cells (e.g. with glutaraldehyde) tc make them
permeable to the glycerol. This should create substantial
apprehension in the mind of the investigator because the
catalogue of artefacts arising from fixation and/or glycerol
treatment is substantial and growing (e.g. Sleytr and Robards
1982 and other references cited above). For some
investigations this might not matter while for others it can be
of paramount importance. For example, although of different
dimensions, both phloem sieve plate pores and plasmodesmata can
be considered to be tubes through which solutions can move at
high velocity. There is ample evidence to believe that phloem
structure can transform extremely rapidly as a result of damage
during sampling and/or fixation and it has been calculated
(Johnson 1978) that even the most rapid fixation may not be
sufficiently fast to immobilise the contents of sieve elements
precisely in their in vivo positions. Similar reasoning can be
applied to plasmodesmata where, unless they are considered to
be totally inert tubes, the artifacts associated with chemical
fixation and cryoprotection (e.g. Hughes and Gunning 1980) must
necessarily limit the ability to draw valid conclusions about
their ultrastructure when chemical treatments are used either
prior to conventional thin-sectioning or before rapid freezing.

     A further example of the problem of rapidly fixing dynamic
intracellular processes has been provided by Robards (1984) in
relation to the nectary trichomes of Abutilon. In this gland
the prenectar passes into the hairs via a modified epidermal
cell, the basal cell; the next most distal cell is referred to
as the stalk cell. Secretion occurs from the cell at the tip of
the trichome. (See Gunning and Hughes, 1976 and Robards 1984
for more details on the biological background to this secretory
system.) It was calculated that it would take approximately 10

minutes for glutaraldehyde to penetrate to the tip cells whereas
rapid freezing could immobilise the whole tip cell within 1.0
millisecond. This is in a system where the volume flow is
sufficient to fill and empty the cells more than once every
minute and fusion of cytoplasmic vesicles or cisternae with cell
membranes is probably occuring at millisecond intervals. There
must be concern that structure revealed *after* chemical fixation
and cryoprotection may be quite different from that portrayed in
cells that have been frozen from the native state.

It is, of course, usually the case that experimental
techniques should be tailored to the material under study
although sometimes the obduracy of the scientific mind seems to
want to combine the most intractable specimens with the most
esoteric methods. Wherever possible it seems sensible to
attempt to improve the chances of success by chosing amenable
(model) systems for investigation. Nowhere is this more true
than in the application of rapid freezing methods. In
selecting the *Abutilon* gland for investigation, the first
criterion was necessarily that the physiological processes were
characteristic of the phenomena that we wished to study.
However, we were also influenced by the fact that the long,
relatively slender, exposed trichomes should freeze very
rapidly and that a high intracellular sugar concentration (0.4
to 0.6M) should contribute at least some natural cryoprotectant
activity.

Our own work now uses almost exclusively unfixed and
uncryoprotected material, preferring to put up with the
difficulties of relatively poor freezing within tissues rather
than the insidious artefacts induced by chemical treatment.

The major classes of cryoprotectants used are the
penetrating (e.g. glycerol) and non-penetrating (e.g.
polyvinylpyrrolidone). In general, plant cells are relatively
impermeable to glycerol (Richter 1968a,b; Rottenberg and
Richter 1969) and, therefore, must be fixed before treatment.
Classically attractive images can be obtained from freeze-
fracture replicas so prepared but the possibility of artefacts,
especially relating to rapid processes, is ever present. The
non-penetrating cryoprotectants (Echlin *et al*. 1977; Franks *et
al*. 1977; Skaer *et al*. 1977, 1978; Skaer 1982) have not, in
our hands proved particularly useful for roots or other highly
vacuolated cells (see Wilson and Robards 1980, 1981). This is
a pity, because, in principle, the virtues of these polymers
offer a means of avoiding many of the difficulties referred to
above.

The best advice to be offered is that different freezing
methods, both with and without cryo-protectants, should be

used when setting out to study a new experimental system.   The
results can then be compared and the investigator can decide
which method is best in relation to its technical complexity.
For example, in a study of the distribution of IMPs in freeze-
fracture replicas of root cell membranes, both treated and
untreated material was used (Robards et al. 1980, 1981).   It can
be seen (Table 3.1) that, while glycerol treatment had little
effect on the measured parameters, PVP-treated tissue gave

Table 3.1

Particle frequencies on plasmalemma fracture faces from
endodermal and cortical cells of Zea mays

Fracture face	Cell type	Cryoprotection treatment		
		Untreated	PVP	Glycerol
		(>8.0nm particles $\mu m^{-2} \pm$ S.E.)		
P-face	Endodermis	2500 ± 110	2985 ± 145	2305 ± 200
	Cortex	1140 ± 65	1690 ± 142	1105 ± 130
E-face	Endodermis	1295 ± 25	1450 ± 30	1240 ± 95
	Cortex	350 ± 20	655 ± 85	415 ± 10

(Taken from Robards et al. 1980.)

slightly higher IMP frequencies which could be attributed to the
shrinkage of cells, and hence possible reduction in membrane
area, caused by this cryoprotectant.   Therefore, although it
could be concluded that fixation and cryoprotection were not
contra-indicated for this work, chemically untreated material
was subsequently used for the rest of the study in the belief
that the results would be more reliable.

3.4 LOW TEMPERATURE SCANNING ELECTRON MICROSCOPY

Once a specimen has been frozen, one of the simplest things
that can be done with it is to observe it, while still frozen,
on the cold stage of a low temperature scanning electron
microscope (LTSEM).   Early work in low temperature scanning
electron microscopy of plant specimens was pioneered by Echlin
(see reviews in Echlin 1978a,b). Apparatus is now commercially
available which makes the process extremely easy. (The
technical aspects of this technique have been fully reviewed in

*Robards and Sleytr, 1985. The first commercially available
system in Europe derived from the work of Robards and Crosby
1979 which led to production of the Emscope SP2000. A recent
detailed review of applications of this method is provided by
Beckett and Read [in press] and an atlas of micrographs of
diverse specimens viewed in an SEM at low temperature has been
published by Wilson and Robards 1984.)*

In brief, the specimen is frozen, possibly fractured
and/or etched, coated with a conducting film and then observed
at low temperature (<-160°C). Because resolution in the SEM is
unlikely to be very high (compared with TEM), the cooling rate
for the specimen (and consequent ice crystal size) is not too
critical and, therefore, freezing in subcooled liquid nitrogen
is the most usual method. It should be borne in mind that the
actual surface revealed by fracturing a frozen specimen is
identical to that used for freeze-fracture replication, the
only differences being the nature of the finally observed
specimen and the resolution of the subsequent viewing methods.

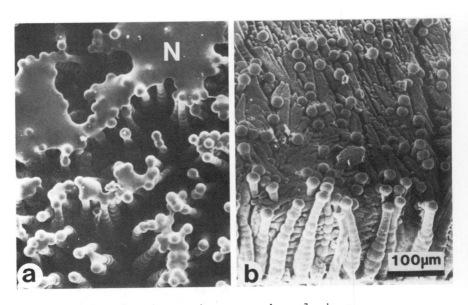

Fig. 3.1. Low temperature scanning electron
micrographs of *Abutilon* nectary hairs.
a) Without pretreatment, showing how the nectar (N)
forms a layer across the tops of the hydrophobic hairs;
b) after treatment with 0.6M sucrose containing a trace
of Triton X-100 which facilitates 'wetting' of the hairs
right down to their bases.

3.4.1 Structural studies

LTSEM can be an extremely quick and useful means of assessing
methods before applying them to the more demanding processes of
freeze-fracture replication. For example, in our work on
Abutilon nectaries we have wanted to prepare trichomes so that
cell ultrastructure can be studied by both freeze-fracture
replication and freeze-substitution. In such difficult tissue
it is often the case that to obtain reasonable amounts of
replica is so time-consuming and demanding that progress is
either very slow or the work is abandoned. LTSEM provides the
means for observing the fracturing behaviour of the material
and, therefore, allows the more rapid development of
preparation/freezing/fracturing techniques that will give a much
higher probability of success when applied to the replication
method. The trichomes of *Abutilon* are covered with a
hydrophobic cuticle. This means that when nectar is secreted
(or when aqueous solutions are applied) the liquid tends to
remain as a superficial layer across the tips of the hairs, the
more basal parts being surrounded by air (Fig. 3.1). In
consequence, when longitudinal fractures are made, the replica
falls to pieces because it is discontinuous. The solution to

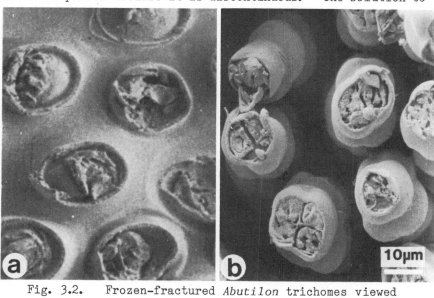

Fig. 3.2. Frozen-fractured *Abutilon* trichomes viewed
by low temperature scanning electron microscopy.
a) Shows how intact hairs can be broken off at the
level of the basal cells (see also Fig. 3.6);
b) illustrates cross-fractures through the mid/apical
region of the hairs and also shows how deep etching can
help to reveal structure within the individual hairs
cells.

this problem was to apply a 0.6M solution of sucrose
(approximate concentration of the nectar), containing a little
Triton X-100, to the hairs. This immediately penetrated
between the hairs, supporting them and allowing good replicas
to be recovered. A similar approach is applicable to other
plant tissues, such as leaves and aerenchymatous roots
(although possible deleterious effects of Triton X-100 should
obviously be taken into account). While attempting to improve
freeze-substitution methods for *Abutilon*, it has been found
helpful to break-off the hairs at their bases. Once again,
LTSEM provides a simple and quick means of seeing exactly where
the fractures occur (Fig. 3.2).

LTSEM has also proved most useful in other botanical
investigations such as the determination of the distribution of
aerenchyma and the presence of water droplets within
aerenchymatous cavities, structural preservation of very
delicate specimens, and retention of wax/lipid components such
as occur on many plant surfaces (Beckett and Read [in press];
Wilson and Robards 1984).

Critical point drying (CPD) has become the method of choice
for most biological SEM preparation schedules but is well known

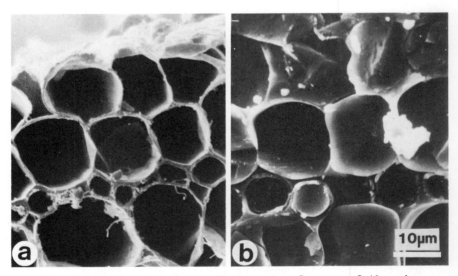

Fig. 3.3. A comparison of the outer layers of the rice
(*Oryza sativa*) root prepared by:
a) critical point drying; and b) low temperature
scanning electron microscopy. Comparison of
measurements (Table 3.2) shows that the critical point
dried specimen is significantly shrunken.

Table 3.2

A comparison of the dimensions of critical point dried
and frozen hydrated peripheral cells of rice (*Oryza sativa*)
as measured from scanning electron micrographs

Cell layer	Dimension	Mean Measurement (µm)		Shrinkage by CPD	
		CPD	LTSEM	µm	%
Epidermis	Tangential	13.9	15.9	2.0	12.6
	Radial	16.1	20.1	4.0	19.7
Hypodermis	Tangential	13.3	15.8	2.5	15.8
	Radial	15.2	18.5	3.3	17.8
Fibrous layer	Tangential	6.2	8.4	2.2	26.2
	Radial	4.6	6.3	1.7	27.0
Cortical first layer	Tangential	17.0	20.9	3.9	18.7
	Radial	17.7	23.9	6.2	25.9

to produce considerable shrinkage (Boyde and Maconnachie 1979).
This can be a serious impediment to quantitative studies and
LTSEM provides a fast, simple and accurate alternative
procedure (Fig. 3.3). Organisms such as fungi, which are
notoriously difficult to prepare by conventional methods, are
also relatively easy to maintain in a life-like state when
frozen and viewed by LTSEM (Beckett and Read, in press).
Similarly, freeze-drying (FD) also has its limitations when
applied to biological tissues. In particular, largely through
inadequate apparatus, freeze-drying can be most inconsistent
and, as with CPD, can lead to distortion of cells (Boyde and
Franc 1981; Campbell and Roach 1983). The necessary use of
solvents prior to CPD inevitably leads to lipid extraction
which, in some circumstances, can be severe (Robards 1978).
Thus, for work concerned with extracellular waxes, freeze-
drying or LTSEM are essential and the latter, in addition,
provide almost perfect structural retention.

3.4.2   X-ray microanalysis

While LTSEM has proved to be a most valuable tool for plant
ultrastructural observations, a major potential contribution
arises from its use for microanalysis. Numerous preparatory
methods for microanalysis of plant cells have been described and
used but none is so simple, so quick, and less likely to

produce artefacts as the analysis of frozen hydrated specimens.
It is not appropriate here to discuss the pros and cons of
frozen hydrated thin sections in comparison with frozen bulk
material; the different approaches have been well discussed
elsewhere (e.g. Echlin and Saubermann 1977; Gupta and Hall
1984; Marshall 1980a; Saubermann and Echlin 1975; Scmlyo and
Silcox 1979). Suffice it to say that very little work has been
done on the cryoultramicrotomy of plant material and even fewer
results have accrued. The techniques are difficult, even for
favourable specimens. Partly for this reason, botanists have
tended to look towards LTSEM for microanalysis (e.g. Echlin *et
al*. 1982), even though there are substantial problems
concerning the quantification of data. However, qualitative
and semi-quantitative data can be obtained with an ease that is
not apparent in other methods. Furthermore, most of the
problems concerning soluble compound diffusibility are removed.

In the *Abutilon* hairs, a matter of some significance is
the distribution of potassium, the level of which is relatively
high in the phloem compared with the nectar from which it is
virtually absent. It is necessary, therefore, to determine
the relative levels of potassium in cells between the phloem and
the tip of the trichomes. To this end, frozen-hydrated
specimens have been coated with aluminium and then analysed
using a low temperature system attached to JEOL JSM50 SEM.   To
make consecutive and comparable analyses along a single hair
(or, indeed, any piece of tissue) it is crucially important
that the microscope operational conditions are maintained
constant and also that the take-off angle is essentially
identical for successive analyses. This was established from
stereo-pair micrographs which were recorded on Polaroid film
from all specimens; only when it was shown that the whole
fracture along the length of a hair would present the same take-
off angle to the X-ray detector was microanalysis undertaken.
In practice, this inevitably meant discarding many, otherwise
well-prepared, specimens.

The potassium counts from different positions along the
trichomes did not vary greatly in any predictable manner (Table
3.3). In some individual trichomes there was a significant
difference ($> x2$) in the potassium counts from the basal cell
compared with relatively fewer counts from the stalk or hair
cells. However, more analyses are required before these
results can be confirmed. Nevertheless, such data do suggest
that the potassium is being withheld from the nectar at a
secretory stage taking place from all trichome cells rather than
at some specific site within a cell or group of cells from which
a discrete symplastic compartment becomes loaded. A
consistent finding was an increase in the level of chlorine
(chloride) apically from the basal cell. The significance of

this is unclear but the important point here is that it is
unlikely that such a clear pattern would have been revealed in
specimens prepared other than for LTSEM.

Table 3.3

X-ray data from frozen-fractured *Abutilon* nectary trichome cells
(Flower at the point of secretion)

	Basal cell[*] (9)	Stalk cell (4)	SC+1 (4)	SC+2 (2)	SC+3 (3)
Phosphorus	7.8 (1.0)	6.3 (0.8)	3.3 (0.4)	4.3 (0.6)	7.2 (0.9)
Chlorine	0.7 (1.0)	2.3 (3.3)	4.7 (6.7)	5.7 (8.^)	10.0 (14.2)
Potassium	12.5 (1.0)	12.5 (1.0)	8.6 (0.7)	11.7 (0.9)	18.7 (1.5)

Data (mean counts) expressed as a percentage of total net counts
for aluminium, phosphorus, chlorine and potassium (figures in
brackets giving ratio against basal cell counts).
SC = Stalk Cell; [*] = number of determinations.
(This work was carried out in collaboration with Dr. K. Oates,
University of Lancaster.)

3.5. FREEZE-SUBSTITUTION

Once a plant specimen has been frozen, the possibilities for
observing or analysing it in a transmission electron microscope
reside in subsequent processing by cryosectioning, freeze-
drying and embedding (not much used by botanists and with few
apparent advantages), freeze-etching (structural studies only),
or freeze-substitution.

The potential advantages of freeze-substitution as a
preparative method have long been appreciated although, again,
the intractability of many plant tissues has frequently
frustrated its used. However, recent work has provided some
strikingly good results (e.g. see Chapter 10 in this Volume) and
we can hope that improved methods, resulting from a better
understanding of the processes involved, will lead to greater
utilisation of the technique.

Most freeze-substitution work has used acetone as the
substituting medium but the results of Humbel and colleagues
(Humbel *et al.* 1983; Humbel and Müller 1984) strongly suggest

that methanol is generally better for structural studies.   It
can dissolve ice more rapidly and at lower temperatures than
acetone and it also continues to act as a solvent even in the
presence of a significant amount of water.   We have produced
acceptable results (Fig. 3.4) using the method recommended by
Humbel *et al.* which use a 'cocktail' of glutaraldehyde, uranyl
acetate and osmium tetroxide dissolved in methanol.   Other
workers have also achieved successful results using either
acetone (e.g. Chapter 10 in this Volume; Dempsey and Bullivant
1976) or, particularly for microanalytical studies, diethyl
ether (Harvey 1980, 1982) has been used.   Freeze-substitution
has also been employed successfully prior to microanalysis of
animal tissues (e.g. Marshall 1980b; Ornberg and Reese 1981).

        Whichever substitution method is used, there is a number
of common difficulties that must be overcome.   The first of
these is that penetration of the substituting medium into plant
tissues at low temperatures is *extremely* slow.   Rates of
substitution have not been measured or calculated for complex
tissues but there is good reason to believe that the major part
of many freeze-substituted specimens has not had the ice
crystals dissolved until warming up to a much higher temperature
than the initial substituting temperature.   As ice crystal
growth is strongly temperature dependent, this leads to the
appearance of very large ice crystals within cells that may well
have been quite well-frozen initially.   Because of these
difficulties, it is critically important to take all possible
steps to make the area of interest as accessible to the
substituting medium as possible.   This is usually quite
straightforward because, as the specimen is deep-frozen, it
can be fractured open under liquid nitrogen before being
transferred into the substituting fluid.

        In our studies on the *Abutilon* hairs, initial freeze-
substitution results were unacceptable.   In view of the
impermeable cuticle overlying all the hairs this was not
surprising.   We therefore adopted the method of breaking off
hairs from the frozen glands by gently striking them with a cold
metal rod while under liquid nitrogen.   This proved to be
highly successful and, in fact, the sites of the fractures
could easily be checked by LTSEM (Fig. 3.2).   The result was a
mixture of isolated hairs, some intact and some broken at
different locations, together with pieces of sepal with exposed
basal cells.   It was found that the cytoplasm of the fractured
cells was extremely well preserved with almost no signs of ice
crystals whatsoever, so confirming that the initial freezing
had been satisfactory.   Interestingly, structural preservation
only two or three cells distant from the fractured surface was
usually poor and ice crystal damage was considerable.   It is
highly improbable that freezing rates were consistently

different over such small distances and, therefore, the
conclusion is that the very small ice crystals in the accessible
cells were fully dissolved at low temperature (-75°C) while
those in more distant cells were not dissolved at low
temperature and grew as the temperature rose until, ultimately,
the substituting medium eventually penetrated to them.
Provided that pieces of specimen that have been broken under
liquid nitrogen can subsequently be recognised, there is little
limitation to the further use of this method.   As an additional
aid to good substitution, it is useful to agitate or rotate the
specimen vial during the substitution process.   Once the tissue
has been fully freeze-susbstituted it is ready for embedding.
A number of authors, including ourselves, have been impressed
with the potential advantages of low temperature embedding using
resins such as the Lowicryl media (Armbruster *et al.* 1982;
Carlemalm 1984; Carlemalm *et al.* 1982)   However, results on
tissues so far, especially from plants, have been extremely
disappointing.   The problem seems largely to be one of

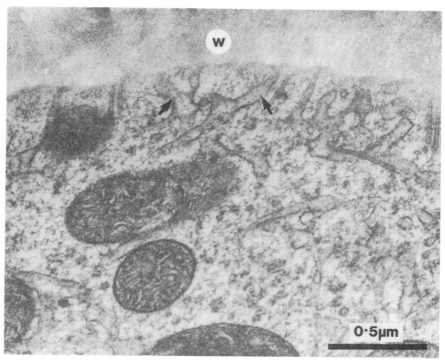

Fig. 3.4. Micrograph of a freeze-substituted *Abutilon*
hair cell from a nectary just prior to the onset of
secretion.   There is little evidence of ice crystal
damage and structural preservation is good. W - Cell
wall; arrows - Secretory reticulum.

inadequate infiltration at low temperature.   If structural
studies only are to be made, then infiltration at around 0°C or
room temperature in a low viscosity resin such as the Spurr
(1969) formulation can be quite satisfactory.   Again, however,
there are some pitfalls.   Because freeze-substitution is an
effective fixation method, many cell membranes retain semi-
permeable properties.   This is little appreciated but the
results of over-rapid exposure to high concentrations of resin
monomer in solvent can lead to dramatic and disastrous withdrawal
of solvent from the specimen (see also McCully and Canny 1985).
This can result in specimens that appear badly fixed or
'plasmolysed'.   With specimens such as the isolated nectary
hairs, it is easy to make experimental observations during
resin infiltration and to see that an initially well preserved
hair can, within a few seconds, be completely spoilt (Fig.
3.5).   The solution is to present the resin, at least
initially, in *very* gently increasing concentration steps so
that the tissue has time to equilibrate without a massive efflux
of solvent.   Alternatively, the solvent in which the resin is
dissolved can be gradually evaporated away by passing a stream
of dry nitrogen gas over the vial in which it is contained.

3.6   FREEZE-FRACTURE REPLICATION

Although freeze-fracturing methods have now been available for
more than twenty years, their contribution towards botanical
knowledge has been spasmodic and limited.   To a considerable
extent this is due to the difficulties of retaining sufficiently
large areas of freeze-fracture replica so that correct cell
identification can be made and so that proper statistical
methods can be applied.

     In some instances (e.g. freeze-fracture of tissues) it can
be extremely difficult (although not, in our experience,
usually impossible) to obtain replicas from chemically untreated
material.   Nevertheless, the very fact that replica retention
is made easier after fixation should give cause for concern that
chemical changes have taken place.   Some workers have used
heavily fixed and cryoprotected specimens simply because they
were unable to obtain replicas from untreated material.
Nevertheless, such a justification has to be viewed critically.
Numerous methods have been described in the literature for
improving the retention of replicas and perseverance here
usually justifies the extra work.   The major problem is
breakage of the replica due to swelling and shrinkage of the
specimen in the cleaning solutions.   Two general approaches
recommend themselves: i) to make a stronger replica so that it
can withstand the mechanical effects; or, ii) to attempt to
minimise dimensional changes to the tissue as it is digested.

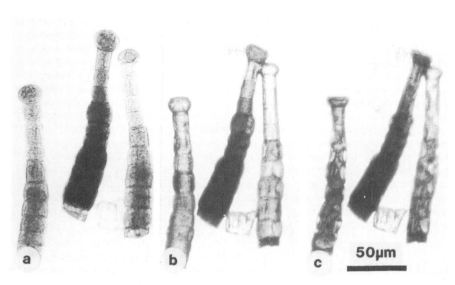

Fig. 3.5.   Micrographs of isolated freeze–substituted
Abutilon nectary hair cells taken at approximately 10
second intervals while exposed to relatively
concentrated resin monomer in acetone.   The initial
appearance (a) is of well–preserved hairs that appear
very similar to the living state.   However, as soon as
the hairs are exposed to the resin (having been
initially in 100% acetone), severe solvent withdrawal
results in rapid distortion and collapse.   Were it not
for the fact that this is so easily seen by light
microscopy, the final ultrastructural appearance could
easily be misinterpreted as damage that had occurred at
some other stage of processing.

In the first category, methods have been described for
strengthening the carbon/platinum replica with, for example,
collodion, naphthalene or vacuum deposited silver (e.g. Bordi
1979; Robards and Umrath 1978; Stolinski *et al.* 1983).   The
most successful method of reducing specimen distortion has, in
our hands, been that of thawing the frozen specimen, with the
replica intact, in cold methanol (see also De Mazière *et al.*
1985).   The technique is to remove the specimen/replica
assembly from the freeze–etching apparatus and *immediately* to
place it under liquid nitrogen onto the surface of frozen
methanol contained in a small plastic vial.   The vial is then
allowed to warm–up slowly so that the nitrogen evaporates and
then the methanol melts, so allowing the (still frozen)
specimen to sink into it.   A substitution/fixation process

occurs, rendering the membranes fully permeable so that, when chemicals such as hypochlorite, sulphuric acid or chromic acid are applied, far less distortion takes place and it is then possible to obtain replicas from otherwise 'impossible' specimens.

Another impediment to the successful collection of large intact replicas from plant tissues can be the presence of air spaces, as in roots or between adjacent superficial structures such as hairs on the surface of plant organs. Our first attempts to freeze-fracture *Abutilon* nectary trichomes were unsuccessful because the replica was not continuous across the air gap between the adjacent hairs. Once again, LTSEM was used in developing a solution of this problem. Observation of secreting trichomes shows that the viscous nectar collects and coalesces over the tips of the hairs but, no doubt due to the hydrophobic cuticle, does not penetrate between them (Fig. 3.1a.). Application of a droplet of 0.6M sucrose containing a trace of Triton X-100 served to 'wet' the hairs over their whole surface (Fig. 3.1b.), so eliminating air cavities and providing a completely solid block for freeze-fracturing and from which intact replicas could be recovered (Fig. 3.6).

Once having obtained a clean replica within which the individual cells can be recognised, interpretation and analysis can proceed. The vast majority of observations from freeze-fracture replicas are made on membrane fracture faces. Features such as intramembrane particle (IMP) frequency, distribution and appearance are recorded. This can pose substantial problems of sampling and statistical analysis. Work in our own laboratory, which has attempted to define the different sources of variation, has shown that the major

----

Fig. 3.6. Freeze-fractured views of *Abutilon* nectary hairs. Fractured hairs viewed by low temperature scanning electron microscopy (a) can be used to assess possible fracturing behaviour preparatory to the more demanding methods of freeze-fracture replication. Moreover, such fractured cells provide suitable specimens for X-ray microanalysis (although, of course, coatings other than gold are then used). A similarly fractured hair is illustrated in (b), but prepared by freeze-fracture replication and showing the basal and stalk cells, together with part of the intervening transverse wall which is shown at high magnification in (c). Using a combination of such low temperature methods it is thus possible to accumulate a range of structural and analytical information from cells that can be considered close to their *in vivo* condition.

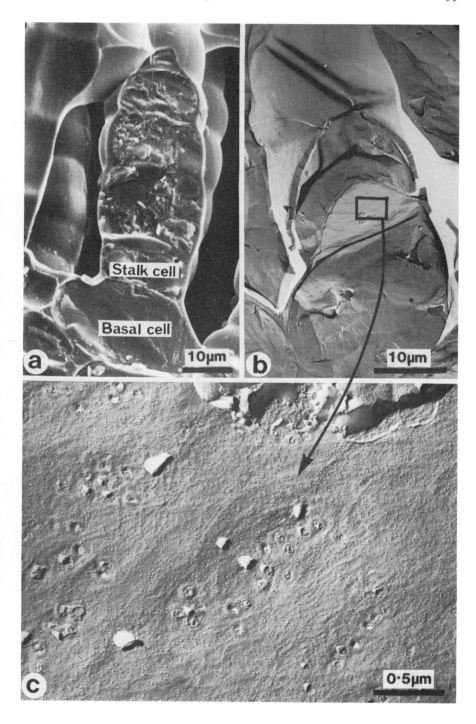

variation arises from differences between individual replicas.
At all events, any comparisons to be made between experimental
and control samples need the most rigorous examination and
quantitative analysis. If this is not done, then it is all
too easy to acquire results that, while having the trappings of
statistical validity, are in fact meaningless.

A number of authors have reported attempts to quantify IMPs
on membrane fracture faces (see, e.g., Duniec et al. 1982;
Robards et al. 1981; Van Winkle and Etman 1981; Weinstein et
al. 1979; Niedermeyer and Wilke 1982). There are significant
problems, not only in the application of statistical methods to
determine the frequency and distribution pattern of IMPs, but
also in relation to the topography of the replica. For
instance, particle frequencies will clearly be inaccurate if
determined from a fracture face that is not normal to the
electron beam. Stereo-pair electron micrographs should be used
to evaluate such geometric features (Steere et al. 1974) so that
surfaces can be 'normalised' to the electron beam before
carrying out quantitative analysis. Weibel et al. (1976) have
also approached the problem of obtaining accurate membrane
surface areas from freeze-fracture micrographs by the use of
stereological methods.

An increasing number of publications has been concerned
with the *pattern* of IMP distribution in membranes. This
relates to interest in determining whether different
experimental treatments (for example, chilling or desiccation
stress) produce IMP-free patches, so reflecting possible phase
transitions in membrane lipids. Some clear results have been
shown from microorganisms (e.g. Furtado et al. 1979; Martin et
al. 1976; Ono and Murata 1982; Tsien and Higgins 1974) but the
situation is far less well resolved in higher plants.
Experiments using chilling temperatures on *Zea mays* and *Hordeum
vulgare* revealed no significant changes in the coefficient of
particle dispersion (Robards and Clarkson 1984; Table 3.4).
Other authors have reported particle-free areas on membranes
(e.g. Toivio-Kinnucan et al. 1981). However, to be sure that
the observed distributions are real and not artefactual and,
still more, to make a proper comparison of results from
different laboratories, requires that experimental and
analytical conditions are very carefully defined and controlled.
Taking up the point made earlier, chemical pretreatments should
certainly be avoided and even the *rate* of cooling should at
least be taken into account in interpreting micrographs (some
estimate of this can be deduced from ice crystal size in the
adjacent cells). Once a replica has been produced, the
quantitative analyses must be applied carefully (see papers
cited above). Furthermore, there should be sufficient
different samples, sufficient different replicas and sufficient

Table 3.4

Particles on marrow root cortical cell membranes
after chilling at 4°C for 2 hours

Temperature and face		Particle diameter (nm)	Frequency (particles μm^{-2})	Coefficient of dispersion
18°C	P	9.6±0.08	1568±84	1.19±0.01
	E	10.4±0.06	327±30	1.06±0.02
4°C	P	9.7±0.06	1545±82	1.21±0.01
	E	9.7±0.07	565±54	1.13±0.02

Means ± standard error; fracture plane approximately 3.0 mm
from root apex.
(Taken from Robards and Clarkson 1984.)

different micrographs to minimise bias arising from inadequate
sampling. Together, these requirements lead to a substantial
amount of work but this is inevitable if significant results are
to be achieved. It is all too easy to be influenced by looking
at a couple of electron micrographs that appear to illustrate
some clear change arising from different treatments; further
analysis all too often shows such variation not to be
significant. For example, as part of a large programme to
investigate the pattern of IMP frequency and distribution in the
membranes of successive cell layers across plant roots, we
recorded and examined many hundreds of micrographs. It was
relatively straightforward to demonstrate that the *mean* IMP
frequency on the endodermal P-faces was significantly higher
than the frequency on the cortical cell P-faces. However, an
attempt to make a more detailed study of variation across the
membranes of individual cells soon demonstrated that the
variation from different areas within the membrane of a single
cell could be as large as that from different cells. This
accentuates the inescapable requirement that sufficient samples
are taken and properly analysed.

3.7 SUMMARY

This paper has attempted to outline some of the low temperature
methods that can now be successfully applied to plant cells and
tissues. Inevitably, it has been impossible to cover all
techniques; it also very much reflects a personal selection of
examples. As in the field of microscopy in general, low
temperature techniques should not be seen as 'competing' with
each other; rather, they can provide complementary information.

Much of the original impetus for using frozen specimens as the starting point for microscopical observation and/or analysis came for the potential avoidance of chemical treatment. Despite this, there is a worrying trend towards the relatively indiscriminate use of chemical pretreatments without sufficient consideration of whether this has any adverse effects on the results obtained. Similarly, the relative difficulty of obtaining freeze-fracture replicas containing large areas of membrane fracture faces sometimes leads to inadequate quantitative assessments of experimental results. As low temperature methods become more widely available and used, it is going to be increasingly important that the whole artefact problem - both in preparation and in interpretation - is constantly taken into consideration. If this is done, then low temperature methods will continue to have a vital role to play in botanical microscopy.

3.8 ACKNOWLEDGEMENTS

In preparing this paper I have really only been the spokesman for other members of my laboratory over a number of years. In particular, I should like to thank Peter Crosby, Meg Stark and Ashley Wilson who have all helped, in various ways, with the work presented here. I am greatly indebted to Dr. Ken Oates (Department of Biological Sciences, University of Lancaster) whose expertise and help made the X-ray microanalytical work reported here possible. My grateful thanks also go to Dr. Eva Kronestedt for help with the manuscript and to Deborah Green for wrestling with many of the problems of camera-ready copy. Much of this work was carried out under grants awarded by the AFRC.

3.9. REFERENCES

Arancia, G., Rosati Valente F. and Trovalusci P. (1979). Effects of glutaraldehyde and glycerol on freeze-fractured *Escherichia coli*. *J. Microscopy* **118**, 161-176.

Armbruster, B.L., Carlemalm E., Chiovetti R., Garavito R.M., Hobot J.A., Kellenberger E. and Villiger W. (1982). Specimen preparation for electron microscopy using low temperature embedding resins. *J. Microscopy* **126**, 77-86.

Beckett, A. and Read, N. (in press). Low temperature scanning electron microscopy. In: *Ultrastructure techniques for microorganisms*. (eds. H.C. Aldrich, W.J. Todd) Plenum Press, New York.

Bordi, C. (1979). Parlodion coating of highly fragile freeze-fracture replicas. *Micron* 10, 139-140.

Boyde, A. and Franc F. (1981). Freeze-drying shrinkage of glutaraldehyde fixed liver. *J. Microscopy* 122, 75-86.

Boyde, A. and Maconnachie E. (1979a). Volume changes during preparation of mouse embryonic tissue for scanning electron microscopy *Scanning* 2, 149-163.

Campbell, G.J. and Roach M.R. (1983). Dimensional changes associated with freeze-drying of the internal elastic lamina from cerebral arteries. *Scanning* 5, 137.

Carlemalm, E. (1984). Low temperature embedding resins. *Proc. 8th Eur. Reg. Conf. Electron Microscopy*, Budapest 3, 1786-1787.

Carlemalm, E., Garavito R.M. and Villiger W. (1982). Resin development for electron microscopy and an analysis of embedding at low temperature. *J. Microscopy* 126, 123-144.

Costello, M.J. (1980). Ultra-rapid freezing of biological samples. *Scanning Electron Microscopy* 2, 361-370.

De Mazière, A.M.G.J., Aertgeerts, P. and Dietrich, W. (1985). A modified cleansing procedure to obtain large freeze-fracture replicas. *J. Microscopy* 137, 185-188.

Dempsey, G.P. and Bullivant S. (1976a). Copper block method for freezing non cryoprotected tissue to produce ice crystal free regions for electron microscopy. I. Evaluation using freeze-substitution, *J. Microscopy* 106, 251-260.

Duniec, J.T., Goodchild, D.J. and Thorne, S.W. (1982). A method of estimating the radial distribution function of protein particles in membranes from freeze-fracture micrographs. *Comput. Biol. Med.* 12, 319-322.

Echlin, P. (1978a). Low temperature biological scanning electron microscopy.   In: *Advanced techniques in biological electron microscopy*. (ed. J.K. Koehler). pp. 89-122. Springer-Verlag, Berlin.

Echlin, P. (1978b). Low temperature scanning electron microscopy: a review. *J. Microscopy* 112, 47-62.

Echlin, P. and Saubermann A.J. (1977) Preparation of biological specimens for X-ray microanalysis. *Scanning Electron Microscopy* 1, 621-638.

Echlin, P., Skaer, H.B., Gardiner, B.O.C., Franks, F. and

Asquith, M.H. (1977). Polymeric cryoprotectants in the
preservation of biological ultrastructure. II. Physiological
effects. *J. Microscopy* <u>110</u>, 239-256.

Echlin, P., Lai C.E. and Hayes T.L. (1982). Low temperature X-
ray microanalysis of the differentiating vascular tissue in root
tips of *Lemna minor*. *J. Microscopy* <u>126</u>, 285-306.

Elder, H.Y., Gray, C.C., Jardine, A.G., Chapman, J.N. and
Biddlecombe, W.H. (1982). Optimum conditions for the
cryoquenching of small tissue blocks in liquid coolants. *J.
Microscopy* <u>126</u>, 45-61.

Fineran, B.A. (1972). Fracture fraces of tonoplast in root tips
after various conditions of pretreatment prior to freeze-
etching. *J. Microscopy* <u>96</u>, 333-342.

Fineran, B.A. (1978). Freeze-etching. In: *Electron microscopy
and cytochemistry of plant cells*. (ed. J. L. Hall). pp. 279-
341. Elsevier/ North-Holland, Amsterdam.

Franks, F., Asquith, M.H., Hammond, C.C., Skaer H. le B. and
Echlin P. (1977). Polymeric cryoprotectants in the preservation
of biological ultrastructure. I. Low temperature states of
aqueous solutions of hydrophilic polymers. *J. Microscopy* <u>110</u>,
223-238.

Furtado, D.,   Williams, W.P., Brian A.P.R. and Quinn P.J.
(1979). Phase separation in membranes of *Anacystis nidulans*
grown at different temperatures. *Biochim. Biophys. Acta* <u>555</u>,
352-357.

Gunning, B.E.S. and Hughes, J.E. (1976). Quantitative
assessment of symplastic transport of pre-nectar into trichomes
of *Abutilon* nectaries. *Aust. J. Plant Physiol.* <u>3</u>, 619-637.

Gupta, B.L. and Hall T.A. (1984). X-ray microanalysis of frozen-
hydrated cryosections: problems and results. *Proc. 8th. Eur.
Reg. Conf. Electron Microscopy, Budapest* <u>3</u>, 1665-1675.

Harvey, D.M.R. (1980). The preparation of botanical samples for
ion localisation studies at subcellular level. *Scanning
Electron Microscopy* <u>2</u>, 409-420.

Harvey, D.M.R. (1982), Freeze-substitution, *J. Microscopy* <u>127</u>,
209-222.

Hasty, D.L. and Hay E.D. (1977). Freeze-fracture studies of the
developing cell surfaces. II. Particle-free membrane blisters on
glutaraldehyde fixed corneal fibroblasts are artefacts. *J. Cell*

Biol. 78, 756-768.

Hay, E.D. and Hasty D.L. (1979).   Extrusion of particle-free membrane blisters during glutaraldehyde fixation. In: *Freeze-fracture: methods, artefacts and interpretations.* (eds. J. E. Rash, C. S. Hudson). pp. 59-66. Raven Press, New York.

Hughes, J.E. and Gunning, B.E.S.   (1980). Glutaraldehyde-induced deposition of callose. *Can. J. Bot.* 58, 250-258.

Humbel, B., Marti T. and Müller M. (1983).   Improved structural preservation by combining freeze-substitution and low temperature embedding.   *Beitr. elektronenmikroskop. Direktabb. Oberfl.* 16, 585-594.

Humbel, B. and Müller M. (1984).   Freeze-substitution and low-temperature embedding. *Proc. 8th Eur. Reg. Conf. Electron Microscopy, Budapest* 3, 1789-1798.

Johnson, R.P.C. (1978).   The microscopy of P-protein filaments in freeze- etched sieve pores, *Planta* 143, 191-205.

Lefort-Tran, M., Gulik, T., Plattner, H., Beissen J. and Weissner W. (1978).   Influence of cryofixation proceedures on organisation and partition of intramembrane particles. *Proc. 9th Int. Congr. Electron Microscopy, Toronto* 2, 146-147.

Marshall, A.T. (1980a).   Quantitative X-ray microanalysis of frozen hydrated bulk biological specimens. *Scanning Electron Microscopy* 2, 335-348.

Marshall, A.T. (1980b).   Freeze-substitution as a preparation technique for biological X-ray microanalysis. *Scanning Electron Microscopy* 2, 395-408.

Martin, C.E., Hiramitsu, K., Kitajima, Y., Nozawa, Y., Skriver L. and Thompson G.A. (1976). Molecular control of membrane properties during acclimation. Fatty acid desaturase regulation of membrane acclimating *Tetrahymena* cells. *Biochemistry* 15, 5218-5227.

McCully, M.E. and Canny, M.J. (1985). The stabilization of labile configurations of plant cytoplasm by freeze-substitution. *J. Microscopy* 139, 27-33.

Niedermeyer, W. and Moor H. (1976).   Effect of glycerol on structure of membranes: a freeze-etch study. *Proc 6th Europ. Congr. Electron Microscopy, Jerusalem* 2, 108-109.

Niedermeyer, W. and Wilke, H. (1982). Quantitative analysis of intramembrane particle (IMP) distribution on biomembranes after freeze-fracture preparation by a computer-based technique. *J. Microscopy* 126, 259-274.

Ono, T.A. and Murata M. (1982). Chilling susceptibility of the blue-green alga *Anacytis nidulans*. III. Lipid phase of cytoplasmic membrane. *Plant Physiol*. 69, 125-129.

Ornberg, R.L. and Reese T.S. (1981). Quick freezing and freeze-substitution for X-ray microanalysis of calcium. In: *Microprobe analysis of biological systems*. (eds. T.H. Hutchinson, A.P. Somlyo). p. 213. Academic Press, London.

Parish, G.R. (1975). Changes of particle frequency in freeze-etched erythrocyte membranes after fixation. *J. Microscopy* 104, 245-256.

Richter, H. (1968a). Die Reaktion hochpermeabler Pflanzenzellen auf die Gefrierschutzstoffe (Glyzerin, Athylenglykol, Dimethylsulfoxid). *Protoplasma* 65, 155-166.

Richter, H. (1968b). Low-temperature resistance of *Campanula*-cells treated with glycerol. *Protoplasma* 66, 63-78.

Robards, A.W. (1978). An introduction to techniques for scanning electron microscopy of plant cells. In: *Electron microscopy and cytochemistry of plant cells*. (ed. J. Hall). pp. 343-415. Elsevier/North Holland, Amsterdam.

Robards, A.W. (1984). Fact or artefact - A cool look at biological electron microscopy. *Proc. Roy. microsc. Soc.* 19, 195-208.

Robards, A.W. and Clarkson, D.T. (1984). Effects of chilling temperatures on root cell membranes as viewed by freeze-fracture electron microscopy. *Protoplasma* 122, 75-85.

Robards, A.W. and Crosby, P. (1979). A comprehensive freezing, fracturing and coating system for low temperature scanning electron microscopy. *Scanning Electron Microscopy* 2, 325-344.

Robards, A.W. and Crosby, P. (1983). Optimisation of plunge freezing: linear relationship between cooling rate and entry velocity into liquid propane. *Cryo-Letters* 4, 23-32.

Robards, A.W. and Umrath, W. (1978). Improved freeze-etching of difficult specimens. *Proc. 9th Int. Congr. Electron Microscopy, Toronto* 2, 138-139.

Robards, A.W., Newman, T.M. and Clarkson, D.T. (1980).

Demonstration of the distinctive nature of the plasmamembrane of the endodermis in roots using freeze-fracture electron microscopy. In: *Plant membrane transport: current conceptual issues*. (eds. W.J. Lucas and J. Dainty). pp. 395-396. Elsevier/North Holland, Amsterdam.

Robards, A.W., Bullock, G.R., Goodall, M.A. and Sibbons, P.D. (1981). Computer assisted analysis of freeze-fractured membranes following exposure to different temperatures. In: *Effects of low temperatures on biological membranes*. (eds. G.J. Morris A. Clarke). pp. 219-238. Academic Press, London.

Rottenberg, W. and Richter, H. (1969). Automatische Glyzerinbehandlung pflanzlicher Dauergewebszellen fur die Gefrierätzung. *Mikroskopie* 25, 313-319.

Saubermann, A.J. and Echlin P. (1975). The preparation examination and analysis of frozen hydrated tissue sections by scanning transmission electron microscopy and X-ray microanalysis. *J. Microscopy* 105, 155-192.

Skaer, H.le B. (1982). Chemical cryoprotection for structural studies. *J. Microscopy* 125, 137-148.

Skaer, H.le B., Franks, F., Asquith M.H. and Echlin P. (1977). Polymeric cryoprotectants in the preservation of biological ultrastructure. III. Morphological aspects, *J. Microscopy* 110, 257-270.

Skaer, H.le B., Franks F. and Echlin P. (1978). Nonpenetrating polymeric cryofixatives for ultrastructural and analytical studies of biological tissues. *Cryobiology* 15, 589-602.

Sleytr, U.B. and Robards, A.W. (1982). Understanding the artefact problem in freeze-fracture replication: a review. *J. Microscopy* 126, 101-122.

Somlyo, A.V. and Silcox, J. (1979). Cryoultramicrotomy for electron probe analysis. In: *Microbeam analysis in biology*. (eds. C.P. Lechene and R.R. Warner). pp. 535-555. Academic Press, New York.

Spurr, A.R. (1969). A low-viscosity epoxy resin embedding medium for electron microscopy. *J. Ultrastruct. Res.* 26, 31-43.

Steere, R.L., Erbe, E.F. and Moseley, J.M. (1974). Importance of stereoscopic methods in the study of biological specimens prepared by cryotechniques. *Proc. 8th Int. Congr. Electron Microscopy, Canberra* 2, 36-37.

Stolinski, C., Gabriel, G. and Martin, B. (1983). Reinforcement and protection with polystyrene of freeze-fracture replicas during thawing and digestion of tissue. *J. Microscopy* <u>132</u>, 149-152.

Toivio-Kinnucan, M.A., Chen, H.-H., Li P.H. and Stushnoff C. (1981). Plasmamembrane alterations in callus tissues of tuber bearing *Solanum* species during cold acclimation. *Plant Physiol.* <u>67</u>, 478-483.

Tsien, H.C. and Higgins M.L. (1974). Effect of temperature on distribution of membrane particles in *Streptococcus faecalis* as seen by the freeze-fracture technique. *J. Bact.* <u>118</u>, 725-734.

Van Winkle, W.B. and Entman, M.L. (1981). Accurate quantitation of surface area and particle density in freeze-fracture replicas. *Micron* <u>12</u>, 259-266.

Weibel, E.R., Losa, G. and Bolender, R.P. (1976). Stereological methods for estimating relative membrane surface area in freeze-fracture preparations of subcellular fractions. *J. Microscopy* <u>107</u>, 255-266.

Weinstein, R.S., Benefiel, D.J. and Pauli, B.U. (1979). Use of computers in the analysis of intramembrane particles. In: *Freeze-fracture: methods, artefacts and interpretations.* (eds. J.E. Rash, C.S. Hudson). pp. 175-183. Raven Press, New York.

Willison, J.H.M. and Brown R.M. (1979). Pretreatment artefacts in plant cells, In: *Freeze-fracture: methods, artefacts and interpretations.* (eds J.E. Rash, C.S. Hudson). pp. 51-57. Raven Press, New York.

Wilson, A.J. and Robards, A.W. (1980). Some limitations of the polymer polyvinylpyrrolidone for the cryoprotection of barley (*Hordeum vulgare*) roots during quench freezing. *Cryo-Letters* <u>1</u>, 416-425.

Wilson A.J. and Robards, A.W. (1981).   Some experiences in the use of a polymeric cryoprotectant in the freezing of plant tissue. *J. Microscopy* <u>125</u>, 287-298.

Wilson, A.J. and Robards, A.W. (1984). *An atlas of low temperature scanning electron microscopy.* pp. 64.   Centre for Cell and Tissue Research,   York. (ISBN 0 9509670 0 9).

4

Stereo-transmission electron microscopy of thick sections and whole cells

Chris Hawes

4.1   INTRODUCTION

4.1.1   The problem

All electron micrographs represent two dimensional projections
of three-dimensional (3-D) structures and in order to obtain
information on the 3-D structure of the specimen special
techniques have to be applied. With scanning electron
microscopy clues about the third dimension are given by
perspective, shadows and the obstruction of features within the
specimen. This situation also applies in part to replicas and
shadow cast specimens observed with the transmission electron
microscope. However, the majority of biological specimens
studied by transmission electron microscopy are thin sections
of resin embedded material and as such simply present the
microscopist with an image of a two dimensional plane sliced
from the specimen. Although stereoscopical information can be
obtained from such thin sections (Willis 1972) this approach
contributes little towards an understanding of the spatial
organisation of the unsectioned specimen as a whole.

It is easy to misinterpret the structure of an organelle
within a cell from the observation of thin sections (Elias 1971
and Fig. 1 Gunning and Steer 1975). The conventional method of
overcoming this problem is to use the technique of serial
sectioning and reconstruction in order to attain some insight
into the third dimension (Atkinson *et al.* 1974, Gaunt and Gaunt
1978, Wright and Moisand 1982). This is a time consuming
technique beset with many technical problems. These start with
the cutting of many sequential sections of equal thickness, an
impossible task according to some authors (Williams and Meek
1965). Serious problems can also arise with thinning of
sections by the electron beam (Willis 1972, Bennet 1974, King

et al. 1980) and in the alignment of tracings from the
micrographs with respect to one another in order to build up a
3-D model of the specimen. Recently computer aided
reconstruction techniques have facilitated the production of
stereo-reconstructions from such serial thin sections,
information being transferred from the micrographs into a
computer via a digitiser tablet (Green *et al.* 1979, Moens and
Moens 1981). However, there are severe restrictions in the
quantity and detail of the information that can be included in
such reconstructions (Hawes 1981) and this, coupled with the
section thickness problems, suggests the accuracy and validity
of much serial section work must be open to question. To
obtain accurate information on the spatial organisation of a
specimen it is therefore desirable to use techniques that
permit the observation of "thick" specimens coupled with the
application of stereo-electron microscopy for the analysis of
the resulting micrographs (Hawes 1981). For large
reconstructions serial thick section techniques may be used
(Crang and Pechak 1978, Peachy 1974).

4.1.2   Historical

The first biological electron microscopists were forced into
the observation of thick specimens due to the lack of suitable
preparative techniques. Indeed one of the very first
biological micrographs was of a 15 µm thick unembedded osmium
impregnated section of a *Drosera intermedia* leaf (Marton 1934),
and Elvers (1943) used thick (1-3 µm) paraffin sections to
study meiosis in various *Lilium* spp. Needless to say the
micrographs from these studies showed very little structural
detail due to excessive section thickness and beam damage. In
1947 van Doorsten *et al.* looked at whole yeast cells with
accelerating voltages of up to 350 kV, and demonstrated
increased penetration of the specimen by the electron beam when
the accelerating voltage was increased. However, considerable
effort was very soon put into the development of preparative
techniques such as resin embedding and ultramicrotomy in order
to facilitate the production of ultrathin sections. For a
number of years the study of thick biological specimens was
restricted to relatively few studies using high voltage
electron microscopes (HVEMs). These started with a study of
hydrated bacteria at 650 and 750 kV within a microscope
environmental chamber in an attempt to observe living organisms
with the HVEM (Dupouy *et al.* 1960). One of the first uses of
the HVEM with plant material was in the observation of the
labarynthine wall ingrowths in xylem transfer cells which were
clearly revealed in a 1 µm thick section (Gunning *et al.* 1970).
Thus the use of higher than conventional accelerating voltages
with the HVEM held great potential for the observation of thick

specimens. When combined with stereo-techniques this provided microscopists with a quick alternative method for extracting three-dimensional information from their specimens.

4.2 MICROSCOPY OF THICK SPECIMENS

For the purpose of this review thin sections will be considered to be less than 0.25 µm in thickness, semi-thick sections between 0.25 and 1 µm and thick sections 1 µm and thicker Conventional transmission electron microscopes can be considered to have accelerating voltages up to 120 kV, medium voltage microscopes usually up to 400 kV and high voltage microscopes 750 kV and above.

Two major factors have to be taken into consideration when attempting microscopy on "thick" specimens. First and foremost is specimen preparation. For instance it is far easier to study a thick specimen that is not embedded in resin as there is less beam/specimen interaction and therefore greater beam penetration can be achieved (see 4.3.3, 4.4, Hawes 1985). In conventionally prepared and stained resin sections the usable section thickness is often limited by an excess of overlapping information in the final image (see 4.3.2a) and not by any constraints imposed by the microscope being used. However by employing selective staining techniques much thicker sections can be used (see 4.3.2b).

Secondly, the accelerating voltage of the microscope is a major factor influencing the maximum thickness of specimen that can be observed for any given resolution. Most reviews on microscopy of thick specimens have concentrated on the use of the HVEM (eg Glauert 1974, Humphreys 1976, King *et al*. 1980, Hawes 1981). These have discussed the advantages to be gained by using very high accelerating voltages and so will only be briefly considered here. At 1000 kV the penetrating power of an electron beam is approximately three times greater than at 100 kV and coupled with the great depth of field of an HVEM the viewing of extremely thick specimens (10 µm) is possible. In microscopy of thick specimens poor resolution caused by chromatic aberration due to energy loss in the electron beam within the specimen is often a major problem. At 1000 kV it has been suggested that chromatic aberration in thick biological specimens is reduced by at least twenty times compared with 100 kV (Humphreys 1976). At such high accelerating voltages there is also less beam damage to the specimen. Compared to a 100 kV electron beam, use of a 1000 kV beam will result in three times less ionisation damage and three times less beam heating for a given specimen thickness (Humphreys 1976). Indeed it was these factors which inspired

the early attempts to observe living material with the HVEM
(Dupouy *et al*. 1960). However, King *et al*. (1980) believe that
radiation damage can be a serious problem at 1000 kV as well as
at 100 kV. Cratering of 10 µm thick sections after irradiation
in the HVEM has been reported (Favard and Carasso 1973),
although it is doubtful that any biologist would find the need
to study sections this thick. However, at high accelerating
voltages thick specimens have better stability under the beam,
an important fact when observing unembedded sections and whole
cells.

The preceding discussion has concentrated on the merits of
using extremely high accelerating voltages for the study of
thick specimens. Indeed there are many cases when use of the
HVEM is vital (see 4.4). However, microscopy on semi-thick and
thick specimens can be carried out with conventional
transmission electron microscopes (Harris and Oparka 1983.
Thiéry and Rambourg 1976), or with the new generation of medium
kV microscopes. With most microscopes a semi-thick section of
0.5 µm can easily be photographed at the lower end of the
magnification range. Most manufacturers are now incorporating
design features into their machines that help overcome some of
the problems of studying thick specimens at low kVs (Fig. 4.2).
Thus the objective lens system in the Philips 410 compensates
for loss of resolution in semi-thick specimens caused by
chromatic aberration as does the energy loss spectrometer in
the Zeiss 902. It is quite feasible to observe a section cut
at 2 µm at low magnifications with a 120 kV microscope but at
higher magnifications considerable instability and poor
resolution will be apparent. With 200-300 kV microscopes
sections as thick as 3 µm can be studied at reasonably high
magnifications with good resolution. In all cases loss of
resolution due to chromatic aberration can be reduced by the
use of as small an objective aperture as is possible, although
this may lead to excessive contrast in the final image.

4.3   THICK SECTIONING

4.3.1 Section cutting and thickness

With one or two exceptions all the microscopy on thick plant
specimens has been on sectioned material. It is important to
remember that the images obtained from microscopy of thick
sections differ considerably from those of conventional thin
section work and therefore it is advisable to carry out both
procedures during any one study. Furthermore, when carrying
out thick section work, a range of thicknesses will have to be
used in order to find those most suitable for extracting
optimum 3-D information from the specimen.

Cutting thick resin sections is an easy procedure. To date the majority of studies have used sections cut in the range 0.5-1.0 μm (see Hawes 1981) although considerably thicker sections can be used in the HVEM (5 μm Hawes 1981, up to 10 μm Marty 1983). Sections adhere readily to coated grids or can be gently heat sealed to uncoated grids (Hawes 1981). They can be stabilised by coating with a thin layer of carbon (Peterson and Ris 1976) and this may prevent some beam damage during microscopy (King *et al.* 1980).

It is difficult to cut sections accurately to a known thickness due to inaccuracies in microtome coarse advance mechanisms plus compression and stretching effects. By re-embedding and cross sectioning sections cut at 1 μm, variation in thickness of between 0.8 μm and 1.2 μm can be demonstrated (Hawes, unpublished data). Beam thinning of the section, is also a problem (see 4.1 Hawes, 1981, King *et al.* 1980) and the thickness of the irradiated area can be measured by stereoscopic techniques (see 4.5). Table 4.1 shows the variation in section thickness that can be measured in Spurr resin sections cut at 1 μm and 2 μm. It is likely that most published micrographs of thick sections give an over estimate of the thickness of the section area exposed to the beam; to avoid this error the thickness of the sections should be measured stereoscopically (see 4.5.3b).

Table 4.1

Stereoscopic measurement of section thickness

Microtome Setting	Section Thickness (μm)		No. of Sections
	Mean	Range	
1 μm	0.64	0.4 - 0.95	23
2 μm	1.33	0.92 - 1.84	14

4.3.2 Staining for thick sectioning

4.3.2a Conventional stains. It is not necessary here to discuss fixation techniques for thick section work as it is usual to employ conventional primary fixatives, although on occasion techniques such as freeze substitution have been used (Howard 1981). However, the application of correct staining procedures is vital for the success of a thick section study. The maximum thickness of section that can be observed is more often than not limited by the staining techniques employed. It is important to remember that a thick section contains many times the quantity of information than does a conventional thin

section. The use of cytoplasmic stains such as uranium and
lead salts will often result in a confused image due to the
quantity of overlapping information and background cloudiness
due to masses of stained ribosomes (Fig. 4.1.a). Most studies
using these stains have therefore been restricted to sections
of 1 μm or less (Cox and Juniper 1973, Coss and Pickett-Heaps
1974, Gunning 1980), but have proved useful in the study of
plant microtubules (Gunning 1980, Palevitz 1981) and the
structure of the spindle (Coss and Pickett-Heaps 1974, Peterson
and Ris 1976).

     Staining thick sections can take upwards of 1 h in each
solution and therefore it is advisable to stain 'en bloc'
before embedding wherever possible. Pesacreta and Parthsarathy
(1984) have developed a method of staining the whole trimmed
block in hot alcoholic uranyl acetate in order to increase the
staining of microfilament bundles for HVEM observation. Hot
alcoholic phosphotungstic acid (Locke and Krishnan 1971) and
potassium permanganate (Fineran *et al.* 1978) have also been
used as stains for thick sections.

4.3.2b Selective staining. To overcome the problems
associated with conventional staining of thick sections it is
advisable to employ selective staining techniques which may
either be performed 'en bloc' or on the section. In this way
only the organelles or components of the cell which are to be
studied are stained and they can then be imaged against an
unstained or lightly stained background. This greatly enhances
the 3-D effect in stereo-micrographs, simplifies their
interpretation and permits the use of very thick sections
(Glauert 1974, Hawes 1981).

     The most popular selective staining technique is that of
osmium impregnation which over recent years has contributed
considerably to our understanding of the structure of the
endomembrane system in plant cells (Fig. 4.2). Material can
either be post-fixed in osmium tetroxide for long periods (Poux
et al. 1973) or rapid impregnation can be achieved by post-
fixation and staining in a mixture of osmium tetroxide and zinc
iodide, the ZIO technique (Figs. 4.1.b - 4.4, Marty 1973,
Gilloteaux and Naud 1979, Harris 1980, Hawes 1981, Hawes *et al.*
1981, Poux 1981, Whatley *et al.* 1982).

     Osmium impregnation results in the selective staining of
the endoplasmic reticulum, nuclear envelope, Golgi dictyosomes
and plastid thylakoid lamellae (Fig. 4.1.b - 4.4). In some
tissues tonoplasts and mitochondria also stain (Hawes 1981).
By a combination of this technique, thick sectioning and
stereo-microscopy it has been shown that in plant cells the
endoplasmic reticulum has two distinct forms, tubular (TER) and

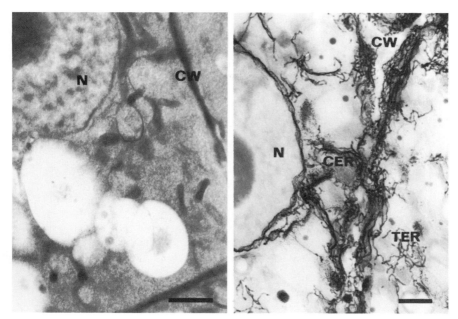

Fig. 4.1.a. Maize root tip meristem cell stained for
1 h in uranyl acetate and 1 h in lead citrate. Compare
with a similar cell, Fig. 4.1.b. from ZIO impregnated
tissue. CER = cisternal ER, CW = cell wall, N =
nucleus, TER = tubular ER. Sections cut at 1 μm,
1000 kV, bar = 2 μm.

cisternal (CER) which interconnect with each other and cannot
be easily distinguished in thin sections (Poux 1981, Harris and
Chrispeels 1980, Marty 1973, 1983, Barlow *et al*. 1984,
Stephenson and Hawes 1985). The relationship between the forms
of ER and the forming or cis face of Golgi dictyosomes has been
investigated in several tissues using thick sections.
Connections between ER and dictyosomes were reported in the
trap glands of *Dionaea muscipula* (Juniper *et al*. 1982) and in
cotyledons of *Vigna radiata* (Harris and Oparka 1983) yet no
connections could be found in developing endosperm of *Triticum
aestivum* (Parker and Hawes 1982) nor in *Zea mays* and *Phaseolus
vulgaris* root tip meristems (Juniper *et al*. 1982). Such
studies have led some authors to propose the existence of a
Golgi-endoplasmic reticulum lysosome (GERL) complex in plant
cells similar to that found in animal tissues (Marty 1978,
Harris and Oparka 1983).

Thick sectioning has shown the endomembrane system to be a
major structural feature of the mitotic apparatus in plant
cells. The spindle is completely surrounded by a mass of
tubular and cisternal membranes some of which traverse the

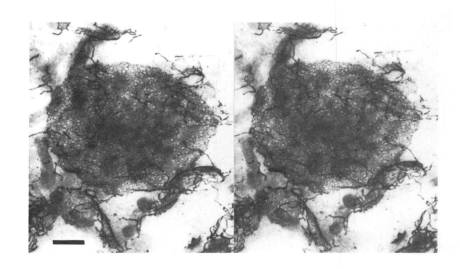

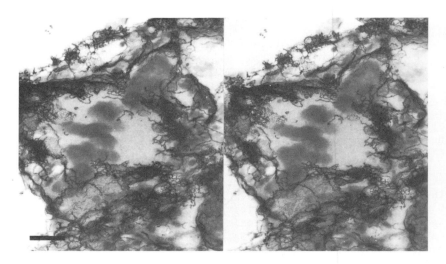

Fig. 4.2. Stereo-pair of an interphase nucleus frcm
a ZIO impregnated maize root tip cell. Note the TER
and CER associated with the nuclear envelope.
Section cut at 1 µm, 100 kV Philips 410, 10° tilt,
bar = 1 µm.
Fig. 4.3. Pea root tip nucleus in metaphse. The
mitotic apparatus is totally enveloped by CER and TER.
Section cut at 2 µm, 900 kV, 10° tilt, bar = 2 µm.

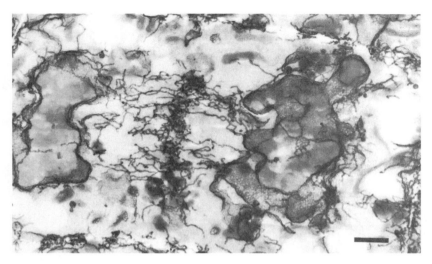

Fig. 4.4. Telophase in a pea root tip meristem. Note
the mass of TER around the cell plate. Section cut
at 1 μm, 900 kV, bar = 2 μm.

mitotic apparatus (Fig. 4.3). TER subsequently forms a complex
three dimensional net around the cell plate and connects the
plate with the telophase nuclei (Fig. 4.4, Hawes *et al.* 1981).

Another selective staining technique which has considerable
potential is the uranyl acetate, copper and lead citrate
(Ua/Cu/Pb) impregnation procedure developed by Thiéry and
Rambourg (1976). When applied to plant tissue, plastid
envelopes and sites of Golgi activity are most heavily stained
(Fig. 4.5.a) and can be easily identified in thick sections
(Fig. 4.5.b). Thus this procedure may be used to compliment
the osmium impregnation techniques (Hawes and Horne 1983).
Selective staining of the plasma membrane can be achieved by
modification of the acidified phosphotungstic acid technique
using sections up to 1 μm in thickness (Hawes 1981).

4.3.2c Enzyme cytochemistry. The use of thick sections to
study the spatial distribution of enzyme activity within plant
cells has yet to be fully exploited. The reaction products of
many enzyme cytochemical staining procedures are so electron
opaque as to be excellent candidates for selective stains.
Cytochrome oxidase activity on the inner mitochondrial
membranes of *Candida utilis* sphaeroplasts has been located with
3-3' diaminobenzidine and osmium tetroxide, and in thick
sections mitochondria were stained selectively (Davison and
Garland 1975). The Gomori technique has been used in
conjunction with thick sectioning to locate acid phosphatases
in TER of the cells of the radicle of *Euphorbia characias*

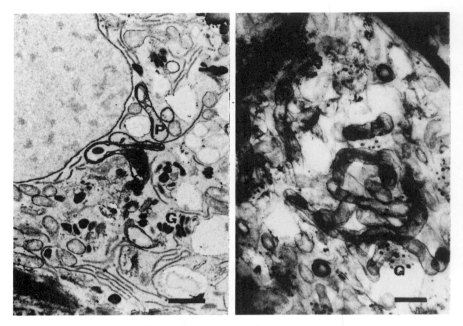

Fig. 4.5.a. Maize root cap cell impregnated by the
Ua/Cu/Pb technique. Plastid envelopes (P) and Golgi
vesicles (G) show the heaviest impregnation. Thin
section, 80 kV, bar = 2 µm. Fig. 4.5.b. Thick section
of a cell stained as in a. showing tubular plastids (P)
and sites of Golgi activity (G). Section cut at 1µm,
900 kV, bar = 1 µm. (From Hawes & Horne 1983, courtesy
of the Societe Francaise de Microscopie Electronique)

(Marty 1978), in the ER of *Cucumis sativus* root tip cells (Poux
et al. 1974) and in the ER of protoxylem cells of *Vigna radiata*
(Pitt and Wakely, in preparation).

4.3.3 Resin-free sections

It is possible to produce sections which are free of any
embedding medium. The most popular method is the polyethylene
glycol (PEG) technique developed by Wolosewick (1980) for the
study of the cytoskeleton and cytoplasmic matrix in animal
tissues (see also 4.4). Tissue is conventionally fixed and
then embedded in a high molecular weight PEG. Semi-thick
sections are cut, but prior to microscopy the PEG is dissolved
out of the sections which are subsequently critical-point dried
and carbon coated. The sections can then be observed without
any further heavy metal staining. Unosmicated material can
also be studied by this technique. The implication is that
electron scattering to produce adequate contrast is generated

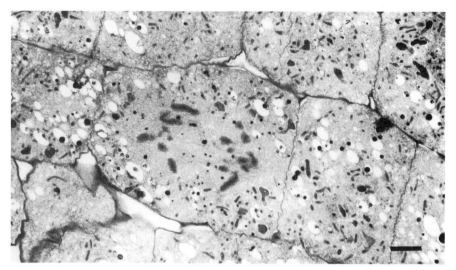

Fig. 4.6. PEG processed section of an unosmicated, unstained maize root tip meristem with an anaphase cell. Section cut at 0.5 μm, 1000 kV, bar = 3 μm.

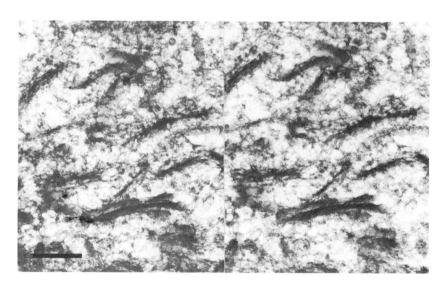

Fig. 4.7. Stereo-pair showing dictyosomes in a PEG processed section of a maize root cap cell. Note the filamentous nature of the cytoplasmic matrix. Section cut at 0.25 μm, 12° tilt, 1000 kV, bar = 0.5 μm. (From Hawes et al 1983, courtesy Springer-Verlag)

in resin-free sections by the specimen itself, whereas in resin embedded material the similar scattering properties of the resin and of the specimen result in low contrast unless heavy metal stains are employed. The PEG technique has recently been adapted for the study of unembedded plant tissues and again unstained and unosmicated specimens can be observed at a range of accelerating voltages (Fig. 4.6, Hawes *et al.* 1983, Hawes and Horne 1985).

In a PEG section all the major organelles can easily be identified. Furthermore a 3-D network of fine filaments 6-18 nm in diameter appears to pervade the cytoplasm and link all the organelles (Fig. 4.7). This filamentous network has been equated with the microtrabecular lattice described in both whole mount preparations of animal cells (see 4.4) and in PEG sections of animal tissues (Wolosewick 1980). The information obtained by this technique supports the contention that the cytoplasmic matrix of plant cells is structured (Hawes *et al.* 1983). That PEG embedding does not disrupt the ultrastructure of the cell has been demonstrated by re-embedding critical-point dried semi-thick sections in resin and thin sectioning them for conventional electron microscopy. All the organelles appeared intact and largely undamaged although the 3-D network of fine filaments could no longer be distinguished, (Hawes *et al.* 1983, Hawes and Horne 1985). Future work using the PEG technique with plant cells will probably concentrate on the use of such embedment-free sections for the localisation of cytoskeletal and other proteins by both immunofluorescence and immunogold labelling techniques.

4.4   WHOLE CELLS

4.4.1   Introduction

Early attempts at transmission electron microscopy of whole plant cells met with very little success even when the high penetrating power of the HVEM was used. The yeast cells observed by van Dorsten *et al.* (1947) at 350 kV and hyphae of *Trichothecium roseum* photographed with a 3 MV HVEM (Dupouy 1973) were both air dried and the considerable disruption of the cytoplasm precluded any meaningful interpretation of the images. Also in whole critical-point dried spores of *Ceratocystis ulmi* very little internal structure was discernable (Harris 1975).

Considerable advances in cell biology have since been made with electron microscopy of whole, fixed and critical-point dried animal cells using both conventional and high voltage electron microscopes (Temmink and Spiele 1980, Porter and

Tucker 1981). The three dimensional organisation of the
cytoskeleton and cytoplasmic matrix has been studied in these
embedment-free systems and interpretation of the results has
aroused considerable controversy. Arguments have revolved
around the possible existence of a structured matrix or
microtrabecular lattice, a system of extremely labile 3-6 nm
protein filaments only visible in resin free-specimens (Schliwa
et al. 1982, Wolosewick and Porter 1979, Porter and Tucker 1981,
Porter and Anderson 1982, Pawley *et al.* 1983).

4.4.2 Whole mount plant cells

Studies on the 3-D organisation of the plant cytoskeleton and
cytoplasmic matrix using whole mount preparations are hampered
by the very nature of plant cells. The cellulose based walls
are electron transparent when unstained and viewed at high
accelerating voltages (Hawes 1981, 1985) but their rigidity
prevents the spreading of cells onto grids. Thus the walls
make the production of whole mount preparations for observation
at conventional accelerating voltages impossible. Even with
the high penetrating power of the HVEM it is hard to find areas
of cytoplasm sufficiently thin (2-3 μm) to allow the structure
to be interpreted by stereo-microscopy. Observations are best
restricted to thin filamentous cells or small cells from
suspension cultures. Microscopy is facilitated if the cells
are vacuolate with only a thin layer of cortical cytoplasm.

   In such a study on whole critical-point dried moss
protonema there was some indication of fine cytoskeletal
elements in the cytoplasm (Cox and Juniper 1983). More
recently an HVEM investigation of whole critical-point dried
cells from suspension cultures of carrot (*Daucus carota* L.)
(Fig. 4.8.a) has revealed complex 3-D network of fine inter-
connecting filaments in the cytoplasmic matrix (Fig. 4.9 and
Hawes 1985). These varied between 5 nm and 18 nm in diameter
and have been compared to the microtrabecular lattice of animal
cells. As with PEG sections re-embedding and thin sectioning
of the dried cells confirmed that the whole mount technique did
not damage cell ultrastructure. Triton extraction of these
carrot cells prior to fixation reveals the cortical system of
microtubules to be in spiralling bands next to the wall (Fig.
4.8.b).

   The wall-free liquid endopsperm cells of *Scadoxus*
(*Haemanthus*) *multiflorus* Bak. have also proved suitable for the
study of the cytoplasmic matrix. When whole mount preparations
were observed with the HVEM a complex three dimensional system
of filaments was again identified in the cytoplasm and these
appeared to interconnect all the organelles (Hawes unpublished).

Fig. 4.8.a. Critical-point dried carrot cell from a suspension culture. Fig. 4.8.b. Similar cells after Triton extraction showing the cortical bands of microtubules. 1000 kV, bar = 4 μm.

Fig. 4.9. Stereo-pair of the cytoplasm of a critical-point dried carrot cell as in Fig. 4.8.a. The cytoplasmic matrix consists of a network of anastomosing filaments. P = plastid, M = mitochondrion, 1000 kV, 1.5^{o} tilt, bar = 0.5 μm. (From Hawes 1985, courtesy European Journal of Cell Biology)

4.4.3 Dry cleaving

A technique that permits the observation of the cytoskeleton in unembedded cells at conventional accelerating voltages without sectioning is that of dry cleaving (Traas 1984). Sheets of cells can be digested out of tissues such as root tip meristems with wall degrading enzymes (Traas 1984) or the walls of cells from suspension culture can be softened enzymically (Hawes 1985). The fixed cells are then attached to poly-L-lysine coated grids, dehydrated and critical-point dried as for whole cell preparations. Grids are then inverted onto Scotch tape and carefully removed. Many of the cells will be fractured leaving areas of wall, plasma membrane and cytoplasm attached to the grid. These chunks of cytoplasm vary in thickness and some may prove too thick for microscopy. However, by this technique the cortical cytoskeleton of microtubules associated with the plasma membrane is revealed and exceedingly long microtubules of up to 20 μm can be measured, an impossibility with thin sections (Traas 1984). Also the network of fine filaments comprising the cytoplasmic matrix has been described both in the dry cleaved cytoplasm of *Lepidium sativum* and *Ceratopteris thaliotroides* root cells (Traas 1984) and in suspension culture cells of carrot (Hawes 1985).

4.5   STEREO-TECHNIQUES

4.5.1 Taking stereo-pairs

At the resolution required for most thick specimen microscopy the depth of field of the objective lens can be several times greater than the thickness of the specimen. Focus is therefore maintained throughout the specimen and the only means of extracting 3-D information is from stereo-pairs of micrographs. This is a simple procedure which involves tilting the specimen between exposures in order to record two micrographs of the specimen showing the same structural details from slightly different viewpoints. When the micrographs are examined with an appropriate stereoscope the illusion of three dimensions will be created due to the introduction of parallax caused by tilting. The parallax is the difference in the position of corresponding image points along an axis at right angles to the tilt axis (Fig. 4.10). If the microscope is not fitted with a eucentric goniometer stage it is necessary to track the image with the specimen translates during the tilting procedure in order to maintain its position on the screen. With thick specimens any magnification changes due to tilting are normally slight and do not have to be corrected (Hawes 1981). If only qualitative stereoscopy is envisaged it is not necessary to record each half of the pair at equal tilt angles either side

C. Hawes

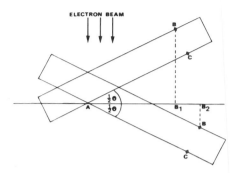

Fig. 4.10. Parallax between two points (A and B) on either surface of a section, induced by tilting, can be measured from the micrographs as the line $AB_2 - AB_1$. Section thickness (BC) can then be calculated as follows:

$$BC = \frac{AB_2 - AB_1}{2M \; Sin\frac{1}{2}\theta}$$

of the normal specimen plane (Hawes 1981) and thus the production of multiple stereo-pairs over a large tilt range can be used to study the specimen from a variety of viewpoints.

The choice of tilt angle is dependent on the final micrograph viewing magnification and the specimen thickness, as parallax (Y) is related to the height difference (h) between vertically separated points, the tilt angle (θ) and the magnification (M) in the equation,

$$Y = 2h.M \; Sin\frac{1}{2}\theta$$

derived by Nankivell (1963). Maximum stereoscopic impression is observed when the parallax between two points is about 5 mm if the prints are viewed at 25 cm from the eyes (Hudson and Makin 1979). Graphs of suggested tilt angles for various specimen thicknesses are given in Fig. 4.11. These graphs should only be used as guidelines for tilt angles as it must be remembered that section thickness cannot be accurately known prior to microscopy (see 4.3.1) and the features of interest may lie in a confined area within the total specimen thickness. Though, tilting to give optimum parallax is not necessary for visualisation of depth or for quantitative stereo-analysis, excessive tilting will make optical fusion of the two halves of the stereo-pair impossible.

4.5.2 Beam-tilt stereoscopy

The problem of choosing suitable areas of the specimen for

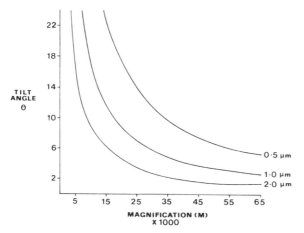

Fig. 4.11. Suggested maximum tilt angles for three
thicknesses of section at different final viewing
magnifications, calculated assuming an optimum parallax
of 3 mm.

stereo-microscopy would be overcome by the development of an on
line stereo-imaging system for the TEM using a double-image-
display system. A pre-requisite of this is the production of
two images by rapidly tilting the electron beam and not the
specimen (Doole *et al.* 1983). Use of the full range of beam
tilt is restricted by the introduction of aberrations in the
objective lens due to the off axis beam. However, preliminary
experiments with the HVEM have indicated that these
aberrations can be corrected by bringing the beam back on axis
below the specimen by the installation of extra coils in the
objective lens between the specimen and the objective apertures
(Doole and Hawes unpublished).

4.5.3 Quantification with thick specimens

4.5.3a Direct measurements. Relatively few attempts have been
made to quantify the information within thick specimens.
Direct measurement techniques are limited to simple counting
procedures (Haskins 1976), to structures which lie parallel to
the section surface (Cox *et al.* 1980) and to spherical
components. It has been demonstrated that the diameter of
secretory granules and peroxisomes is more accurately measured
from thick rather than from thin sections (Ohno *et al.* 1978).
Morphometry on the large branched mitochondria of fibroblasts
has been carried out using whole cell preparations. The number
per cell, their length, width and number of branches were
recorded (Goldstein *et al.* 1984). However, such measurements

take no account of foreshortening effects on the length of the
organelle caused by it passing obliquely through the depth of
the specimen. For accurate measurements account has to be
taken of the distribution of the components in the third
dimension, that is by the use of stereometry.

4.5.3b Stereometry. The position of structures within a thick
specimen can be calculated from parallax measurements (see
Fig. 4.10). Three criteria have to be fulfilled before stereo-
analysis can be undertaken; 1) the magnification of both
halves of the stereo-pair must be identical, 2) the tilt angles
must be known and 3) the micrographs must be aligned correctly
with respect to the axis of tilt i.e. at right angles to the
parallax axis. If the microscope does not have a rotation-free
imaging system for each magnification it will be necessary to
calibrate the rotation of the image on photographic plate with
respect to the specimen. Once the parallax between structures
or points of interest has been measured, height differences can
be calculated using Nankivell's (1963) equation. However, if
the specimen is not tilted an equal number of degrees about the
horizontal plane, errors are introduced which result in
artificially high parallax readings and these have to be
compensated for (Hawes 1981).

     The simplest way to measure parallax is with a ruler to
find the displacement along the axis at right angles to the
tilt axis, between the same points on both micrographs. This
approach can be facilitated by the use of a digitiser tablet
interfaced to a microcomputer (Niederman et al. 1983). Such a
system has the advantage that stereo-vision is not a
prequisite for parallax measurement and X, Y and Z
co-ordinates of structures can be quickly obtained.
Micrographs have to be aligned on the tablet with respect only
to the tilt axis. The major problem with this approach is that
the parallax to be measured is often very small (i.e. 1 mm or
less on the prints) and therefore any slight displacement of
the pen or cursor along the X axis can give rise to large
errors in measurement.

     The most popular method of parallax determination is by the
use of measuring mirror stereoscopes or stereometers (Thomas
et al. 1974, Gaunt and Gaunt 1978, Bauer and Exner 1981, Hawes
1981). In one such system two stereoscopically fused light
spots are introduced into the optional paths to the two
micrographs. Movement of one of the light spots along the
X-axis with a micrometer changes the apparent Z position of the
fused spots within the stereo-field. Thus a spot of light can
be made to land on different objects within the stereo-field
and the parallax can be determined from the micrometer reading.
To facilitate measurements a displacement transducer mounted on

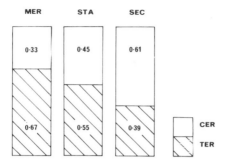

Fig. 4.12. Proportions of CER and TER in meristem,
starch and secretory cells of the maize root cap
measured by point-counting within stereo-micrographs
of thick sections. (Courtesy of J. Stephenson)

the micrometer can be interfaced via an A/D converter to a
microcomputer (Hawes and Doole unpublished, Fink and Faust
1977). After appropriate programming height measurements in
microns can be read directly and stored on disc (Hawes and
Doole unpublished). X and Y co-ordinates may be obtained by
manually gridding the micrographs or from a digitiser tablet
mounted under the platen of the stereoscope (Bauer and Exner
1981).

    Considerable practice is needed before a measuring
stereoscope can be used with any degree of accuracy and
multiple readings are always needed for each point measured.
However, once X, Y and Z co-ordinates are obtained, measurements
and computer reconstructions of structures within the stereo-
field can easily be achieved (Bauer and Exner 1981, Niederman
et al. 1983).

4.5.3c Stereology on thick sections. With the aid of a
measuring stereoscope sampling grids can be used in any given
plane within a thick section. Thomas *et al.*(1974) described
the 'floating test plane' where identical grids were
superimposed on the stereo-pair. When viewed with the
stereoscope horizontal movement of one of the grids varied the
depth of the fused grid plane within the stereo-field.
However, with biological specimens it was found difficult to
sample with either grids of dots for point counting or lines
for intersect counting (Stephenson and Hawes 1985). A
modification of the technique uses a dot grid on one half of
the stereo-pair to provide the X,Y position of the sample
points. Each point is then lowered to a pre-determined
position into the interior of the stereo-field using the
stereoscope floating spot as the sample point. The changing

proportions of TER and CER in the differentiating cells of the
maize root cap could be quantified by this method (Fig. 4.12)
whereas sampling techniques on thin sections could only produce
information on the total ER in the cells (Stephenson and Hawes
1985).

4.6 SUMMARY

The observation of thick specimens by both conventional and
high voltage electron microscopy overcomes many of the problems
in the interpretation of the 3-D organisation of plant cell
cytoplasm. Thick sections can be stained with conventional
heavy metal salts although the final image may appear confused
due to excessive overlapping information. Selective staining
techniques such as osmium impregnation greatly enhance the 3-D
effect in stereo-micrographs and facilitates interpretation of
the micrographs. Semi-thick resin-free sections can be
observed without recourse to heavy metal staining. The high
accelerating voltage of the HVEM allows the study of the
cytoskeleton and cytoplasmic in whole mount preparations of
critical-point dried cells.

    Tilt angles used for the taking of stereo-micrographs
depend on both the specimen thickness and the final viewing
magnification. Quantitative data can be extracted from
micrographs of thick specimens by 1) direct measurement of
spherical components or structures running parallel to the
section surface, 2) by stereometric analysis and 3) by
stereological techniques.

4.7 REFERENCES

Atkinson, A.W. Jr., John, P.C.L. and Gunning, B.E.S. (1974).
The growth and division of the single mitochondrion and other
organelles during the cell cycle of *Chlorella* studied by
qauntitative stereology and three dimensional reconstruction.
Protoplasma 81, 77-109.

Barlow, P.W., Hawes, C.R. and Horne, J.C. (1984). Structure of
the amyloplasts and endoplasmic reticulum in the root caps of
Lepidium sativan and *Zea mays* observed after selective membrane
staining and by high voltage electron microscopy. *Planta* 160,
363-371.

Bauer, B. and Exner, H.E. (1981). Quantification and
reconstruction of spatial objects and surfaces by computer-
aided stereometry. *Proc. 3rd Europ. Symp. stereol. Stereol.
Iugosl.* 3, suppl. (1), 255-262.

Bennet, P.M. (1974). Decrease in section thickness on exposure to the electron beam; the use of tilted sections in estimating the amount of shrinkage. *J. Cell Sci.* 15, 693-701.

Coss, R.A. and Pickett-Heaps, J.D. (1974). The effects of isopropyl-N-phenyl carbamate on the green alga *Oedogonium cardiacum*. I. Cell division. *J. Cell Biol.* 63, 84-98.

Cox, G. and Juniper, B.E. (1973). The application of stereo-micrography in the high voltage electron microscope to studies of cell wall structure and deposition. *J. Microscopy* 97 29-40.

Cox, G. and Juniper, B.E. (1983). High voltage electron microscopy of whole, critical point dried plant cells.   Fine cytoskeletal elements in the moss *Bryum tenuisetum*. *Protoplasma* 115, 70-80.

Cox, G., Morgan, K.J., Sanders, F., Nickolds, C. and Tinker, P.B. (1980). Translocation and transfer of nutrients in vesicular-arbuscular mycorrhizas. III Polyphosphate granules and phosphorus translation. *New Phytol.* 84, 649-659.

Crang, R.E. and Pechak, D.G. (1978). Serial section reconstruction of the black yeast *Aureobasidium pullulans* by means of high voltage electron microscopy. *Protoplasma* 96, 225-234.

Davison, M.T. and Garland, P.B. (1975). Mitochondrial structure studied by high voltage electron microscopy of thick sections of *Candida utilis*. *J. Gen. Micro.* 91, 127-138.

Doole, R.C., Dowell, W.C.T., Hawes, C.R. and Küstner, G. (1983). Beam tilt stereoscopy at 1 MeV. *Optik* 64, 259-262.

Dupouy, G., Perrier, F. and Durrieu, L. (1960). Microscopie électronique. L'observation de la matière vivante au moyer d'un microscope électronique functionnant sous très haute tension. *C.r. hebd Seanc. Acad. Sci. Paris* 251, 2836-2841.

Dupouy, G. (1973). Performance and applications of the Toulouse 3 million volt electron microscope. *J. Microsc.* 97, 3-28.

Elias, H. (1971). Three dimensional structure identified from single sections. *Science* 174, 993-1000.

Elvers, I. (1943). On an application of the electron microscope to plant cytology. *Acta Horti Bergiani* 13, 149-245.

Favard, P. and Carasso, N. (1972). The preparation and
observation of thick biological specimens in the high voltage
electron microscope. *J. Microsc.* 97, 59-81.

Fineran, B.A., Juniper, B.E. and Bullock, S. (1978).
Gramiferous tracheary elements in the hasutorium of the
Santalaceae. *Planta* 141, 29-32.

Finke, H.S. and Faust, H.W. (1977). Use of the G-2 stereocord
in practical planning work. *Vermessungswesen und Raumordnung*
39, 3-18.

Gaunt, W.A. and Gaunt, P.N. (1978). Stereo-photography for
reconstruction. In: *Three dimensional reconstruction in
biology*, pp. 88-131. Pitman Limited.

Gilloteaux, J. and Naud, J. (1979). The zinc iodide-osmium
tetroxide staining-fixative of Maillet. Nature of the
precipitate studied by x-ray microanalysis and detection of
Ca^{2+} affinity subcellular sites in a tonic smooth muscle.
Histochemistry 63, 227-243.

Glauert, A.M. (1974). The high voltage electron microscope in
biology. *J. Cell Biol.* 63, 717-748.

Goldstein, S., Moerman, E.J. and Porter, K. (1984). High
voltage electron microscopy of human diploid fibroblasts during
ageing *in vitro*. Morphometric analysis of mitochondria. *Exp.
Cell Res.* 154, 101-111.

Green, R.J., Perkions, W.J., Piper, E.A. and Stenning, B.T.
(1979). The transfer of selected image data to a computer
using a conductive tablet. *J. biomed. Engng.* 1, 240-246.

Gunning, B.E.S. (1980). Spatial and temporal regulation of
nucleating sites for arrays of cortical microtubules in root
tip cells of the water fern *Azolla pinnata*. *European J. Cell
Biol.* 25, 53-65.

Gunning, B.E.S., Pate, J.S. and Green, L.W. (1970). Transfer
cells in the vascular system of stems. Taxonomy, association
with nodes, and structure. *Protoplasma* 71, 147-171.

Gunning, B.E.S. and Steer, M.W. (1975). *Ultrastructure and the
biology of plant cells*. Edward Arnold, London.

Haskins, E.F. (1976). High voltage electron microscopical
analysis of chromosomal number in the slime mould *Achinostelum
minutum* de Bary. *Chromosa* 56, 95-100.

Harris, J.L. (1975). Some three dimensional aspects of *Ceratocystis ulmi* as observed by high voltage electron microscopy. *Mycologia* <u>67</u>, 332-341.

Harris, N. and Chrispeels, M.J. (1980). The endoplasmic reticulum of mung-bean cotyledons: quantitative morphology of cisternal and tubular ER during seedling growth. *Planta* <u>148</u>, 293-303.

Harris, N. and Oparka, K.J. (1983). Connections between dictyosomes, ER and GERL in cotyledons of mung bean (*Vigna radiata* L.). *Protoplasma* <u>114</u>, 93-102.

Hawes, C.R. (1981). Applications of high voltage electron microscopy to botanical ultrastructure. *Micron*, <u>12</u>, 227-257.

Hawes, C.R. (1985). Conventional and high voltage electron microscopy of the cytoskeleton and cytoplasmic matrix of carrot (*Daucus carota* L.) cells grown in suspension culture. *Europ. J. Cell Biol.* In press.

Hawes, C.R., Juniper, B.E. and Horne, J.C. (1981). Low and high voltage electron microscopy of mitosis and cytokinesis in maize roots. *Planta* <u>152</u>, 397-407.

Hawes, C.R., Juniper, B.E. and Horne, J.C. (1983). Electron microscopy of resin-free sections of plant cells. *Protoplasma* <u>115</u>, 88-93.

Hawes, C.R. and Horne, J.C. (1983). Staining plant cells for thick sectioning: Uranyl acetate, copper and lead citrate impregnation. *Biol. Cell* <u>48</u>, 207-210.

Hawes, C.R. and Horne, J.C. (1985). Polyethylene glycol embedding of plant tissues for transmission electron microscopy. *J. Microsc.* <u>137</u>, 35-45.

Howard, R.J. (1981). Ultrastructural analysis of hyphal tip cell growth in fungi: Spitzenkörper, cytoskeleton and endomembranes after freeze substitution. *J. Cell Sci.* <u>48</u>, 89-103.

Humphreys, C. (1976). High voltage electron microscopy. In: *Principles and techniques of electron microscopy. Biological applications* Vol. 6. (ed M.A. Hayat). pp. 1-39. van Nostrand Reinhold Company.

Hudson, B. and Makin, J. (1970). The optimum tilt angle for electron stereo-microscopy. *J. Phys.* E. <u>3</u>, 311.

Juniper, B.E., Hawes, C.R. and Horne, J.C. (1982). The relationships between the dictyosomes and the forms of endoplasmic reticulum in plant cells with different export programmes. *Bot. Gaz.* 143, 135-145.

King, M.V., Parsons, D.F., Turner, J.N., Chang, B.B. and Ratkowski, A.J. (1980). Progress in applying the high-voltage electron microscope to biological research. *Cell Biophys.* 2, 1-95.

Locke, M., and Krishnan. (1971). Hot alcoholic phosphotungstic acid and uranyl acetate as routine stains for thick and thin sections. *J. Cell Biol.* 50, 550-557.

Marton, L. (1934). Electron microscopy of biological objects. *Nature* 133, 911.

Marty, F. (1973). Sites réactifs à l'iodure de zinc tétroxide d'osmium dans les cellules de la racine d'*Euphorbia characias* *C.R. hebd. Seanc, Acad. Sci. Paris* 227(D), 1317-1320.

Marty, F. (1978). Cytochemical studies on GERL, provacuoles and vacuoles in root meristematic cells of *Euphorbia*. *Proc. Natl. Acad. Sci. USA* 75, 852-856.

Marty, F. (1983). Microscopie électronique à haute tension de l'appareil provacuolaire dans les cellules méristématiques des racines de radis (*Raphanus sativus* L.). *Ann. Sci. Nature. Bot. Paris* 13, 245-260.

Moens, P.B. and Moens, T. (1981). Computer measurements and graphics of three-dimensional cellular ultrastructure. *J. Ultrastructure Res.* 75, 131-141.

Nankivell, J.F. (1963). The theory of electron stereo microscopy. *Optik.* 20, 171-198.

Niederman, R., Amrein, P.C. and Hartwig, J. (1983). Three-dimensional structure of actin filaments and of an actin gel made with actin-binding protein. *J. Cell Biol.* 96, 1400-1413.

Ohno, S., Yoshida, K., Murata, F. and Nagata, T. (1978). Morphometry of cell organelles on thick sections by high voltage electron microscopy. *Electron Microscopy 1978*. (ed. J.M. Sturgess). 2, 14-15.

Palevitz, B.A. (1981). Microtubules and possible microtubule nucleation centres in the cortex of stomatal cells as visualized by high voltage electron microscopy. *Protoplasma* 107, 115-125.

Parker, M.L. and Hawes, C.R. (1982). The golgi apparatus in developing endosperm of wheat (*Triticum aestivum*). *Planta* 154 277-283.

Pawley, J., Ris, H. and Neuberger, D. (1983). Recent projects at the Madison H.V.E.M. *Proc. 41st Ann. Meet. of E.M. Soc. of Am.* (ed. V.W. Bailey). 706-707.

Peachey, L.D., Damsky, C.H. and Veen, A. (1974). Computer assisted three-dimensional reconstruction from high-voltage electron micrographs of serial slices of biological material *Proc. 8th Int. Congress on E.M., Canberra, Aust.* 1 330-331.

Pesacreta, T.C. and Parthsarathy, M.V. (1984). Improved staining of microfilament bundles in plant cells for high voltage electron microsciopy. *J. Microscopy* 133, 73-77.

Peterson, J.B. and Ris, H. (1976). Electron microscopic study of the spindle and chromosome movement in the yeast *Saccharomyces cerevisiae*. *J. Cell Sci.* 22, 219-242.

Porter, K.R. and Anderson, K.L. (1982). The structure of the cytoplasmic matrix preserved by freeze-drying and freeze substitution. *Europ. J. Cell Biol.* 29, 83-96.

Porter, K.R. and Tucker, J.B. (1981). The Ground substance of the Living Cell. *Scient. Am.* 244(B), 41-51.

Poux, N. (1981). Analyse de la cellule végétale par microscopie électronique à haut voltage. *Bull. Soc. bot. Fr. 128, Actual. bot.* 1, 7-18.

Poux, N., Favard, P. and Carasso, N. (1974). Étude en microscopie électronique haute tension de l'appareil vacuolaire dans les cellules méristématiques de racines de concombre. *J. Microscopie* 21, 173-180.

Schliwa, M., van Blerkom, J. and Pryzwansky, K.B. (1982). Structural organisation of the cytoplasm. *Cold Spring Harbor symposium* XLVI, 51-67.

Stephenson, J.L.M. and Hawes, C.R. (1985). Stereology and stereoscopy of endoplasmic reticulum in the root cap cells of maize. *Submitted to Protoplasma.*

Temmink, J.H.M. and Spiele, H. (1980). Different cytoskeletal domains in murine fibroblasts. *J. Cell Sci.* 41, 19-32.

Thiéry, G. and Rambourg, A. (1976). A new staining technique for studying thick sections in the electron microscope. *J. Microscopie Biol. Cell* 26, 103-106.

Thomas, L.E., Lentz, S. and Fisher, R.M. (1974). Stereoscopic methods in the H.V.E.M. *In High Voltage Electron Microscopy. Proc. 3rd Int. Conf.* pp. 255-259.

Traas, J.A. (1984). Visualisation of the membrane bound cytoskeleton and coated pits of plant cells by dry cleaving *Protoplasma* 115, 81-87.

van Dorsten, A.C., Oosterkamp, W.J. and Le Poole, J.B. (1947). An experimental electron microscope for 400 kilovolts. *Phillips Technical Rev.* 9, 193-201.

Whatley, J.M., Hawes, C.R., Horne, J.C. and Kerr, J.D.A. (1982). The establishment of the plastid thylakoid system. *New Phytol.* 90, 619-629.

Williams, M.A. and Meek, G.A. (1965). Studies on thickness variation in ultrathin sections for electron microscopy. *J. Roy. Microsc. Soc.* 85, 337-352.

Willis, R.A. (1972). Optimization of stereoscopic and high angle tilting procedures for biological thin sections. *J. Microscopy* 98, 379-395.

Wolosewick, J.J. (1980). The application of polyethylene glycol to electron microscopy. *J. Cell Biol.* 86, 675-681.

Wolosewick, J.J. and Porter, K.R. (1979). Microtrabecular lattice of the cytoplasmic ground substance; Artifact or reality. *J. Cell Biol.* 82, 113-139.

Wright, M. and Moisand, A. (1982). Spatial relationships between centrioles and the centrosphere in monasters induced by taxol in *Physarum polycephalum* amoebae. *Protoplasma* 113, 69-79.

5

Quantitative morphological analysis of botanical micrographs

L. G. Briarty

5.1 INTRODUCTION

To anyone who has observed cytoplasmic streaming taking place in a vacuolate plant cell, or who has watched some of the elegant time-lapse filmed studies of fertilisation or of cell division in endosperm, the possibility of being able to record these phenomena in quantitative terms must seem remote. When the inevitable distortions brought about during fixation and processing for light or electron microscopy are also considered, it must appear even less likely that data from any quantitative analysis of such material represents the situation to be found in the living cell.

While there exists, as we shall see, a wide range of techniques and procedures for obtaining accurate quantitative information from images such as micrographs, it must be appreciated that in reality most of the attempts to use the techniques for obtaining data on living plant cell systems are compromises. The compromise exists at three levels; at the lowest lies the acceptance that cell structure, following fixation and processing, is unlikely in many cases to be exactly representative of the living state (Mersey and McCully 1978). The second compromise is the assumption that the cell parameters of importance are both measurable and measured. It is tempting to regard numerical data as being somehow more real than the images from which they have been derived; certainly they are more easily handled, but cannot represent more than a sub-set of all the information really present, since they only represent what has actually been looked for. The third compromise concerns the representativeness of the samples actually studied; while large samples may be theoretically desirable they may not be

achievable in practice. With these provisos, then, the careful
application of quantitative analytical procedures can provide
data which will increase our understanding of plant cell
structure and function.

5.2 METHODS FOR QUANTITATION

5.2.1 Approaches to quantitation

In attempting to measure cellular structure from sectioned
material, the investigator is immediately faced with having to
take one of two contrasting approaches. These are either to
study in detail the structure of one or a few associated cells,
taking accurate measurements which may be true only for the cell
or cells measured, or to sample a large cell population in a
systematic way, and to estimate average values for parameters
which may not actually be the case for any of the individual
cells making up the population. These approaches are typified
by, on the one hand, serial sectioning followed by
three-dimensional reconstruction and observation and measurement
of the model produced, and on the other sterological analysis of
sections from suitably randomised samples, calculating
parameters from measured estimators and the known relationships
between their appearance in a two-dimensional section and their
occurrence in a three-dimensional volume.

5.2.2 Reconstruction from sections

5.2.2a Serial sectioning. Serial sectioning and
reconstruction provides the simplest, and in some case the only
way, of determining some features of cell structure. The shapes
of cells themselves and of individual structures within them can
be accurately determined, as can the spatial relationships
between them and their contained structures. While sterological
analysis is suitable for determining bulk parameters concerning
volume and surface relationships, a knowledge of shape is
necessary when using sterological procedures for determining
number (see below). Features such as the connectivity of the
chondriome, for example, can be determined only from serial
sections through cells, and the nature of contact areas betwen
cells or organelles is most easily appreciated from such
preparations.

Techniques for the production of serial sections and for
maintaining their correct registration have been described in
detail in their book by Gaunt and Gaunt (1978), and some of the
problems of interpretation of three-dimensional structures are

discussed by Steer (1981).

Serial sectioning can be useful both at light and electron microscope levels, depending on the size of the objects studied. This latter is important in determining section thickness, and the spacing between sections in cases where not all adjacent sections are used in the reconstruction. (Zilles, Schleicher and Pehlemann 1982). While fixation, embedding and physical sectioning are most frequently used, there are alternative non-destructive procedures that are available in some cases. Optical sectioning is possible on fairly transparent live specimens, while the exciting possibility of obtaining clear, contrasty optical sections from translucent specimens is provided by the tandem scanning reflected light microscope (TSRLM)(Petran, Hadravsky, Benes, Kucera and Boyde 1985).

Reconstruction of the form of the specimen may be undertaken in a number of ways. Perhaps the simplest is a straightforward graphic procedure where details of the sections are drawn superimposed; this may be improved by drawing the individual sections in pairs, with the spacing of one set staggered so that parallax between them is introduced and the images may be viewed stereoscopically (Yamada and Yoshida 1972). Alternatively models may be constructed from the section information using transparent or opaque sheet material, depending on the degree of internal complexity that is required to be resolved.

Until recently these were the only techniques available to most workers. With the advent of microcomputers and sophisticated graphics software, however, a range of programs has been developed which will allow the storage and display of serial section information, with the possibilities of rotation of the reconstructed image so that it can be viewed from any direction, or used to produce stereo pairs.

There are many examples of such computer reconstructions in botanical microscopy. Stevens (1977) used such a procedure to study the number of mitochondria and volume of the chondriome during the cell cycle of *Saccharomyces cerevisiae*, demonstrating a cyclic process of fragmentation and fusion, and the presence of a single giant mitochondrion. Atkinson, Gunning and John (1974) carried out a similar analysis of *Chlorella* throughout the cell cycle and in addition used stereological procedures. They were able to show growth of most organelles during cell growth, changes in nuclear surface area and volume, and breaking up of the mitochondrical reticulum on cell division, and subsequent fusion of the fragments; *Euglena* was similarly described by Pelligrini (1976) and Pelligrini and Pelligrini (1976). In a careful study of pollen organisation Russell (1984)

has described the complex, three-dimensional relationships of
the sperm cells of *Plumbago*, showing that the two cells differ
in size and in their content of plastids and mitochondria, and
discussing the implications that this has on the identity of the
sperm which fuses with the egg. At a higher level of
magnification Kristen (1984) has described the structure of the
dictyosome in cultured sycamore cells, showing the connections
between cisternae and associated elements.

Absolute measurements of the volumes of cellular
components can easily be made from serial sections if the
magnification and the section thickness and/or spacing is known;
the volume is simply the sum of the individual 'slice' volumes
(areas x thickness) with a correction factor if necessary to
allow for the perimeters being cut at an angle. Schotz (1972)
describes a structural and volumetric analysis of *Chlamydomonas*
using this approach, Gaffal (1983) has followed the dynamics of
the nuclear envelope in the phytoflagellate *Polytoma papillatum*,
and Russell (1984) has measured the cell and organelle volumes
in *Plumbago* sperm.

In most cases data for this type of reconstruction are
saved by tracing the section outlines using a digitizer pad.
This sends the data to its host computer as a stream of x,y
co-ordinates which are stored and used to produce a composite
image. Many programs have been published for the computerized
production of reconstructions from serial sections (e.g. Tipper
(1977), Veen and Peachey (1977), Dykes and Afshar (1982) and
Briarty and Jenkins (1984)), and there is a bibliography of
serial section software in Howard and Eins (1984). Ferwerda
(1982) gives comprehensive information on viewing stereo images.

Once the digitized section data are saved on a file,
measurements in three dimensions may be made from them (Moens
and Moens 1981); alternatively stereo pairs may be produced from
the stored images (Briarty, Fischer and Jenkins 1982) and then
these used to calculate dimensions from parallax distances
following photogrammetric procedures (Smith 1984).

5.2.2b Digitized data recording. The digitizer pad itself is
a useful tool for measurement of distance, length, area and
other parameters from microscope images, in combination with a
microcomputer and suitable software. Some software is available
commercially (e.g. 'Dimensions', A.W. Agar), the principles of
digitizer measurement and some algorithms are given by Moss
(1981), Santi, Fryhofer and Hansen (1980), Coster and Chermant
(1980) and Chermant and Coster (1981), and a suitable system is
described by Peachey (1981,1982). Mize (1984) reviews
planimetric and reconstruction procedures using computers.

While the device can give rapid and accurate measurements under suitable conditions, it has been pointed out by Cornelisse and van den Berg (1984) that in certain situations overestimation of distance can occur. This results from the fact that the co-ordinates from which the final value is calculated are situated on a square lattice and are sampled at a finite rate; any line at an angle to the lattice will be measured as a series of steps, the intervals depending on the angle of the line with respect to the lattice, the rate at which the cursor is moved, and the rate at which the co-ordinates are sampled. The overestimate could be as much as 41%, and they suggest techniques to minimise such errors.

5.2.2c Reconstruction from semi-thin sections. Where objects are too large to be included in the thickness of a single semi-thin (i.e. 1 μm) section, high voltage electron microscopy of such sections can be used to yield high resolution data for subsequent reconstructions. Such an approach is most useful when trying to obtain three dimensional information on structures too small to be clearly resolved by light microscopy, and yet too large to be conveniently contained within a reasonable number of conventional thin sections (Peachey, 1975). Calveyrac and Lefort-Tran (1976) used such an approach to understand chondriome and plastidome organisation in *Euglena* ; both are present as networks in the non-dividing cell, and using 0.5 or 1.0 μm thick sections they were able to section through complete cells using 15 to 20 sections.

5.3   STEREOLOGICAL APPROACHES

The general principles of stereological analysis of sectioned material by point and intersection counting are well established, and their application to the study of biological material has been developed over the past twenty or so years. Methods for estimating volume, surface and numerical density (volume of a component per unit volume, VV, surface area, SV, and number per unit volume, NV, respectively) as well as some other parameters are clearly set out by Weibel, Kistler and Scherle (1966) and Freere and Weibel (1967). Steer (1981) surveys some of the practical problems of stereological analysis giving a number of worked examples, and Weibel's (1979,1980) two-volume work deals comprehensively with both practical methods and the theoretical bases for their development, giving a bibliography of earlier work. Many general presentations of the methods have appeared in the literature (e.g. Chayes 1965; Briarty 1975; Williams (1977) and James 1977)), and Toth (1982) gives an introduction to stereology with particualar reference to botanical applications.

5.3.1 Problems peculiar to plant tissues

The greater part of stereological analyses of biological
material has been carried out on animal tissues, and in many
cases on tissues which are present in large volumes and are
relatively homogeneous such as lung or liver. The relatively
small absolute volumes of many plant tissues, together with a
frequently high degree of anisotropy and inhomogeneity, and the
frequent difficulty of defining cell shape and volume
characteristics, provide a set of problems for which a range of
solutions has been developed.

5.3.1a Specific volume definition. When the comparison of
stereologically derived parameters from different samples is
necessary the data for comparison must be of the same type; in
the case of volume, surface and numerical density it is often
most logical to present these on a per cell basis, rather than
simply as a percentage volume of tissue (see also 5.3.1e below).
For this to be done a knowledge of the size of the cells in
question is required. In cases where cell shape is regular and
geometric a value for cell volume may be obtained from
measurements on transverse and longitudinal sections;
alternatively the cells may be separated by maceration in acid
(Lowenberg 1955) or in a suitable wall-degrading enzyme, and
individual cells measured directly.

     Where cell shape is not easily defined, but cell number
per unit volume of tissue (numerical density) is known,
measurements from sections can yield data on mean cell volume;
if the volume density of tissue occupied by cells is calculated,
then the mean cell volume will be given by $(1/NV).VV$.
Alternatively, if the cells are uninucleate, the nuclei are
spherical, and the volume density of nuclei (VVnuc) is known,
then measurements of nuclear profile diameters on sections can
be used to estimate the mean nuclear diameter, and hence the
nuclear volume, Vnuc. The cell volume is then Vnuc/VVnuc. (for
details see Steer 1981). Since the possibility of a structure's
appearance in a section is related both to the number of
structures per unit volume and to their shape and size
distributions, estimates of numerical density are not easy.
Weibel (1979) discusses the approaches to such analyses, and
some empirical equations for calculating NV for variously shaped
objects are given by Durand and Warren (1981).

5.3.1b The cell population under investigation. The basis of
sterological sampling lies in ensuring that the samples are
representative of the cell population as a whole, and that all
cells have an equal chance of being sampled. The corollary of
this is that unless all the sampled cells represent only a
single cell population then any calculated parameters will

represent an average of all the cell populations, and thus may
have no real existence. While this is probably not of major
importance in studying differentiated tissues, in cases where
the object of study is undergoing major change, or where
recognition of a cell type is ambiguous, it becomes significant
(Fritsch 1982). Thus a study of a population of dividing cells
will be of limited value unless division in the system is
synchronous; the data will represent only a ficticious average
cell whose structural parameters will depend on the changes that
occur during the cell cycle and the proportion of the cells at
each stage. A number of studies of developing plant cells appear
to have been carried out without appreciating this point (see
below).

5.3.1c Anisotropy. Tissue anistropy is perhaps the most
difficult feature of plant material that has to be taken into
account when sampling for stereological analysis. While in most
cases volume densities may be estimated without error, accurate
estimation of other parameters requires a rigorous approach.

The higher plant leaf is perhaps one of the most
profitable organs on which to carry out structural analyses
since there is a wealth of biochemical and physiological data
with which to relate such information. While knowledge of
surface parameters, such as internal cell wall area exposed to
air space and cell contact areas, can be directly related to
physiological and gas exchange data, many of the problems of
sampling in an anistropic organ are posed by such a study. The
basic stereological relationships require that the specimen
should be isotropic and sectioned at random. While volume
density estimates of anisotropic structures are independent of
the sectioning angle (Weibel, 1979), it is clear that surface
density estimates, made from intersection counts with test
lines, will be influenced by the section plane's relationship to
any organised structure. Thus the intersection counts from a
transverse section of leaf palisade tissue will be very
different from counts made from a paradermal section through the
same tissue, and if the test system is composed of straight
lines then the counts will be further influenced by its position
in relation to the section.

The leaf exhibits a number of anisotropies; the different
tissues are present in graduated layers, vascular traces run
through them in a non-random way, and within the various layers
the cells are oriented, quite strongly in the case of the
palisade tissue and epidermes and less so in the case of the
mesophyll.

A first approach to the analysis of anisotropic material
is to measure the degree of anisotropy that is present, and then

to make specific allowance for it in subsequent calculations.
Where anisotropy is suspected it can be tested for quite simply
by rotating a rectilinear test system in 5 or 10 degree
increments over the sample, and counting the intersections made
with the relevant interface at each position. A rose diagram, a
plot of number of intersections radially for each angle, will
indicate, by its deviation from a circle, the direction(s) and
degree of anistropy. Where the anisotropy is regular and
specifically oriented, surface estimates may be made by taking
intersection counts with the test system parallel and
perpendicular to the direction of orientation (Underwood 1970).

Several different approaches have been used to derive
accurate estimates of structural parameters of anisotropic leaf
tissues. The simplest way to ensure that the test system used
for measuring intersects with a random sample of cells at random
angles is to randomise the leaf cells before measurement. This
may be done by isolating them enzymically, and then fixing,
embedding and sectioning the entire macerate, as has been
suggested by Morris and Thain (1983) and Thain (1983). They also
proposed a rather more sophisticated approach to leaf cell
surface density measurements on transverse sections of leaf
mesophyll, and present simple correction factors, based on
direct measurements or intercept counts, which may be used to
take into account the organisation in this tissue. Parkhurst
(1982) has also discussed the problems of surface and volume
density measurement on anisotropic structures such as wood and
leaf, and suggests a cylindricity parameter, k, for correcting
surface density measurements on leaf palisade cells, as well as
a test system of randomly oriented lines.

Where a tissue is organised in layers, with perhaps
variation within the layers, test areas should be positioned so
that such variation is taken into account. The analysis of
structures whose organisation is longitudinal (e.g. wood,
vascular bundles) is discussed by Thain (1983), Mathieu,
Cruz-Orive, Hoppeler and Weibel, and Weibel (1972), and a
discussion of different test systems and their application to
the analysis of transfer cells in vascular tissue is given by
Gunning, Pate, Minchin and Marks (1974).

5.3.1d Isolated cell measurements. Stereological analysis of
a population of isolated cells, algal unicells for example,
appears at first sight much simpler than that of an organised
tissue; sections from a pellet prepared from a single culture
will often give a sufficiently large sample size that is
representative of the whole population.

There are a number of precautions that should however be
borne in mind in such a study. The first of these relates to the

state of the cell population overall; unless all the cells are
in the same stable state, or are synchronised in development,
then the derived parameters will be averages of all the states
in the population (Fritsch 1982), unless there is some
morphometric marker to allow cells in a particular state to be
unambiguously sampled. However, the choice of such a marker, say
a nucleus having a particular form, involves choosing sections
which include that marker, thus biasing the sample. In the case
where the marker is the nucleus then the result will be a
systematic over-estimation of nuclear volume density and
underestimation of cell surface area.(Fritsch 1982). If some
details of the shape and eccentricity of the cell and its
contained nucleus are known,it is possible to correct such
volume and surface data (Cruz-Orive 1976a,b).

Sections taken from a pelleted population of free cells
should of course be representative of the whole population,
taking into account any sedimentation or separating-out of
different cells during the preparation. If a normal square test
area is used for making counts then the result will be an
over-estimation of peripheral structures in the cell, since the
counting frame will cut off peripheral samples from profiles
which it does not completely cover. In order to correct this
Fischer (1977) suggests a counting frame containing an inner
sampling area, inset by the radius of the cells under study;
only profiles of cells whose centres lie within the inner area
are counted.

5.3.1e Sampling. The choice of sample size and sampling
approach determines both the effort required and the validity of
the results of any stereological analysis. Choice of the basic
sample size for volumetric analysis by point counting is fairly
straightforward and depends on the volume density of the
component phase being measured (based on a small pilot survey)
and the degree of error which can be tolerated; Weibel (1979)
gives a formula and nomogram for deriving the required number of
test points. Having determined the test point number their
spacing (d) should be such that at least one (and ideally not
more than one) point falls over each profile of the phase being
measured, in general this is achieved by a point spacing on a
square lattice of $d=\sqrt{a}$, where a is the mean profile area. In
most plant material, however, interest will be centred not just
on one cell component but on several, and the point number and
spacing will thus be in many cases a compromise.

Information on point number and spacing can be used to
calculate the basic representative sample area, and this may
consist of one or more micrographs depending upon magnification;
usually images will be square to fit a coherent test system
format (Weibel 1979, appendix 3).

In many cases a study of cell structure will involve
components occupying both large and small volume densities in
the tissue; thus a study of leaf cells might require data on
volume densities of intercellular space and of microbody
content, components occupying perhaps 40% and 0.5% of the
reference phase (the tissue volume) respectively. In such a case
an analysis designed to estimate microbody volume density would
be wasteful of effort if at the same time used to assess air
spaces; a much lower magnification and point density would be
adequate for the latter. To obtain the greatest accuracy the
measured phase should occupy a large proportion of the reference
phase, and the magnification should be as low as possible while
still allowing recognition of the measured phase (Cruz-Orive,
Gehr, Muller and Weibel 1980).

To fulfil these conflicting requirements a sequential,
multi-level sampling strategy is required (Cruz-Orive and Weibel
1981). The sample is photographed at several levels of
magnification; at each level the phases which can be easily and
accurately measured are estimated, and then one of these
measured phases (the one containing further phases to be
estimated) becomes the reference phase at the next highest
magnification level. Thus a study to determine the composition
of a vacuolate tissue such as apple fruit might begin with
fairly low power light micrographs; these could probably be
resolved into air spaces, walls, cytoplasm (including
organelles) and nuclei, and the volume density of each of these
determined by point counting with respect to the total tissue
volume – the reference phase in this case. A second higher level
of magnification would be necessary to distinguish the
organelles within the cytoplasm; here the sample micrographs
would be taken from the cytoplasm alone, since higher resolution
of the other tissue components would not be necessary, and the
resulting volume densities of components expressed per unit
volume of cytoplasm. Since the proportion of cytoplasm within
the tissue has already been estimated, specific tissue
parameters in real units may be calculated for all the cell
components measured. Such a process may be continued at higher
levels of magnification if necessary to estimate structural
parameters of organelles. An example of such an approach is
presented in Briarty 1980a.

Such a sampling method requires random placing of the test
system over the sample; where the sample phase under study is
relatively small (as in the second level above), then efficient
random sampling requires a systematic procedure such as the
surface area weighted quadrat (SAWQ) method described by
Cruz-Orive and Weibel (1981).

When sampling tissue sections with a conventional Weibel-type test system for estimating volume density, the normal procedure is to calculate for each sample the necessary parameters as quotients (e.g. VV vacuole = points over vacuole / total point number), and then to average the sample quotients so that a mean and a standard error can be calculated for the parameter to check the adequacy of the sample size and the degree of variance in the samples.

Where the total point number is the same for each sample this is quite suitable. However, Mayhew and Cruz-Orive (1974) have pointed out that where the sample size is not the same in each case, such as when counting on isolated cells, or small phases in a cell, then it is more accurate to calculate the sample parameters as quotients of the means (per sample) of the points over the phase in question, and the total point number (e.g. VV nucleus = mean of points over nucleus/mean of total points per sample). Though a standard error cannot be calculated in this case, Weibel (1979,p.138) describes the application of the standard error of the ratio estimate to determine parameter errors so calculated.

There is a great deal of discussion in the stereological literature of sampling procedures, and of the balance of economy against sampling efficiency (e.g. Gundersen and Osterby 1980, Kroustrup 1981). It has been clearly shown, at least for animal tissues, that of the total variance in a stereological analysis the majority results from the biological variation between organisms (Gundersen and Osterby 1981). Thus, in a pilot experiment on a drug effect on rat kidney basement membrane structure using 10 animals, 3 tissue blocks per animal, 9 pictures per block and 16 intercept measurements per picture, the contributions to the total variance at each step were as follows:

Animals	70%
Blocks	19%
Fields	8%
Intercepts	3%

Clearly the easiest way to increase accuracy is to increase the sample size at the lowest level, to investigate more organisms. This is neatly summed up by Weibel (quoted in Gundersen and Osterby 1981) as "Do more less well!". More organisms measured at a lower degree of accuracy will give a more representative result than a smaller number sampled more accurately. Similar results are presented by Gupta, Mayhew, Bedi, Sharma and White (1983). Thus studies where the basic sample level is perhaps a single organism are of questionable value.

It is important that all samples for an analysis should be processed in an identical way so as not to introduce further variation; some effects of different fixation and embedding procedures on tissue parameters are described by Mathieu, Cruz-Orive, Hoppeler and Weibel (1983), Loud, Barany and Pack (1965), and Bertram and Bolender (1983). Mathieu and Costello (1983) point out that the same fixative solution may produce different effects in specimens of the same material but in different physiological states.

Having calculated structural parameters for a cell or tissue they need to be expressed in a way which is both meaningful, and which allows their comparison with other (biochemical or physiological) data on the tissue in question. Volume, surface and numerical densities are calculated on the basis of total tissue volume; these may be converted to specific units, say mitochondrial volume per cell, if the cell volume is known and other compartments which do not contain the structure in question can be accounted for.

The use of the cell as a reference volume is adequate in many cases. However, it assumes a constancy of size and form for the average cell, and knowledge of cell volume which is not always easy to obtain. Bolender (1978) shows very clearly that direct comparison of volume densities as a percent volume of a tissue can be ambiguous, since a change in the relative volume of any one compartment produces an opposite change in relative volumes of other compartments, while their absolute volumes may remain constant. Any change in cell volume by division will involve changes in absolute volumes of components per cell while their volume density overall in the tissue may remain constant. Bolender (1978) suggests that in some cases, where comparisons with biochemical data are being made, instead of a reference volume such as the cell being used, data should be related to a reference area. Bolender (1983) discusses the problems of integrating stereological and biochemical data, while problems of changing reference volumes are also discussed by Osterby and Gundersen (1980).

5.3.1f Automatic and semi-automatic methods. While the use of automatic image analysers and semi-automated tracing devices might at first seem attractive as means of speeding up stereological analyses, for most straightforward analyses of biological specimens they are of limited value. While such devices can give rapid and accurate measurements of distances, areas and numerical counts on individual images, it has been

shown that for comparable precision a point counting procedure
may be up to eleven times faster than a tracing device (Mathieu,
Hoppeler and Weibel 1980). Since the overall error is determined
largely by variation betwen the sampling units (see 5.3.1e
above), precise sampling at the individual level is of little
relevance (Gundersen, Boysen and Reith 1981, Mathieu,
Cruz-Orive, Hoppeler and Weibel 1981).

Fully automated image analysis is of value only when the
various object phases can be clearly and unambiguously
identified by the instrument. The human eye is still better at
recognising cell organelles in sections than computer software,
however there is some work reported on automatic recognition of
mitochondrial membranes (Favre, Keller, Mathieu and Weibel
1980).

5.3.2 Software for stereology

Data collection for sterological analysis is a tedious procedure
and can be considerably eased by using a suitably programmed
microcomputer on which subsequent parameter calculation and
statistical analysis can also be performed. As in most
microcomputer applications absence of standardisation makes the
transfer and distribution of such programs difficult, with the
result that a number of workers have developed different
programs to carry out similar functions.

Howard and Eins (1984) have summarised in detail the
published software available up to mid-1984. They list programs
under the headings of classical point counting methods, size
distribution analysis, serial section and 3-D reconstruction,
statistics, educational aids, image analysis and special
applications; the programs are also clasified according to both
language and machine make. Since this review two further
programs have appeared; Vodopich and Moore (1984) describe a
FORTRAN IV program, MORPHO, for point counting stereology
calculation and data analysis, while Larsen, Pederson,
Pentcheff, Bertram and Bolender (1983) detail a BASIC program,
designed for the Tektronix 4050 computer, which carries out
similar data collection, storage and analysis. Hoppeler,
Mathieu, Bretz, Kramer and Weibel (1980) discuss the use of
small computer systems in stereology.

Although interest in this type of software is restricted
to the scientific community, the increased sophistication of
"off-the-shelf" business software is worthy of consideration.
The "spreadsheet" type of program, normally used for financial
analysis and projection, has some application in stereological
analysis; here data are entered into a matrix and processed by

the program through a sequence of user-defined formulae. Such
programs have potential, particularly where data gathered by an
experimenter's program can be fed into and processed by such a
spreadsheet. Not all such programs will accept data from a file,
hjowever, and where this is possible, care is needed to ensure
that a compatible file structure can be produced.

While it is perhaps not realistic to suggest a single
standard for stereology software, as Howard and Eins (1984)
argue it would be helpful for communicating and adapting
programs if in-house software packages could be written in a
limited number of language dialects (PASCAL AND Microsoft BASIC
are proposed), and designed so that machine-dependent routines
are kept separate from the main part of the algorithm.

5.4 QUANTITATIVE STUDIES ON PLANT CELLS

Over the last five or so years the number of published papers
presenting stereological analyses of plant cells has grown
rapidly. Many of the studies are analyses of a single cell type
or organ, either at a fixed point in time, or throughout a
developmental sequence; others concentrate on the development of
an organelle or a group of organelles. In many cases the results
can be related to already available biochemical data, in others
where there are no biochemical data available the changes in
structure allow predictions as to the system's biochemistry.

5.4.1a <u>Nuclear structure</u>. Changes in the nucleus during
development have been noted by a number of workers; the ratio of
dispersed to condensed chromatin has been shown to increase in
the shoot apical meristem of mustard immediately after
transition to flowering, and increases in nucleolar size and
vacuolation indicate a synthesis of ribrosomes in evoked
meristems (Havelange and Bernier 1974). In a similar situation
in *Xanthium* there was a change in nuclear structure (Havelange
1980). Reynolds (1984) demonstrated an increase in dispersed
chromatin in pollen grains of *Hyoscyamus niger* when they were
induced to undergo embryogenic development. In a general study
of nuclear structure in various tissues of a number of unrelated
species, however, Nagl, Cabriol, Lahr, Greulach and Ohliger
(1983) showed that the proprtion of condensed chromatin is
species-specific but not tissue-specific or function-related.

Changes in nuclear volume, and in the ratio of cell to
nuclear volume, have been presented in a number of developmental
studies. In the developing wheat endosperm for example the cell
volume:nuclear volume ratio varies from 23.7 to 49.0 during
reserve deposition (Briarty, Hughes and Evers 1979) and the
nuclear volume increases without any DNA increase, whereas in

the developing cotyledon of *Phaseolus* during reserve deposition
(Briarty 1980b) there is a DNA increase of 3.2 times, and a cell
volume increase of 29 times; the cell volume:nuclear volume
ratio increases from 17 to 26. Dean and Leach (1982) have
demonstrated a relationship between plastid number and ploidy
level and ribulose bisphosphate carboxylase activity.

5.4.1b Plastids. Quantitative data on plastid structure and
development are fairly widely available. The behaviour of
plastids under a gravitational stimulus has been studied in both
normal roots and in those which show little or no graviresponse
in order better to understand their role in this phenomenon.
Comparisons of columella statocytes in agravitropic mutants of
pea and *Arabidopsis* with those in normal plants (Olsen and
Iversen 1980; Olsen, Mirza, Maher and Iversen 1984) have shown,
by comparisons of organelle distributions in different cell
quadrants, that in the mutants the redistribution of plastids
after geostimulation may be affected by a different distribution
of rough endoplasmic reticulum, so that the rate at which they
fall through the cytoplasm is slowed.

Distribution of statocyte organelles has also been
investigated by Moore (1983), Ransom and Moore (1984) and Stoker
and Moore (1984) in attempting to understand graviresponse.
While it is generally accepted that amyloplast sedimentation in
statocytes is a primary response to the gravitational stimulus,
these studies show that in *Zea mays* , *Phaseolus vulgaris* and
Helianthus annuus the distribution of all the organelles is
asymmetric, not simply that of the statoliths; when the
"diluting effect" of plastid redistribution is taken into
account asymmetry of the other organelles remains evident (Moore
1983). Hensel and Sievers (1980) measured the effects of
prolonged klinostat rotation on statocytes, finding that ER
cisternal length was increased and the starch content of the
amyloplasts decreased; the symmetry of the statenchyma was lost,
and graviresponse delayed in rotated roots.

Roots differ in their graviresponse, that of primary roots
being greater than laterals. Stoker and Moore (1984) have shown
that in *H. annuus* this response is not a reflection of a
difference in cell ultrastructure in the statocytes of the two
types of root, since they posses a similar volume density of
organelles. However the total volume of columella tissue is less
in laterals than in primary roots (Moore and Pasieniuk, 1984a
and b) and the response may be related to the total volume of
columella tissue (Moore 1984) or to the number of amyloplasts
per cell (Ransom and Moore 1983). Other quantitative studies on
ultrastructural changes during root cap differentiation have been
carried out by Clowes and Juniper (1964), Juniper and Clowes
(1965), Moore and McClelen (1983) and Moore and Coe (1984); the

latter suggest that the changes associated with cellular
differentiation in the root cap are organelle-specific,
different organelles changing at different rates with the
development of particular functions. The effect of IAA, as a
growth substance likely to be involved in graviresponse, on oat
coleoptile cells was investigated by Shen-Miller and Gawlik
(1977); they found an apparent decrease in the numbers of
mitochondria and microbodies (but not of plastids) in the
expanding subapical cells, but no effect on the contents of the
apical cells, and suggested that the decrease in number was in
fact a decrease in volume.

Sampling accurately small volumes of tissue, such as
differentiating regions of perhaps only a few cells in the root
cap or root tip is not easy; it is difficult to ensure that like
areas are being compared with like. To simplify comparisons in
such situations Darbelly and Perbal (1984a and b) have devised a
semi-quantitative procedure for describing cell structure. For
each organelle one or more characteristics are scored on a scale
of 0 to 3, and the cell is then described by means of a formula
containing all the scores. Comparison of cell types is then made
simply by comparing the formulae; common values cancel out
leaving the essential differences. Used to compare the structure
of horizontally placed *Lens* root cortex cells undergoing
gravitropic bending, the technique indicated that the cells
undergoing extension in the upper part of the root were
differentiating faster than controls, while in the lower part
though cell length was similar to that of the controls, the cell
contents were in a more juvenile state. This was interpreted as
indicating a downward movement of inhibitor substance, with a
lower concentration in the upper part of the root than in the
equivalent part of the controls, and higher concentration in the
lower part. Though this method is not entirely objective, it is
as useful as, and perhaps less misleading than a "statistical"
method where sampling may not be entirely objective.

A number of studies providing data on plastid number per
cell give an insight into plastid division. In analysing
developing wheat endosperm Briarty, Hughes and Evers (1979)
showed that plastid division and cell division were not
synchronised, plastid division stopping before cell division so
that the mean plastid number per cell dropped. The final
composition of the endosperm in terms of the two types of starch
granules present has been described (Hughes and Briarty 1976),
as has the development of small starch granules in germinating
legume seeds (Briarty and Pearce 1982).

Clauhs and Grun (1977), studying areas of organelles in
cell sections rather than deriving stereological parameters,
followed the fate of plastids and mitochondria in maturing

pollen cells of *Solanum*; they showed a loss of plastids in the generative cell which they attributed to a lethality factor.

Fagerberg (1984a), studying developing *H.annuus* leaves, suggested that organelle replication is independent of cell division; Thomas and Rose (1983) also showed a lack of co-ordination between cell and plastid division during the early divisions of tobacco mesophyll protoplasts. In *Euglena* Orcival-Lafout, Pineau, Ledoigt and Calveyrac (1972) demonstrated changes in the chloroplast during the cell cycle; before cell division the plastid breaks down into small units which are shared between the daughter cells and subsequently fuse.

A number of studies have related physiological changes to quantitative changes in plastid ultrastructure. Fagerberg and Dawes (1977) studying wound regeneration in *Sargassum* showed that thylakoid surface density increased, perhaps giving a capability for a higher photosynthetic rate, and in the same species Fagerberg, Moon and Truby (1979) showed that rates of photosynthesis and respiration in blade and stipe tissues could be related directly to membrane surface area of plastids and mitochondria on a tissue basis. The effect of low light levels on plastid structure in *H.annuus* was to increase the surface to volume ratio of chloroplast membranes (Fagerberg 1984b), while Berlin, Quinsberry, Bailey, Woodworth and McMichael (1982) showed that the effect of water stress on cotton leaf cells was to produce smaller chloroplasts with increased grana and stroma lamellae surfaces; however stress also reduced leaf size and chloroplast volume on a per leaf basis. Santos and Salema (1983) demonstrated the occurrence of diurnal changes in chloroplast tubules of the CAM plant *Sedum telephium*. Jellings and Leech (1982) point out the importance of assessing the proportions of different cell types in interpreting biochemical data on whole leaf preparations.

Davydenko, Palilova and Shchetinina (1984) used 12 stereologically derived chloroplast parameters in a study of the plasmagenes of *Aegilops* and *Triticum*; they found significant differences between most of the characters in the two genera, and suggested that there is a wide range of variability for the plasmagenes determining chloroplast structure.

5.4.1c Mitochondria. One of the areas in which quantitative measurement of these organelles has been useful is the better understanding the chondriome genome. Bendich and Gauriloff (1984), studying the relationship between genome size and morphology in cucurbit mitochondria, concluded that the average 2N/4N cell probably contains 200 to 300 genomes distributed among a larger number of organelles. They further suggested that

an individual mitochondrion might not have a whole genome, and
that this might be related to the observed coalescence of these
organelles. Davydenko (1984) and Davydenko and Palilova (1984)
carried out comparative studies on chloroplast and mitochondrial
ultrastructure in a range of alloplasmic wheats, and found
significant differences in mitochondrial structure in the
alloplasmic lines; however a classification based on
mitochondrial structure did not agree with the phenotypic
classification. They concluded that cytoplasms differ in the
plasmagenes responsible for their mitochondrial structure.

At the level of organelle biochemistry, John, McCullough,
Atkinson, Forde and Gunning (1973) and Forde, Gunning and John
(1976), studying mitochondrial development through the cell
cycle in *Chlorella*, found that while growth of the inner
membrane paralleled overall cell growth, the synthesis of
respiratory enzymes was not similarly co-ordinated, indicating a
change in enzyme density of the membranes during the cell cycle.
Opik (1973), too, noted that in rice coleoptiles, differences in
respiratory metabolism were not parelleled by changes in
mitochondrial structure.

Physiological effects on mitochondria have been noted by
Hajibagheri, Hall and Flowers (1984) where growth of *Suaeda
maritima* in increased salinity levels produced significant
increases in mitochondrial volume and surface density. Fagerberg
(1984a), from an analysis of developing *H.annuus* palisade cells,
noted a decreasing mitochondrial role as the cells matured.

Reconstructions demonstrate a single mitochondria in
Euglena cells (Pelligrini 1976; Pelligrini and Pelligrini 1976;
Calveyrac and Lefort-Tran 1976) and in *Pleurochrysis* (Beech and
Wetherbee 1984), and differing numbers of the organelles in the
two sperm cells of *Plumbago zeylandica* (Russell 1984). Lyndon
and Robertson (1976) following the apical ultrastructure of pea
shoots, noted that during cell enlargement to form pith the
mitochondrial density increased, it remained the same in cells
involved in leaf initiation, and mitochondria were fewer in
cells involved in axillary development.

5.4.1d Dictyosomes and membrane turnover. Some significant
work on rates of membrane production by dictyosomes has been
carried out by Picton and Steer, using growing pollen tubes as
an experimental system. They determined rates for vesicle
production and membrane turnover, measuring the rates of vesicle
accumulation after cytochalasin d treatment (1981), subsequently
showing (1983) that the vesicle production rate was in excess of
that required for cell membrane extension, and that

recycling was probably occurring; the role of calcium in vesicle fusion and tip extension was also investigated (1985). Cunninghame and Hall (1985) studied the effects of IAA on the dictyosomes of pea stem epidermal cells, showing that it produced a short-term increase in their volume density in the cells, while Quaite, Parker and Steer (1983) showed that, in IAA treated *Avena* coleoptile segments undergoing cell extension, there were sufficient dictyosomes present for their vesicle production to sustain the observed rates of plasma membrane extension.

Turnover rates of dictyosomes are reported by Schnepf and Busch (1976) to be 30m in slime secreting glands of *Mimulus*. Mauseth (1980a) gives data on mucilage cells in *Opuntia*, and Ryter and de-Chastellier (1977) calculated membrane turnover rates in *Dictyostelium*; Moore and Coe (1984) give data on dictyosome content of *Cucurbita* root cap cells during development, while Moore (1982) notes an increase in dictyosome volume density as being one of the main ultrastructural changes in wound response in *Sedum*. Forde and Steer (1976) suggest a role for stereological analysis of membrane types in interpretation of lipid analyses of tissues, and Blank, Zaar and Kleinig (1977) give data on the frequency of membrane systems in cultured carrot cells.

5.4.1e The vacuome. Many of the references cited elsewhere in the review give data on changes in the vacuome. More specifically Moore (1982) notes a decrease in volume density and breakdown to smaller units in the vacuole of *Sedum* as a wound response, and Atkinson, Gunning and John (1974) report a similar splitting (and consequent increase in S/V ratio) in *Chorella* . Gaudinet, Ripoll, Thellier and Kramer (1984) used stereologically derived data on vacuoles in a discussion of active and passive transport, and Owens and Poole (1979) and Speiss and Seitz (1975) give data on vacuole and cytoplasm volumes in cell cultures. Ryter and de Chastellier (1977) and Schaap (1983) present developmental changes in autophagic vacuoles in *Dictyostelium*, while Mittelstadt and Muller-Stoll (1984) used planimetry to show a reduction in the vacuome of bark parenchyma cells following cold treatment.

5.4.1f Meristems. Stereological analyses of dividing cell populations, as noted above, are not easy to interpret and should be approached with caution. In a series of studies on shoot apices in the Cactaceae, Mauseth and co-workers show that ultrastructural changes in the germinating apex are zone-specific (Mauseth 1980b), that the structure of each zone is constant and stable (Mauseth 1981a, b) and that the induction of leaves or spines causes specific changes in organelle relationships and not simply in cellular growth rate (Mauseth

1982a). Comparison of a range of cactus and other apices showed
no common apex ultrastructure but similarities between species
having similar growth rates (Mauseth 1982b, c; 1984). There is
evidence presented for concentric morphogenic fields in the
shoot apical meristem (Mauseth 1982d), and for a constancy
between the relative proportions of the zones in apices of
different sizes (Mauseth and Niklas 1979; Niklas and Mauseth
1981). Orr (1981) showed similarly that during the transition to
flowering in *Brassica*, while there was an increase in cell
number in all the apical zones, their relative volumes remained
constant. Havelange and Bernier (1974) and Havelange, Bernier
and Jacqmard (1974) described the quantitative ultrastructural
changes during the transition to flowering in *Sinapis alba*
apices, Lyndon and Robertson (1976) similarly described changes
during the plastochron in pea, and Cecich and Miksche (1970)
studied changes in white spruce apices after gamma irradiation.

5.4.1g The leaf. General data on the proportions of tissue
types in leaves are given by Pazourek (1973; 1977), Dengler and
MacKay (1975) and Jellings and Leech (1982), while Smith (1974)
gives data on cotyledon intercellular space volume. Some
quantitative effects on leaf structure of water stress, nutrient
deficiency and petroleum contamination are given by Berlin,
Quinsberry, Bailey, Woodworth and McMichael (1982), Pazourek
(1981) and Natr and Pazourek (1982) respectively. Data on the
effects of shading on leaf cell structure are presented by
Fagerberg (1984b), who also disucsses the implications on
productivity of organelle development (Fagerberg 1984a).

      Relations between photosynthetic pathway and quantitative
leaf anatomy are discussed by Hattersley (1984), and Kunce,
Trelease and Doman (1984) followed glyoxysome changes in
germinating cotton seed. Parkhurst (1982) describes techniques
for stereological analysis of leaf tissues.

5.4.1h Wood. Point counting techniques have been proposed as
an alternative to destructive techniques for wood analysis
(Quirk 1984), and such analyses have been used to determine cell
changes which can be related to growth conditions (Maeglin and
Quirk 1984; Jagels and Dyer 1983) and used to predict mechanical
properties (Beery, Ifju and McLain 1983; Ifju 1983) and for
identification (Steele, Ifju and Johnson 1976a, b). Automatic
image analysis procedures have been developed for wood anatomy
quantification (Ilic and Hillis 1983).

5.4.1i General. There remain a few quantitative analyses of
plant cell structure which do not fall into the above
categories. Knights, Davey and Lucas (1982) determined the
changes taking place during spore germination of *Puccinia
graminis*, and Fagerberg and Mims followed the same process in
Didymium iridis. Mauseth (1982e) determined that the secretion

process in nectaries of *Ancistrocactus* was an eccrine process, based on an analysis of vesicle changes. Trachtenberg and Mayer (1981a, b) followed changes in the mucilage cells of *Opuntia*. Isaac, Briarty and Peberdy (1979) determined the sequence of hyphal breakdown during protoplast formation in *Aspergillus*.

Analyses of nitrogen fixing nodules, relating structure and function, have been carried out by Briarty (1978) and Wheeler, McLaughlin and Steele (1981). Pazourek (1975) described gradients of cell size and shape parameters across the tuber in potato. Walles and Rowley (1982) present some data on pollen development in *Pinus* .

5.5 CONCLUSIONS

From the data presented above two roles emerge for quantitative information on plant cell ultrastructure. The first of these is in supplementing data from other fields, so that a clearer understanding of cell organisation and function can be obtained. In much of the work cited above biochemical and ultrastructural changes have been well correlated, and together yield new information.

The second role is a development of this first one; information on ultrastructural change might be used to predict biochemical change where information on the latter is, using current technology, impossible to obtain. Where such change is taking place in a tissue that cannot be isolated for biochemical analysis, because the relevant cells are too small, or too dispersed, or otherwise inaccessible, or perhaps where inaccessibility of the plant material results from its being part of an unmanned space flight experiment for example, analysis of its structure may be the only way to obtain information on the material.

5.6 SUMMARY

The increasing application of stereological techniques to the analysis of plant cell structure has resulted in a growing literature of quantitative data. This paper surveys the various approaches to three dimensional reconstruction and stereological analysis that are in use, highlighting some of the particular problems posed by the structure of plant tissues, and reviews some of the computer software that has been developed to assist such analyses.

The contribution of quantitative data on plant cell structure to our understanding of plant function and physiology is reviewed.

112   L.G. Briarty

Thanks are due to the SERC and to the National Westminster Bank
Research Fund for financial support in this work.

Atkinson, A.W., Gunning, B.E.S. and John, P.C.L. (1974). Growth
and division of the single mitochondrion and other organelles
during the cell cycle of *Chlorella* and other algae, studied by
quantitative stereology and three-dimensional reconstruction.
Protoplasma 81, 77-109

Beery, W.H., Ifju, G. and McLain, T.E. (1983). Quantitative wood
anatomy-relating anatomy to transverse tensile strength. *Wood
Fiber Sci.* 15, 395-407.

Bendich, A.J. and Gauriloff, L.P. (1984). Morphometric analysis
of cucurbit mitochondria:the relationship between chondriome
volume and DNA content. *Protoplasma* 119, 1-7.

Berlin, J., Quinsberry, J.E., Bailey, F., Woodworth, M. and
McMichael, B.L. (1982). Effect of water stress on cotton leaves.
I An electron microscopic stereological study of the palisade
cells. *Pl. Physiol., Lancaster* 70, 238-243.

Bertram, J.F. and Bolender, R.P. (1983). The effects of specimen
preparation and tissue composition on stereological estimates.
Acta Stereologica 2 (suppl.1). Proc. 6th. Int. Cong. for
Stereol., Gainesville, Florida 1983, 281-284.

Blank, W., Zaar, K. and Kleinig, H. (1977). Morphometric
measurements of *Daucus carota* suspension culture cells. *Planta*
137, 85-87.

Bolender, R.P. (1978). Correlation of morphometry and stereology
with biochemical analysis of cell fractions. *Int. Rev. Cytol.*
55, 247-289.

Bolender, R.P. (1983). Integrating methods, a key for
stereology. *Acta Stereologica* 2 (Suppl.1) Proc. 6th. Int.
Cong. for Stereol. Gainesville, Florida 1983,, 131-138.

Briarty, L.G. (1975). Stereology: methods for quantitative light
and electron microscopy. *Sci. Prog. (Oxf.)* 62, 1-32.

Briarty, L.G. (1978). The development of root nodule xylem
transfer cells in *Trifolium repens*. *J. exp. Bot.* 29, 735-747.

Briarty, L.G. (1980a). Stereological analysis of cotyledon cell development in *Phaseolus*. I Analysis of a cell model. *J. exp. Bot.* 31, 1379-1386.

Briarty, L.G. (1980b). Stereological analysis of cotyledon development in Phaseolus. II The developing cotyledon. *J. exp. Bot.* 31, 1387-1398.

Briarty, L.G., Fischer, P.J. and Jenkins, P.H. (1982). Microscopy, morphology and microcomputers. *Acta Stereologica* 1, 227-234.

Briarty, L.G., Hughes, C.E. and Evers, A.D. (1979). The developing endosperm of wheat-a stereological analysis. *Ann. Bot.* 44, 641-658.

Briarty, L.G. and Jenkins, P.H. (1984). GRIDSS: an integrated suite of microcomputer programs for three dimensional graphical reconstruction from serial sections. *J. Microsc.* 134, 121-124.

Briarty, L.G. and Pearce, N.M. (1982). Starch granule production during germination in legumes. *J. exp. Bot.* 33, 506-510.

Calveyrac, R. and Lefort-Tran, M. (1976). Organisation spatiale des chloroplastes chez *Euglena* a l'aide de coupes sériees semi-fines. *Protoplasma* 89, 353-358.

Cecich, R.A. and Miksche, J.P. (1970). The response of white spruce (*Picea glauca* (Moench) Voss) shoot apices to exposures of chronic gamma radiation. *Radiat. Bot.* 10, 457-467.

Chayes, F. (1965). Determination of relative volume by sectional analysis *Lab. Invest.* 14, 987-995.

Chermant, J.L. and Coster, M. (1981). Image analysis using a table digitizer: II-Particle by particle analysis, and a study of the anisotropy. *Pract. Metall.* 18, 392-408.

Clauhs, R.P. and Grun, P. (1977). Changes in plastid and mitochondrion content during maturation of generative cells of *Solanum* (Solanaceae). *Am. J. Bot.* 64, 377-383.

Clowes, F.A.L. and Juniper, B.E. (1964). The fine structure of the quiescent centre and neighbouring tissues in root meristems. *J. exp. Bot.* 15, 622-630.

Cornelisse, J.T.W.A. and Van den Berg, T.J.T.P. (1984). Profile boundary lengths can be overestimated by as much as 41% when using a digitizer tablet. *J. Microsc.* 136, 341-344.

Coster, M. and Chermant, J.L. (1980). Image analysis using a table digitizer: I-Programming fundamental quantities in logical analysis. *Pract. Metall.* 17, 178-191.

Cruz-Orive, L.-M. (1976a). Correction of stereological parameters from biased samples on nucleated particle phases. I Nuclear volume fraction. *J. Microsc.* 106, 1-18.

Cruz-Orive, L.-M. (1976b). Correction of stereological parameters from biased samples on nucleated particle phases. II Specific surface areas. *J. Microsc.* 106, 19-32.

Cruz-Orive, L.-M. and Weibel, E.R. (1981). Sampling design for stereology. *J. Microsc.* 122, 235-257.

Cruz-Orive, L.-M., Gehr, P., Muller, A. and Weibel, E.R. (1980). Sampling designs for stereology *Mikroskopie* 37, 149-155.

Cruz-Orive, L.-M. and Weibel, E.R. (1981). Sampling Designs for Stereology *J. Microsc.* 122, 235-257.

Cunninghame, M.E. and Hall, J.L. (1985). A quantitative stereological analysis of the effect of indoleacetic acid on the dictyosomes in pea stem epidermal cells. *Protoplasma* 125, 230-234.

Darbelly, N. and Perbal, G. (1984a). Modèle de differenciation des cellules corticales de la racine de lentille. *Biol. Cell.* 50, 87-92.

Darbelly, N. and Perbal, G. (1984b). Gravité et differenciation des cellules corticales dans la racine de lentille. *Biol. Cell.* 50, 93-98.

Davydenko, O.G. (1984). Divergence of the mitochondrial and chloroplast structures between *Aegilops* and *Triticum*.1.Morphometric analysis of the mitochondrial ultrastructure in eleven alloplasmic wheats. *Sov. Genet.* 20, 235-241.

Davydenko, O.G. and Palilova, A.N. (1984). Divergence of the mitochondrial and chloroplast structures between *Aegilops* and *Triticum*.2.Characteristics of 22 cytoplasms on the basis of mitochondrial structure. *Sov. Genet.* 20, 347-360.

Davydenko, O.G., Palilova, A.N. and Shchetinina, M.I. (1984). Divergence of the mitochondrial and chloroplast structures between *Aegilops* and *Triticum*.3.Morphometric analysis of chloroplast ultrastructure. *Sov. Genet.* 20, 353-360.

Dean, C. and Leech, R.M. (1982). Genome expression during normal leaf development 2.Direct correlation between ribulose bisphosphate carboxylase content and nuclear ploidy in a polyploid series of wheat *Pl. Physiol., Lancaster* 70, 1605-1608.

Dengler, N.G. and MacKay, L.B. (1975). The leaf anatomy of beech *Fagus grandifolia.Can. J. Bot.*53, 2202-2211.

Durand, M.C. and Warren, R. (1981). Empirical equations for the estimation of number per unit volume of particulate systems. *Stereol. Iugosl.* 3, 109-114.

Dykes, E. and Afshar, F. (1982). Computer generated three dimensional reconstruction from serial sections. *Acta Stereologica* 1, 289-296.

Fagerberg, W.R. (1984a). Cytological changes in palisade cells of developing sunflower leaves. *Protoplasma* 119, 21-30.

Fagerberg, W.R. (1984b). A stereological analysis of the effect of shading on the structure of palisade cells from mature leaves of *Helianthus annuus.*|*Am. J. Bot.*71, 28.

Fagerberg, W.R. and Dawes, C.J. (1977). Studies on *Sargassum*. II Quantitative ultrastructural changes in differentiated stipe cells during wound regeneration and regrowth. *Protoplasma* 92, 211-227.

Fagerberg, W.R. and Mims, C.W. (1984). A stereological analysis of cytological change during spore maturation in *Didymium Iridis. Tex. J. Sci.*35, 343-344.

Fagerberg, W.R., Moon, R. and Truby, E. (1979). Studies on *Sargassum*.III. A quantitative ultrastructural and correlated physiological study of the blade and stipe organs of S. filipendula. *Protoplasma* 99, 247-261.

Favre, A., Keller, H.J., Mathieu, O. and Weibel, E.R. (1980). The use of local operators for pattern recognition on electron micrographs. *Mikroskopie* 37 suppl. (Proc. 5th. Int. Cong. for Stereology, Salzburg, 1979), 437-443.

Ferwerda, J.G. (1982). *The world of 3-D*. Netherlands Society for Stereo Photography, Borger, Netherlands. 1982.

Fischer, W.M. (1977). Zu Erzeugung morphometrischer Zufallsstichproben bei Einzellern. *Mikroskopie* 33, 111-115.

Forde, B.G., Gunning, B.E.S. and John, P.C.L. (1976). Synthesis

of the inner mitochondrial membrane and the intercalation of
enzymes during the cell cycle of *Chlorella*. *J. Cell Sci.* 21,
329–340.

Forde, J. and Steer, M.W. (1976). The use of quantitative
electron microscopy in the study of lipid composition of
membranes. *J. exp. Bot.* 27, 1137–1141.

Freere, R.H. and Weibel, E.R. (1967). Stereologic techniques in
microscopy *J. Roy. microsc. Soc.* 87, 25–34.

Fritsch, R.S. (1982). Ultrastructural stereology on natural cell
suspensions. *Acta Stereol.* 1, 123–128 (Proc. 3rd. Eur. Symp.
Stereol.).

Gaffal, K.P., Wolf, K.W. and Schneider, G.J. (1983).
Morphometric and chronobiological studies on the dynamics of the
nuclear envelope and the nucleolus during mitosis of the
colourless phytoflagellate *Polytoma papillatum*. *Protoplasma*
118, 19–35.

Gaudinet, A., Ripoll, C., Thellier, M. and Kramer, D. (1984).
Morphometric study of *Lemna gibba* in relation to the use of
compartmental analysis and the flux-ratio equation in higher
plant cells. *Physiol. Plant.* 60, 493–501.

Gaunt, P. .N. and Gaunt, W. .A. (1978). *Three dimensional
reconstruction in biology*. Pitman, Tunbridge Wells.

Gundersen, H.J.G., Boysen, M. and Reith, A. (1981).
Digitizer–tablet or point counting in biomorphometry. *Stereol.
Iugosl.* 3, 205–210.

Gundersen, H.J.G. and Osterby, R. (1980). Sampling efficiency
and biological variation in stereology. *Mikroskopie* 37,
143–148.

Gundersen, H.J.G. and Osterby, R. (1981). Optimizing sample
efficiency of stereological studies in biology: or "Do more less
well!". *J. Microsc.* 121, 65–73.

Gunning, B.E.S.G., Pate, J.S., Minchin, F.R. and Marks, I.
(1974). Quantitative aspects of transfer cell structure in
relation to vein loading in leaves and solute transport in
legume nodules. *Symp. Soc. exp. Biol.* 28 *Transport at the
Cellular Level*. Cambridge University Press, Cambridge.

Gupta, M., Mayhew, T.M., Bedi, K.S., Sharma, A.K. and White, F.H. (1983). Inter-animal variation and its influence on the overall precision of morphometric estimates based on nested sampling designs. *J. Microsc.* 131, 147-154.

Hajibagheri, M.A., Hall, J.L. and Flowers, T.J. (1984). Stereological analysis of leaf cells of the halophyte *Suaeda maritima* (L.)Dum *J. exp. Bot.* 35, 1547-1557.

Hattersley, P.W. (1984). Characterization of C4 type leaf anatomy in grasses (Poaceae.). Mesophyll:bundle sheath area ratios. *Ann. Bot.* 53, 163-179.

Havelange, A. (1980). The quantitative ultrastructure of the meristematic cells of *Xanthium strumarium* during the transition to flowering. *Am. J. Bot.* 67, 1171-1178.

Havelange, A. and Bernier, G. (1974a). Descriptive and quantitative study of ultrastructural changes in the apical meristem of mustard in the transition to flowering. I The cell and nucleus. *J. Cell Sci.* 15, 633-644.

Havelange, A., Bernier, G. and Jacqmard, A. (1974). Descriptive and quantitative study of ultrastructural changes in the apical meristem of mustard in transition to flowering. II The cytoplasm, mitochondria and protoplasts. *J. Cell Sci.* 16, 421-432.

Hensel, W. and Sievers, A. (1980). Effects of prolonged omnilateral gravistimulation on the ultrastructure of statocytes and on the graviresponse of roots. *Planta* 150, 338-346.

Hoppeler, H., Mathieu, O., Bretz, R., Krauer, R. and Weibel, E.R. (1980). The use of small computer systems for Stereology. *Mikroskopie* 37, 408-412.

Howard, V. and Eins, S. (1984). Software Solutions to Problems in Stereology *Acta Stereologica* 3, 139-158.

Hughes, C.E. and Briarty, L.G. (1976). Stereological analysis of the contribution made to mature wheat endosperm starch by large and small granules. *Die Starke* 10, 336-337.

Ifju, G. (1983). Quantitative wood anatomy-certain geometrical-statistical relationships. *Wood Fiber Sci.* 15, 326-337.

Ilic, J. and Hillis, W.E. (1983). Video image processor for wood anatomical quantification. *Holzforschung* 37, 47-50.

118 L.G. Briarty

Isaac, S., Briarty, L.G. and Peberdy, J.F. (1979). The stereology of protoplasts from *Aspergillus nidulans*. In: *Advances in Protoplast Research,* Proc. 5th. Int. Protoplast Symp., Szeged 1979, ed.L. Ferenczy and G.L. Farkas, 213-219.

Jagels, R. and Dyer, M.V. (1983). Morphometric analysis applied to wood structure.1.Cross-sectional cell shape and area change in red spruce. *Wood Fiber Sci.* 15, 376-386.

James, N.T. (1977). Stereology, in: *Analytical and Quantitative Methods in Microscopy,* ed. G.A.Meek and H.Y.Elder. Cambridge University Press, Cambridge.

Jellings, A.J. and Leech, R.M. (1982). The importance of quantitative anatomy in the interpretation of whole leaf biochemistry in species of Triticum,Hordeum and Avena. *New Phytol.* 92, 39-48.

John, P.C.L., McCullough, W., Atkinson, A.W.Jr., Forde, B.C. and Gunning, B.E.S. (1973). The cell cycle in Chlorella. In: *The Cell Cycle in Development and Differentiation.* Ed. M. Balls and F.S. Billett, Cambridge University Press, Cambridge, pp.61-76.

Juniper, B.E. and Clowes, F.A.L. (1965). Cytoplasmic organelles and cell growth in root caps. *Nature* 208, 864-865.

Knights, I.K., Davey, M.R. and Lucas, J.A. (1982). The ultrastructure of dormant, germinating, and photo-inhibited uredospores of the rust fungus *Puccinia graminis* f.sp. tritici. *Protoplasma* 113, 57-68.

Kristen, U., Lockhausen, J. and Robinson, D.G. (1984). Three-dimensional reconstruction of a dictyosome using serial sections. *J. exp. Bot.* 35, 1113-1118.

Kroustrup, J.P., Osterby, R. and Gundersen, H.J.G. (1981). Sampling problems in a heterogeneous organ. *Stereol. Iugosl.* 3, 345-350.

Kunce, C.M., Trelease, R.N. and Doman, D.C. (1984). Ontogeny of glyoxysomes in maturing and germinated cotton seeds-a morphometric analysis. *Planta* 161, 156-164.

Larsen, M.P., Pedersen, E.A., Pentcheff, N.D., Bertram, J.F. and Bolender, R.P. (1983). PCS-I and PCS-II: point counting stereology programs. *Acta Stereologica* 2 (Suppl.1). Proc. 6th. Int. Cong. for Stereol., Gainesville, Florida. 1983, 95-98.

Loud, A.V., Barany, W.C. and Pack, B.A. (1965). Quantitative
evaluation of cytoplasmic stuctures in electron micrographs.
Lab. Invest. 14, 996-1008.

Lowenberg, J.R. (1955). The development of bean seeds. *Pl.
Physiol., Lancaster* 30, 244-250.

Lyndon, R.F. and Robertson, E.S. (1976). The quantitative
ultrastructure of the pea shoot apex in relation to leaf
initiation. *Protoplasma* 87, 387-402.

Maeglin, R.R. and Quirk, J.T. (1984). Tissue proportions and
cell dimensions for red and white oak groups. *Can. J. For. Res.*
14, 101-106.

Mathieu, O. and Costello, M.L. (1983). Differential effect of
glutaraldehyde fixative osmolarity on the ultrastructure of
normal and atherosclerotic pigeon aorta. *Acta Stereologica* 2
(suppl.1) Proc. 6th. Int. Cong. for Stereol., Gainesville,
Florida, 1983, 277-280.

Mathieu, O., Cruz-Orive, L.-M., Hoppeler, H. and Weibel, E.R.
(1981). Measuring error and sample variation in stereology:
comparison of the efficiency of various methods for planar image
analysis. *J. Microsc.* 121, 75-88.

Mathieu, O., Cruz-Orive, L.-M., Hoppeler, H. and Weibel, E.R.
(1983). Estimating length density and quantifying anisotropy in
skeletal muscle capillaries. *J. Microsc.* 131, 131-146.

Mathieu, O., Hoppeler, H. and Weibel, E.R. (1980). Evaluation of
tracing device as compared to standard point-counting.
Mikroskopie 37, 413-414.

Mauseth, J.D. (1980a). A stereological morphometric study of the
ultrastructure of mucilage cells in *Opuntia polycantha*
(Cactaceae). *Bot. Gaz.* 141, 374-378.

Mauseth, J.D. (1980b). A morphometric study of the
ultrastructure of *Echinocereus engelmannii* (Cactaceae). I Shoot
apical meristems at germination. *Am. J. Bot.* 67, 173-181.

Mauseth, J.D. (1981a). A morphometric study of the
ultrastructure of *Echinocereus engelmannii* (Cactaceae).II The
mature, zonate shoot apical meristem. *Am. J. Bot.* 68, 96-100.

Mauseth, J.D. (1981b). A morphometric study of the
ultrastructure of *Echinocereus engelmannii* (Cactaceae). III
Subapical and mature tissues. *Am. J. Bot.* 64, 531-534.

Mauseth, J.D. (1982a). A morphometric study of the ultrastructure of *Echinocereus engelmannii* (Cactaceae). IV Leaf and spine primordia *Am. J. Bot.* 69, 546-550.

Mauseth, J.D. (1982b). A morphometric study of the ultrastructure of *Echinocereus engelmannii* (Cactaceae). V Comparison with the shoot apical meristems of *Trichocereus pachanoi*. *Am. J. Bot.* 69, 551-555.

Mauseth, J.D. (1982c). A morphometric study of the ultrastructure of *Echinocereus engelmannii* (Cactaceae). VI The individualised ultrastructures of diverse types of meristems. *Am. J. Bot.* 69, 1524-1526.

Mauseth, J.D. (1982d). A morphometric study of the ultrastructure of *Echinocereus engelmannii* (Cactaceae). VII Homogeneity of zones in the shoot apical meristem. *Am. J. Bot.* 69, 1527-1529.

Mauseth, J.D. (1982e). Development and ultrastructure of extrafloral nectaries in *Ancistrocactus scheeri* (Cactaceae). *Bot. Gaz.* 143, 273-277.

Mauseth, J.D. (1984). Effect of growth rate, morphogenetic activity and phylogeny on shoot apical ultrastructure in *Opuntia polycantha*. *Am. J. Bot.* 71, 1283-1292.

Mauseth, J.D. and Niklas, K.J. (1979). Constancy of relative volumes in zones in shoot apical meristems in Cactaceae: implications concerning meristem size, shape and metabolism. *Am. J. Bot.* 66, 933-939.

Mayhew, T.M. and Cruz-Orive, L.-M. (1974). Caveat on the use of the Delesse principle of areal analysis for estimating component volume densities. *J. Microsc.* 102, 195-207.

Mersey, B. and McCully, M.E. (1978). Monitoring the course of fixation of plant cells. *J. Microsc.* 114, 49-76.

Mittelstadt, H. and Muller-Stoll, W.R. (1984). Changes of the volume of vacuomes in the parenchyma cells of bark after frost damage. *Flora* 175, 231-242.

Mize, R.R. (1984). Computer applications in cell and neurobiology:a review. *Int. Rev. Cytol.* 90, 83-124.

Moens, P.B. and Moens, T. (1981). Computer measurements and graphics of three-dimensional cellular ultrastructure. *J. Ult. Res.* 75, 131-141.

Moore, R. (1982). Studies of vegetative compatibility-incompatibility in higher plants. 5 A morphometric analysis of the development of a compatible and an incompatible graft. *Can. J. Bot.* 60, 2780-2787.

Moore, R. (1983). A morphometric analysis of the ultrastructure of columella statocytes in primary roots of *Zea mays* L. *Ann. Bot.* 51, 771-778.

Moore, R. (1984). Cellular volume and tissue partitioning in caps of primary roots of *Zea mays*. *Am. J. Bot.* 71, 1452-1454.

Moore, R. and Coe, R. (1984). A morphometric analysis of cellular differentiation in root caps of *Cucurbita pepo*. *Pl. Cell Rep.* 3, 98-101.

Moore, R. and McClelen, C.E. (1983). A morphometric analysis of cellular differentiation in the root cap of *Zea mays*. *Am. J. Bot.* 70, 611-617.

Moore, R. and Pasieniuk, J. (1984a). Structure of columella cells in primary and lateral roots of *Ricinus communis* (Euphorbiaceae). *Ann. Bot.* 53, 715-726.

Moore, R. and Pasieniuk, J. (1984b). Graviresponsiveness and the development of columella tissues in primary and lateral roots of *Ricinus communis*. *Pl. Physiol., Lancaster* 74, 529-533.

Morris, P. and Thain, J.F. (1983). Improved methods for the measurement of total cell surface area in leaf mesophyll tissue. *J. exp. Bot.* 34, 95-98.

Moss, V.A. (1981). Computer-linked planimetry-some hints and algorithms. *Proc. R. microsc. Soc.* 16, 120-124.

Nagl, W., Cabirol, H., Lahr, C., Greulach, H. and Ohliger, H.M. (1983). Nuclear ultrastructure: morphometry of nuclei from various tissues of *Cucurbita, Melandrium, Phaseolus, Tradescantia* and *Vicia*. *Protoplasma* 115, 59-64.

Niklas, K.J. and Mauseth, J.D. (1981). Relationships among shoot apical meristem ontogenic features in *Trichocereus pachanoi* and *Melocactus matanzanus* (Cactaceae). *Am. J. Bot.* 68, 101-106.

Olsen, G.M. and Iversen, T.-H. (1980). Ultrastructure and movements of cell structures in normal pea and an ageotropic mutant. *Physiol. Plant.* <u>50</u>, 275-284.

Olsen, G.M., Miza, J.I., Maher, E.P. and Iversen, T.-H. (1984). Ultrastructure and movements of cell organelles in the root cap of agravitropic mutants and normal seedlings of *Arabidopsis thaliana.* *Physiol. Plant.*   <u>60</u>, 523-531.

Opik, H. (1973). Effect of anaerobiosis on respiratory rate, cytochrome oxidase activity and mitochondrial structures in coleoptiles of rice (*Oryza sativa* L.). *J. Cell Sci.* <u>12</u>, 725-739.

Orcival-Lafout, A.M., Pineau, B., Ledoigt, G. and Calveyrac, R. (1972). Evolution cyclique des chloroplastes dans une culture synchrone *d'Euglena gracilis* 'Z'. Etude stéréologique. *Can. J. Bot.* <u>50</u>, 1503-1508.

Orr, A.R. (1981). A quantitative study of cellular events in the shoot apical meristems of *Brassica campestris* (Cruciferae) during transition from vegetative to reproductive condition. *Am. J. Bot.* <u>68</u>, 17-23.

Osterby, R. and Gundersen, H.J.G. (1980). Stereology in experimental biology: sampling universe, practical sampling procedures, and the biological interpretation of structural quantities. *Mikroskopie* <u>37</u>, 161-164.

Owens, T. and Poole, R.J. (1979). Regulation of cytoplasmic and vacuolar volumes by plant cells in suspension culture. *Pl. Physiol., Lancaster* <u>64</u>, 900-904.

Parkhurst, D.F. (1982). Stereological methods for measuring internal leaf structure variables. *Am. J. Bot.* <u>69</u>, 31-39.

Pazourek, J. (1973). The amounts of different tissues in the leaf of *Acorus calamus* L.. IBP/PT-PP Report No.3.Trebon, 147-149.

Pazourek, J. (1975). Transverse anatomical gradients in potato tubers. *Biol. Plantarum* <u>17</u>, 263-267.

Pazourek, J. (1977). The volumes of anatomical components in leaves of *Typha angustifolia* L.and *Typha latifolia* L.. *Biol. Plantarum* <u>19</u>, 129-135.

Pazourek, J. (1982). The effect of aircraft petroleum on the anatomy of the wheat leaf. *Arch. Phytopathol. u. Pflanzenschutz* <u>18</u>, 59-62.

Pazourek, J. and Natr, L. (1981). Changes in the anatomical

structure of the first two leaves of barley caused by the absence of nitrogen or phosphorus in the nutrient medium. *Biol. Plantarum* 23, 296-301.

Peachey, L.D. (1975). Three-dimensional reconstruction by high voltage electron microscopy of biological specimens. *Proc. 33rd. Ann. Mtg. E.M.S.A.* 288-289.

Peachey, L.D. (1981). A relatively inexpensive microprocessor-linked digital planimeter for electron microscopic morphometry. *39th. Ann. Proc. E.M.S.A.*, Atlanta, Georgia, ed. G.W.Bailey.

Peachey, L.D. (1982). A simple digital morphometry system for electron microscopy. *Ultramicroscopy* 8, 253-262.

Pelligrini, M. (1976). Présence d'une mitochondrie unique et de proplastes isolées chez l'*Euglena gracilis* 'Z', en culture synchrone heterotrophe, a l'obscurité. *C. r. hebd. Seanc. Acad. Sci., Paris, ser. D.* 283, 911-913.

Pelligrini, M. and Pelligrini, L. (1976). Continuité mitochondriale et discontinuité plastidale chez l'*Euglena gracilis* 'Z'. *C. r. hebd. Seanc. Acad. Sci., Paris, ser. D.* 282, 357-360.

Petran, M., Hadravsky, M., Benes, J., Kucera, R. and Boyde, A. (1985). The tandem scanning reflected light microscope. Part I-the principle and its design. *Proc. R. microsc. Soc.* 20, 125-129.

Picton, J.M. and Steer, M.W. (1981). Determination of secretory vesicle production rates by dictyosomes in pollen tubes of *Tradescantia* using cytochalasin d. *J. Cell Sci.* 49, 261-272.

Picton, J.M. and Steer, M.W. (1983). Membrane recycling and the control of secretory activity in pollen tubes. *J. Cell Sci.* 63, 303-310.

Picton, J.M. and Steer, M.W. (1985). The effects of ruthenium red, lanthanum, fluorescein isothiocyanate and trifluoperazine on vesicle transport,vesicle fusion and tip extension in pollen tubes. *Planta* 163, 20-26.

Quaite, E., Parker, R.E. and Steer, M.W. (1983). Plant cell extension: structural implications for the origin of the plasma membrane. *Pl. Cell and Env.* 6, 429-434.

Quirk, J.T. (1984). Cell-wall density of Douglas-fir by two optometric methods. *Wood Fiber Sci.* 16, 224-236.

Ransom, J.S. and Moore, R. (1983). Geoperception in primary and lateral roots of *Phaseolus vulgaris* (Fabaceae). 1.Structure of columella cells. *Am. J. Bot.* 70, 1048-1056.

Ransom, J.S. and Moore, R. (1984). Geoperception in primary and lateral roots of *Phaseolus vulgaris* (Fabaceae). 2.Intracellular distribution of organelles in columella cells. *Can. J. Bot.* 62, 1090-1094.

Reynolds, T.L. (1984). An ultrastructural and stereological analysis of pollen grains of *Hyoscyamus niger* during normal Ontogeny and induced embryogenic development. *Am. J. Bot.* . 7, 490-504.

Russell, S.D. (1984). Ultrastructure of the sperm of *Plumbago zeylandica* . II Quantitative cytology and three-dimensional organisation. *Planta* 162, 385-391.

Ryter, A. and de-Chastellier, C. (1977). Morphometric and cytochemical studies of *Dictyostelium discoideum* in vegetative phase: digestive system and membrane turnover. *J. Cell Biol.* 75, 200-217.

Santi, P.A., Fryhofer, J. and Hansen, G. (1980). Electronic planimetry. *Byte* 5, 114-122.

Santos, I. and Salema, R. (1983). Stereological study of the variation of chloroplast tubules and volume in the CAM plant *Sedum telephium*. *Z. Pflanzenphysiol.* 113, 29-37.

Schaap, P. (1983). Quantitative analysis of the spatial distribution of ultrastructural differentiation markers during development of *Dictyostelium discoideum*. *Roux's Arch. Dev. Biol.* 192, 86-94.

Schnepf, E. and Busch, J. (1976). Morphology and kinetics of slime secretion in glands of *Mimulus telingii*. *Z. fur Pflanzenphysiol.* 79, 62-71.

Schotz, F., Bathelet, H., Arnold, C.-.G. and Schimmer, C. (1972). Die Architektur und Organisation der *Chlamydomonas* -Zelle. Ergebnisse der Elektronenmikroskopie von Serienschnitten und der daraus resultierenden dreidimensionalen Rekonstruktion. *Protoplasma* 75, 229-254.

Shen-Miller, J. and Gawlik, S.R. (1977). Effects of indoleacetic acid on the quantity of mitochondria, microbodies, and plastids in the apical and expanding cells of dark-grown oat coleoptiles. *Pl. Physiol., Lancaster* 60, 323-328.

Smith, C.W. (1984). 3-D or not 3-D? *New Scient.* <u>102</u>, 40-44.

Smith, D.L. (1974). A histological and histochemical study of the cotyledons of *Phaseolus vulgaris* during germination. *Protoplasma* <u>79</u>, 41-57.

Spiess, E. and Seitz, U. (1975). Quantitative determination of growth-dependent changes in the ultrastructure of freely suspended callus cells of parsley (*Petroselinum crispum* (Mill.) A.W.Hill). *Ber. dt. bot. Ges.* <u>88</u>, 319-328.

Steele, J.H., Ifju, G. and Johnson, J. (1976a). Application of stereological techniques to the quantitative characterisation of wood. *Proc. 4th | Int. Cong. for Stereol.* , Gaithersburg, 1975. National Bureau of Standards Special Publication 431, pp.245-256.

Steele, J.H., Ifju, G. and Johnson, J.A. (1976b). Quantitative characterisation of wood microstructure. *J. Microsc.* <u>107</u>, 297-311.

Steer, M. .W. (1981). *Understanding Cell Structure.* Cambridge University Press, Cambridge.

Stevens, B.J. (1977). Variation in number and volume of the mitochondrion in yeast according to growth conditions. A study based on serial sectioning and computer graphics reconstruction. *Biol. Cell.* <u>28</u>, 37-56.

Stoker, R. and Moore, R. (1984). Structure of columella cells in primary and lateral roots of *Helianthus annuus* Compositae. *New Phytol.* <u>97</u>, 205-212.

Thain, J.F. (1983). Curvature correction factors in the measurement of cell surface areas in plant tissues. *J. Exp. Bot.* <u>34</u>, 87-94.

Thomas, M.R. and Rose, R.J. (1983). Plastid number and plastid structural changes associated with tobacco mesophyll protoplast culture and plant regeneration. *Planta* <u>158</u>, 329-338.

Tipper, J.C. (1977). A method and FORTRAN program for the computerized reconstruction of three dimensional objects from serial sections. *Computers and Geosciences* <u>3</u>, 579-599.

Toth, R. (1982). An introduction to morphometric cytology and its application to botanical research. *Am. J. Bot.* <u>69</u>, 1694-1706.

Trachtenberg, S. and Mayer, A.M. (1981). A stereological

analysis of the succulent tissue of *Opuntia ficus-indica* (L)
Mill. I. Development of mucilage cells. *J. Exp. Bot.* 32,
1091-1103.

Trachtenberg, S. and Mayer, A.M. (1981). A stereological
analysis of the succulent tissue of *Opuntia ficus-indica* (L)
Mill. II. Ultrastructural development of the mucilage cells *J.
Exp. Bot.* 32, 1105-1113.

Underwood, E.E. (1970).   *Quantitative Stereology*, Addison-Wesley,
Reading, Mass..

Veen, A. and Peachey, L.D. (1977). TROTS: a computer graphics
system for three-dimensional reconstruction from serial
sections.   *Comput. and Graphics* 2, 135-150.

Vodopich, D.S. and Moore, R. (1984). A computer program to
facilitate morphometric analyses of cellular ultrastructure.
Texas Society for Electron Microscopy Journal   15, 9-10.

Walles, B. and Rowley, J.R. (1982). Cell differentiation in
microsporangia of *Pinus sylvestris* with special attention to the
tapetum.1.The pre- and early-meiotic periods. *Nord. J. Bot.* 2,
53-70.

Weibel, E.R. (1972). A stereological method for estimating
volume and surface of sarcoplasmic reticulum. *J. Microsc.* 95,
229-242.

Weibel, E.R. (1979). *Stereological Methods Vol. I Practical
Methods for Biological Morphometry*. Academic Press, London, New
York.

Weibel, E.R. (1980). *Stereological Methods Vol. II Theoretical
Foundations* . Academic Press London, New York.

Weibel, E.R. and Bolender, R.P. (1973). Stereological techniques
for electron microscopic morphometry, in: *Principles and
Techniques of Electron Microscopy, Biological Applications,
Vol.3*. ed.M.A. Hayat, pp. 237-296. Van Nostrand-Reinhold
Company, New York.

Weibel, E.R., Kistler, G.S. and Scherle, W.F. (1966). Practical
stereological methods for morphometric cytology. *J. Cell Biol.*
30, 23-38.

Wheeler, C.T., McLaughlin, M.E. and Steele, P. (1981). A
comparison of symbiotic nitrogen fixation in Scotland in *Alnus
glutinosa* and *Alnus rubra*. *Pl. Soil* 61, 169-188.

Williams, M.A. (1977). Quantitative methods in biology, in: *Practical Methods in Electron Microscopy, Vol.6,* ed. A.M.Glauert, part II. North-Holland Publishing Company, Amsterdam.

Yamada, M. and Yoshida, S. (1972) Graphic reconstruction of serial sections. *J. Microsc.* 95, 249-256.

Zilles, K., Schleicher, A. and Pehlemann, F.W. (1982). How many sections must be measured in order to reconstruct the volume of a structure using serial sections? *Microscopica Acta* 86, 339-346.

6

Vesicle dynamics

M. W. Steer

6.1 INTRODUCTION

Vesicles may be defined as carriers of macromolecules both between cell compartments, and between the cell interior and the external environment. They range from small, protein (clathrin) bound vesicles, coated vesicles, to large, membrane-bound secretory vesicles. The latter may contain anything from solutes and simple peptides up to complex macromolecules (e.g. polysaccharides, enzymes) which may be assembled into massive structures (e.g. algal scales).

Implicit in every discussion of vesicles is the assumption that they move about in the cell, yet it is only possible to directly visualise this movement in a few exceptional cases. The circumstantial evidence is overwhelming. Vesicles are found in morphological situations corresponding to a cycle of formation, transport and fusion between different membrane systems. The direction of these movements has been established with the aid of radio- and immuno-labels in a number of cell types (Farquar and Palade 1981). The picture that emerges is superficially complete, yet it leaves many key questions unanswered. Here we will be mainly concerned with attempts to provide some quantitative measure of the processes. For example, what is the rate of vesicle formation, and the rate of their movement through the cytoplasm and fusion with the cell surface? Answers to these questions inevitably serve to raise further questions about the quantitative utilisation of the vesicle containers and their contents, and about the processes governing their formation. The main thrust of this work has been to examine a relatively accessible vesicle journey, that from the Golgi apparatus to the cell surface.

The investigation of secretory functions provides some

classic examples of the integration of structural and biochemical analyses of cellular activities. Unequivocal evidence was obtained for the movement of proteins, synthesised on the endoplasmic reticulum, into the cisternae of the Golgi apparatus, and thence into secretory vesicles (Farquar and Palade 1981). While the cells used for these investigations are somewhat atypical, requiring an external signal to trigger vesicle fusion with the cell surface and release of the contents to the exterior, there is no reason to suppose that the secretion process is fundamentally different in cells exhibiting continuous, constitutive, secretory activity. Many subsequent studies have confirmed this basic route from the Golgi apparatus to the cell surface.

     The size of the secretory vesicles precludes accurate quantification of their rates of formation, transport and fusion by direct observation. This has led to the adoption of various strategies to provide an estimate of these processes based on combining measurements made on the living system and on micrographs of preserved cells. This chapter will concentrate on reviewing these procedures as applied to plant cells and then explore the wider implications of the results obtained to date, incorporating ideas and results from the literature on animal cells.

6.2 METHODS OF ESTIMATING VESICLE FORMATION RATES

6.2.1 Indirect method

The term "indirect method" is here applied to those methods which are based on observations of events at or beyond the cell surface to establish the time dimension of events taking place inside the cell. Schnepf (1961, 1974) made the first attempt to glean information of the dynamic events surrounding secretion in a study of *Drosophyllum* mucilage glands that paved the way for all subsequent studies. He recorded the rate of increase in size of the surface droplet on the gland head and calculated a mean rate of volume flow from each cell. Electron micrographs revealed the presence of secretory vesicles in the cytoplasm, and an estimate was made of their diameter, and hence the volume of a typical vesicle was obtained. The number of vesicle volumes required to account for the observed rate of accumulation on the gland head could then be calculated. The key assumption made in this work was that a unit of product outside the cell (in this case mucilage) is derived from a similar unit of product in the secretory vesicles in the cytoplasm.

This method was applied to vesicle production in pollen tubes of *Lilium longiflorum* (van der Woude *et al.* 1971). The rate of tip growth of the cylindrical pollen tubes was established and electron microscopy used to determine the tube diameter, wall thickness and mean vesicle diameter. It was then possible to calculate the rate of addition of wall material (volume) and compare this with the volume of a single vesicle to give a vesicle production per tube. In addition the authors extended their calculations to include the vesicle membranes and found that the rate of membrane addition determined from the above rate of vesicle production approximately balanced the requirement for new membrane by the extending tube.

More recent work has demonstrated that the assumptions underlying the indirect method only apply to a few secretory cells under special circumstances. In general neither the product volume nor the membrane area flow to the cell surface can be equated with subsequent increases in secretory volume flow or membrane area detected at or beyond the cell surface. Highly concentrated secretory products appear to induce an osmotic flow of water from the cell, substantially increasing the volume flow of liquid from the cell surface, leading to overestimates of vesicle formation rates. On the other hand plasma membrane area increases are small compared with the input from vesicle membranes, leading to underestimates of vesicle flows by the indirect method.

6.2.2 Direct method

The term 'direct method' will be used for any system that involves estimating time-dependent changes in vesicle numbers within cells. This method was devized by Picton and Steer (1981), it involves quantifying the rise in vesicle numbers following cytochalasin inhibition of secretion. It had previously been established that cytochalasins cause an increase in vesicle density (number per unit volume of cytoplasm), presumably due to interference with a microfilament-based vesicle transport system (Mollenhauer and Morre 1976; Pope *et al.* 1979). Picton and Steer (1981) exposed pollen tubes to minimal levels of cytochalasin D for short, fixed periods of time (5 and 10 minutes) and showed that the increase in vesicle numbers in the cytoplasm with time was linear. They made the assumption that, in the presence of these low levels of inhibitor, only vesicle transport was inhibited while vesicle formation continued normally. Rates of vesicle accumulation were considered to be equivalent to rates of vesicle formation. The values obtained for pollen tubes of *Tradescantia virginiana* were considerably higher than those obtained by the indirect method for the larger pollen tubes of *Lilium longiflorum* (van der Woude *et al.* 1971).

Data given by Cope and Williams (1981), for reformation of
secretory granules in parotid acinar cells following stimulation
and granule release, can be used as a further direct method of
estimating vesicle formation rates. This involves assuming
that granules do not discharge, nor undergo fusion with each
other, during the peak period of vesicle formation. An increase
of 325 granules per cell was recorded over a four hour period,
corresponding to the formation of 1.35 granules per cell per
minute, each 0.575 μm in diameter.

The work of Shannon (1984) revealed an important limitation
to the cytochalasin method. It was shown in both maize root cap
cells and pollen tubes that cytochalasin progressively inhibits
vesicle formation (Shannon *et al*. 1984). Averaging vesicle
accumulations over long periods of time (> 10 minutes) leads to
a severe underestimate of the true vesicle production rate.
There were no differences between the effects of cytochalasin D
(at 0.3 μg ml^{-1}) and cytochalasin B (at 0.5 μg ml^{-1}). The
implications of these results for our understanding of
secretory processes will be discussed later. Here it is
necessary to emphasise that this limitation restricts the use
of the method to systems, such as pollen tubes, involving
relatively high vesicle production rates per unit volume of
cytoplasm. This is because small increases in vesicle numbers
are difficult to detect against the statistical noise
associated with the method used for estimation of vesicle
densities. Even in favourable tissues it is likely that some
reduction in formation rate will occur during short exposures
to the drug.

The estimation of vesicle densities for the direct method
was considered by Picton (1981). The method developed by Rose
(1980) for the conversion of vesicle profile densities in
micrographs to vesicle densities within the cytoplasm was
found to be the most versatile. It copes with the twin
problems of section thickness and a distribution of vesicle
sizes. Typically the method involves scanning successive
micrographs (each a statistically independent sample),
recording the area of cytoplasm sampled and the number and size
distribution of the vesicle profiles. Approximately 1,000
profiles are measured for each treatment, this number is
necessary to provide an adequate sample of the smaller
profile-size classes. The information on vesicle profile, size
and number, and the section thickness, are used to provide
estimates of vesicle sizes and densities within the cytoplasm
using the conversion matrix given by Rose (1980). A FORTRAN
program for the execution of this conversion is available from
the present author, the tables given in Steer (1981) are less
versatile.

In practice the direct method involves estimating vesicle
densities in untreated cells, solvent-control cells and
cytochalasin-treated cells and then calculating vesicle
accumulation rates based on the differences. It is important
to ensure that the starting material is as uniform as possible,
so that meaningful results can be obtained. Typically 3-5
independent organisms are used to provide each treatment mean,
so that at least 10,000 vesicles have to be measured and counted
to provide a single estimate of vesicle production. A Kontron
Videoplan Image Analyser is used to collect, store and classify
vesicle profiles, and can be programmed to carry out "Rose
analyses" on the results. While some care is taken to avoid
chemical fixatives that lead to swelling of vesicles (using
PIPES rather than phosphate buffers), it has been shown that
freeze-substitution probably provides a better preservation of
the vesicles' size and shape (Howard and Aist 1979; Howard 1981).
This does not, of course, affect the vesicle numbers, but it
does mean that the values derived for vesicle surface area and
volume may be overestimates of the real values. When these
overestimated values for vesicle size are combined with the
underestimated values for vesicle formation rate the resultant
rates of membrane and product delivery probably give a
reasonable estimate of the true cellular activity. Comparative
studies, in which differences between cell treatments are being
examined, will be subject to less uncertainty.

6.3 VESICLE DYNAMICS

6.3.1 Vesicle formation

The previous section reviewed the methods available for
estimating rates of vesicle formation. Here the results of
these investigations will be reviewed and their significance
for the functioning of dictyosomes assessed. Table 15.1
summarises the data that has been published up to the present
time.

There is considerable variation in rates of vesicle
formation between different cell types. This is true whether
the results are expressed in terms of vesicles formed per unit
of cytoplasm, or as vesicles per dictyosome. Large dictyosomes
produce large vesicles less frequently, so this disparity is
reduced when the rates of vesicle membrane formation are
expressed per unit area of a single dictyosome cisterna. For
example in pollen tubes this is 0.3 μm^2 μm^{-2} min^{-1} and in root
cap cells 0.15 μm^2 μm^{-2} min^{-1}. This suggests that there is some
constancy of synthetic capacity between dictyosome cisternal
membranes of different plant cell types.

Table 6.1

Vesicle formation rates in plants

Tissue	Rate (no. min^{-1})	Reference
Drosophyllum mucilage glands*	3 dictyosome^{-1}	Schnepf 1961
Lilium pollen tubes*	2,000 tube^{-1}	van der Woude and Morre 1968
	1,000 tube^{-1}	Morre and van der Woude 1974
*Tradescantia*** pollen tubes	5,388 tube^{-1} 2 μm^{-3} cytoplasm 2.35 dictyosome^{-1}	Picton and Steer 1981
	4,000 tube^{-1} 1.3 μm^{-3} cytoplasm 1.0 dictyosome^{-1}	Picton and Steer 1983
Aptenia ovaries**	1.06 dictyosome^{-1}	Kristen and Lockhausen 1983
Zea root caps**	0.039 μm^{-3} cytoplasm 0.39 dictyosome^{-1}	Shannon and Steer 1984

---

*   Indirect method
** Direct method

The flow of membrane through the dictyosomes can be assessed from the data on vesicle production. This is usually expressed as the turnover time taken to completely replace a single cisterna, or a whole dictyosome, assuming that the membrane of each cisterna is incorporated into the vesicles without any change to its surface area (**Table 6.2**). Estimates of this turnover time are also available from the kinetics of radio-label displacement and other methods; these are included in Table 15.2 for comparison. Such calculations are intended only as indicators of the flow. A great deal depends on the exact nature of the processes leading to vesicle formation and dictyosome membrane renewal. If vesicle formation is restricted to membrane at the edges of cisternae, then these will have to be replaced at a much higher rate than the average values calculated above.

Some studies have suggested that there is a direct relationship between dictyosome cisternal number and vesicle

Table 6.2

Cisternal turnover time

Tissue	Time (min.)	Author

Stereological analysis

Vacuolaria contractile vacuole	1	Schnepf and Koch 1966
Pleurochrysis scale formation	1-2	Brown 1969
Monarda hydathode	1	Heinrich 1973
Tradescantia pollen tubes	10* 23*	Picton and Steer 1981 Picton and Steer 1983
Aptenia ovary glands	7.3	Kristen and Lockhausen 1983
Zea root cap	6.5	Shannon and Steer 1984

Other methods

Pisum Cell wall polysaccharide	1-2	Eisinger and Ray 1972
Zea Cell wall polysaccharide	2-5	Bowles and Northcote 1974
Daucus suspension culture	2-4	Morre *et al.* 1983

---

* Calculated from authors' data

formation rate. Low rates of production are accompanied by increases in the number of cisternae in each stack, which are subsequently 'used up' when secretion resumes (Schnepf 1961; Hawkins 1974; Ueda and Noguchi 1976; Tsekos 1981). Alternatively it has been suggested that small dictyosomes are inactive (Pellegrini 1976).

There is no information on the membrane budding process

that results in the formation of a vesicle, although there are
some suggestions that this may be brought about by the
cytoskeleton. The process of vesicle formation is dependent on
continued protein synthesis (Chrispeels 1976). In pollen tubes
(Picton and Steer 1983b) cycloheximide-inhibition effects
suggest that there is a specific requirement for new protein to
form vesicles..

Control of dictyosome activity has been investigated in
several plant cell types. In *Dionaea* digestive glands definite
cycles of activity have been detected (Schwab *et al*. 1969; Henry
and Steer 1985), which are dependent on cycles of feeding and
trap closure. This appears to be the only documented case of a
plant Golgi system that is under a specific control.
Kwiatkowska and Maszewski (1979) interpreted changes in numbers
of vesicles surrounding each dictyosome as changes in activity
during the cell cycle of antheridial filaments of **Chara**. An
attempt to demonstrate specific dictyosome controls in
Tradescantia pollen tubes revealed that vesicles are produced at
similar rates regardless of tube growth rate (Picton and Steer
1983a), a finding that is consistent with the evolutionary
pressures on pollen tubes. The activities of many plant
secretory systems appear to be geared to general processes of
cell differentiation, so it is difficult to find cases where
specific dictyosome controls can be investigated. One
potentially useful system may be the extension growth of cells
in coleoptiles or seedling hypocotyls. These are dependent on
dictyosome-produced wall polysaccharides which appear to be laid
down strictly in response to the extension process.

Attempts to identify specific inhibitors of vesicle
formation have met with little success. Monensin, a sodium-
specific ionophore, has severely disruptive effects on
dictyosome cisternae, leading to an inhibition of secretion
(Tartakoff and Vassali 1977; Robinson 1981; Mollenhauer *et al*.
1982,1983; Morre *et al*. 1983). In animals, tunicamycin has been
shown to inhibit glycosylation of proteins (Hickman *et al*. 1977),
while in plants it inhibits the related process of lipid-linked
polysaccharide formation (Ericson *et al*. 1977). Both of these
are product-related events, not directly connected with vesicle
formation. Testing such inhibitors against standard plant
secretory systems has not been successful (Shannon and Steer
1984b). There is an urgent need for the identification of a
plant secretory system that produces a readily-characterised
product (enzyme, storage protein?), that is accessible to
applied chemicals and can be easily processed for microscopy.
The suspension cultures of oat cells described by Conrad *et al*.
(1982) are too slow-growing to be of value for such work.

Determination of the rates of vesicle formation from

dictyosomes provides information of importance to considerations
of membrane flow into the dictyosomes and between the individual
cisternae. Research on plant cells has yielded a fair measure
of disagreement over the role of the endoplasmic reticulum in
maintaining supplies of membrane to the cisternal stack
(Dauwalder and Whaley 1982; Juniper *et al.* 1982; Parker and
Hawes 1982; Robinson and Kristen 1982; Shannon *et al.* 1982;
Harris and Oparka 1983), as suggested in the endomembrane
hypothesis of Morre and others (Morré and Mollenhauer 1974;
Morré *et al.* 1979). This proposes that there is a concomitant
flow of vesicular membrane from the endoplasmic reticulum to the
Golgi, to balance the formation of secretory vesicles. In a
steady state situation, it is possible to calculate the
frequency with which such transitional vesicles should be
encountered in thin sections. Similarly if the Rothman (1981)
model of intracisternal exchange is correct, then it should be
possible to observe small vesicles in the neighbourhood of the
cisternal stacks at the expected frequency. It is immediately
clear that if these proposed vesicles are smaller than secretory
vesicles, then they would have to be correspondingly more
numerous to cope with the same membrane flow. Smaller vesicles
will occur less frequently in thin sections than secretory
vesicles, because of their smaller diameter. They may also be
less frequently encountered because of their shorter path
lengths, and hence shorter journey times. The value for the
rate of secretory vesicle formation in outer root cap cells
has been used to estimate the expected frequency of small
vesicle profiles in thin sections, assuming that all the
membrane leaving the dictyosome during secretion is arriving in
the form of small vesicles (Shannon and Steer 1984a). The
expected frequency is about eight times greater than that of
the secretory vesicle profiles, assuming the same rate of
transport and a path length one-tenth of that of the secretory
vesicles. In fact the observed numerical density of small
vesicle profiles is much lower than that of secretory vesicles.
This suggests that in maize root caps movement to, and between,
dictyosome cisternae is not mediated by vesicles.

6.3.2 Vesicle transport

Movement of secretory vesicles to the cell surface appears to be
a microfilament-dependent pathway, based on numerous
observations that cytochalasins inhibit secretion and lead to
vesicle accumulation in the cytoplasm (e.g. Mollenhauer and
Moore 1976; Pope *et al.* 1979; Picton and Steer 1981). Taken at
the simplest level we can envisage some sort of association
between the vesicle membrane and one or more microfilaments
that results in vesicle transport to the surface. There is
evidence for such an association in animal cells, isolated

granules bind actin and form arrays of actin filaments (Wilkins and Lin 1981). In pollen tubes, vesicles synthesised in the presence of cycloheximide accumulate in the cytoplasm and are not transported (Picton and Steer 1983b). One explanation would be that the vesicle membranes are imperfect, lacking a specific protein required for association with the transport system. The transport system itself could be poisoned by cycloheximide, but this is not supported by the continuation of another transport system, cytoplasmic streaming. Microfilaments are known to interact with the plasma membrane of animal cells via specific membrane-anchorage proteins and other cytoskeletal elements (Isenberg *et al*. 1982; Stossel 1982; Tsukita *et al*. 1981). It follows from the signal hypothesis for the insertion of proteins into membranes (Sabatini *et al*. 1982), that vesicles will already possess these anchorage proteins on release from the dictyosomes. Vesicle transport could then be regarded as part of a sorting process whereby the cytoskeleton transports anchorage proteins to its interface with the cell surface. Whatever the mechanism of transport, it is quite clear that it is separate from the process of vesicle fusion, vesicles are merely brought into close proximity to the cell surface.

Under steady-state secretory conditions it is possible to calculate a residence time for the vesicles in the cytoplasm from the rate of vesicle formation (or discharge) and the numerical density of vesicles in transit through the cytoplasm at any instant in time (as determined by glutaraldehyde fixation). This assumes that all vesicles leaving the dictyosomes behave in the same way, taking a similar time to reach the cell surface and discharge. Such assumptions are clearly invalid where dictyosomes lie at different distances from the surface. **Table 6.3** summarises the information that is available on residence times available from microscopic methods, the value for *Drosophyllum* is probably much lower than reality, due to the use of the indirect method for determining vesicle discharge rates. The remaining values of 15-30 minutes are higher than those estimated from the kinetics of radio-isotope secretion, also given in **Table 6.3**. It may be that stereology overestimates the residence time due to slight vesicle accumulation during fixation while radio-isotope labelling underestimates the time due to incorporation of label directly into vesicles.

The range of values presented above provide information on the rate at which vesicles migrate through the cytoplasm. Many secretory cells have relatively small vacuoles, and so the dictyosomes may lie at some distance from the surface. Nevertheless few cells have dictyosomes that are further than 10-20 μm from the surface, so that residence times of the order of 10-20 minutes imply that transport occurs at no more

Table 6.3

Cytoplasmic residence times for vesicles

Tissue	Time (min.)	Author
Stereological analysis		
Drosophyllum mucilage glands	2.6	Schnepf 1974
Mimulus leaf trichomes	32	Schnepf and Busch 1976
Tradescantia pollen tubes	15*	Picton and Steer 1983a
Zea root caps	18*	Shannon and Steer 1984a
Radioisotope kinetics		
Zea root caps	2-3	Bowles and Northcote 1974
Daucus cell wall protein	4-8	Gardiner and Chrispeels 1975
Pisum cell wall polysaccharides	3-4	Robinson *et al.* 1976

---

* Calculated from authors' data

than 1 µm min^{-1}. In pollen tubes this method of calculating residence times is even less satisfactory, since there is a flow of vesicles through the main vesicle-producing regions to the apex. Using the relationship for flow in a pipe (Picton *et al.* 1983), a pollen tube delivering 4,000 vesicles min^{-1} to the apex, would have a vesicle transport velocity of about 4 µm min^{-1}. Any mechanism proposed for the movement of vesicles would have to generate transport at this velocity through cytoplasm with a viscosity of 5-20 mPa s (Steer *et al.* 1984).

There are suggestions that secretory vesicles enlarge after leaving the dictyosomes (Morré *et al.* 1971), and these are supported by observations that coated vesicles fuse with the secretory vesicles (Steer 1977; Picton and Steer 1983b).

Against this, stereological analysis has failed to reveal a
significant increase in size. Comparison of vesicle populations
in control cells and cytochalasin-inhibited cells, where
vesicles are retained for abnormally long times in the cytoplasm,
and vesicles at dictyosomes with those at the discharge point
have all failed to find a significant or consistent effect
(personal communications from Picton and Shannon). This does
not invalidate the original observations, a small increase in
diameter would provide a significant additional volume. For
example, increasing the diameter from 100 nm to 125 nm almost
doubles the volume of the vesicle.

6.3.3 Vesicle fusion

6.3.3a Fusion events. Plant cells differ significantly from
animal cells in that there are two alternative destinations for
secretory vesicles, in addition to the normal exocytosis
through the plasma membrane. These are the cell plate, at
cytokinesis, and the tonoplast membrane, allowing discharge into
the vacuole. Intracellular routing of secretory vesicles must
therefore involve additional sets of controls on the transport
system and possibly on the fusion properties of the vesicles.
In dividing cells the vesicles are brought together under the
control of a microtubule-based phragmoplast, and then undergo
fusion with each other. This self-fusion can be observed in
normal secretory cells if the vesicles become closely packed
together (e.g. Picton and Steer 1983c; 1985). Fusion with the
tonoplast may be a regular feature of plant cell behaviour, it
occurs particularly in the formation of protein bodies in
developing seeds (Bechtel and Gaines 1982; Chrispeels 1983;
Herman and Shannon 1984; Nieden *et al*. 1984). It is of great
interest that this fusion site can be switched from the
tonoplast to the plasma membrane by treatment with ionophores
(Craig and Goodchild 1984), there is no obvious explanation
for this effect.

The molecular events surrounding vesicle fusion have been
extensively studied in artificial vesicle systems (Wilschut *et.
al*. 1980; Fraley *et al*. 1980), in suspensions of isolated
secretory vesicles (Dahl *et al*. 1979) and in actively secreting
cells (Lawson and Raff 1979). The main conclusions from this
work are that vesicles possess one or more docking sites that
interact specifically with docking rossettes on the plasmamembrane. This is accompanied by a migration of many plasma
membrane proteins away from the prospective fusion site.
Fusion is a Ca^{2+} dependent process involving both lipid and
protein components in the respective membranes. Fusion
results in the incorporation of the vesicle membrane into the
plasma membrane continuum, with the inner vesicle membrane

face exposed to the exterior, and the expelling of vesicle contents. Some loss of membrane lipids and proteins to the exterior may also occur during membrane re-organisation.

Few of these events have been investigated in plant cells, mainly due to the refractory nature of plant cell secretory systems. Most operate continuously, with the possible exception of glands such as those found in *Dionaea,* in which regulation is imposed at the level of vesicle formation rather than fusion. Some pollen grains store large numbers of vesicles which only fuse with the plasma membrane on germination (Heslop-Harrison and Heslop-Harrison 1982). The presence of the cell wall makes it difficult to examine cell surface events in plants. Dependence on Ca^{2+} ions has been demonstrated in a number of plant systems (Reiss and Herth 1979; Baydoun and Northcote 1980; Picton and Steer 1983c), and serves to complicate investigations into plant cell extension (Brummell and Hall 1983). Parallel studies to those on animal cells have demonstrated a dependence on external $[Ca^{2+}]$ and show a similar response to the calcium ionophore A23187 (Reiss and Herth 1979; Picton and Steer 1983c). An attempt to use an inhibitor of calcium uptake by endoplasmic reticulum revealed that fluorescein isothiocyanate is a potent inhibitor of vesicle fusion in pollen tubes (Picton and Steer 1985), so that vesicles accumulate at the cell surface. It has not been possible to extend these observations to another plant secretory system, the *Zea* root cap (Shannon, personal communication). Genetic control of fusion has been demonstrated in *Tetrahymena* cells where mutants, unable to secrete, have been shown to lack the plasma membrane docking rosettes (Orias *et al.* 1983), either as a direct or an indirect result of the lesion.

6.3.3b Product release.
Delivery of a secretory product to the cell surface is one of the main end results of the secretory process. In plants these products consist of wall carbohydrates, wall proteins (enzymes and extensin) and slimes and mucilages. As discussed earlier the relationship between product release by the vesicles and the flow of secretion from the cell surface is complicated by accompanying osmotic flows of water. This probably occurs in *Drosophyllum* (Schnepf and Busch 1976) and has been demonstrated in maize root caps where the flow of vesicle product represents less than 0.1% of the final volume of slime (Shannon and Steer 1984a). The reverse situation appears to apply in the case of cell wall polysaccharides. Apparently these are assembled and incorporated into the growing wall more efficiently than they are packaged into secretory vesicles. In pollen tubes there is about a ten-fold increase in the packing density as the carbohydrates are incorporated into the wall at the tube tip (Steer and Picton 1984).

example Shannon and Steer (1984a) estimated that, in secretory
root cap cells, an area equivalent to the cell plasma membrane
is delivered every 10 minutes. Membrane retrieval via
endocytosed vesicles should yield about 8 times more small
vesicle profiles per unit area of micrograph than secretory
vesicles. Such large numbers of small vesicles have not been
found. Similarly, the high rate of membrane recycling found to
occur in slow growing pollen tubes is not accompanied by the
appearance of small vesicles that might form part of a
recycling route.

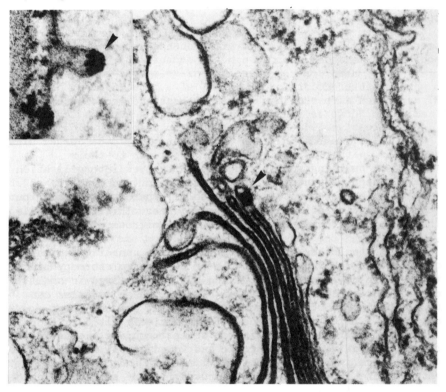

Fig. 6.1.   Maize root cap cells following exposure
to lead salts. Endocytosis of the lead salts (inset,
arrow) is followed by appearance of dense accumulations
in dictyosome cisternae (arrow). By kind permission
of Hübner *et al.* 1986. X 93,500. Inset X 144,500.

The evidence for endocytosis in plant cells is still
controversial. Reviewing earlier literature Cram (1980) found
that the evidence was weak and advanced a number of theoretical
reasons against the uptake of ions by this route. The arguments
concerning excess turgor pressure, generated by the inward
volume flow of water in the vesicles, do not apply to cells in

which there is an outward flow of water and osmoticum in secretory vesicles. It has been demonstrated that endocytosis can occur in plant cells under experimental conditions (Joachim and Robinson 1984; Tanchak *et al.* 1984). Of greater significance is the recent finding that endocytosis occurs in intact tissues. This has been demonstrated by Hübner *et al.* (1986) using lead salts to follow the internalisation of plasma membrane in maize root cap cells. This is followed by the appearance of similar electron-opaque bodies in the dictyosome cisternae **(Figure 6.1)**. In addition the dry-cleave method of preparing fragments of plant cell surfaces for microscopy (Traas 1984) has shown that coated pits are a regular feature of the plant cell surface (Traas 1984; Hawes 1985). If these are involved in membrane recycling then it should be possible to demonstrate an increase in their number and activity with increasing rates of secretion.

6.4 SUMMARY

Vesicles are the visible manifestation of part of the membrane and product traffic within the cell cytoplasm. In the absence of efficient techniques for the quantitative biochemical separation of plant cell membrane systems (Robinson 1984), microscopy provides the main tool for investigating this traffic. Much valuable information of a qualitative nature has been obtained by the examination and interpretation of micrographs from preserved cells, but this approach provides no information about the rates of flow or balances between different routes. The methods available for providing such quantitative information have been assessed and it is concluded that even the successful ones are of limited application. The prospects for future improvements in these methods are not good, and developments aimed at providing such information directly from living cells have met with only limited success (Steer *et al.* 1985). The available quantitative information has been considered in relation to the three main events in vesicle-mediated flows: vesicle formation, vesicle transport and vesicle fusion. Secretory vesicles *en route* from the Golgi apparatus to the plasma membrane are the easiest vesicle population to identify and quantitate. This migration is driven by the cytoskeleton, but the precise details are unknown at present. At the plasma membrane, a requirement for calcium in the vesicle fusion process has been clearly established in plants, as it has already been for animal cells. It is probable that other aspects of the fusion process in plants are similar to those in animals. Quantitation of the secretory part of the membrane flow pathway, from Golgi to plasma membrane, provides a measure of the flow that must occur elsewhere in the system, from endoplasmic reticulum to Golgi and recycling from the

plasma membrane. At present there is no quantitative
microscopic information to support the involvement of vesicles
in these other routes. It is possible that secretory cells have
a finite membrane resource, either in the form of macromolecules
or of assembled bilayers, that is continuously in circulation
between the endo- and plasma membrane systems. Interruption of
one part of the cycle would then lead to a halt in the
circulation. Inhibition of secretory vesicle transport by
cytochalasin leads to the progressive cessation of vesicle
formation, perhaps as additional membrane becomes trapped in the
form of accumulating vesicles. Further information on vesicle
dynamics in plant cells could be obtained using the existing
microscopic methods in conjunction with suitable experimental
systems.

6.5 REFERENCES

Baydoun, E.A.-H and Northcote, D.H. (1980). Measurement and characteristics of fusion of isolated membrane fractions from maize root tips. *J. Cell Sci.* 45, 169-186.

Bechtel, D.B. and Gaines, R.L. (1982). The presence of protease-digestible material in Golgi vesicles during endosperm development of selected cereals. *Am. J. Bot.* 69, 880-4.

Bowles, D.J. and Northcote, D.H. (1974). The amounts and rates of export of polysaccharide found within the membrane system of maize root cells. *Biochem. J.* 142, 139-44.

Brett, J.G. and Godman, G.C. (1984). Membrane cycling and macrovacuolation under the influence of cytochalasin: kinetic and morphometric studies. *Tissue and Cell* 16, 325-35.

Brummell, D.A. and Hall, J.L. (1983). Regulation of cell wall synthesis by auxin and fusicoccin in different tissues of pea stem segments. *Physiol. Plant.* 59, 627-34.

Chrispeels, M.J. (1976). Biosynthesis, intracellular transport, and secretion of extracellular macromolecules. *Ann. Rev. Plant Physiol.* 27, 19-38.

Chrispeels, M.J. (1983). The Golgi apparatus mediates the transport of phytohemagglutinin to the protein bodies in bean cotyledons. *Planta* 158, 140-51.

Conrad, P.A., Binari, L.L.W. and Racusen, R.H. (1982). Rapidly-secreting, cultured oat cells serve as a model system for the study of cellular exocytosis. Characterisation of cells and isolated secretory vesicles. *Protoplasma* 112, 196-204.

Cope, G.H. (1983). Exocrine glands and protein secretion: a stereological viewpoint. *J. Microsc.* 131, 187-202.

Cope, G.H. and Williams, MA. (1981). Secretion granule formation in the rabbit parotid gland after isoprenaline-induced secretion. Stereological reconstructions of granule populations. *Anat. Rec.* 199, 377-387.

Craig, S. and Goodchild, D.J. (1984). Golgi-mediated vicilin accumulation in pea cotyledon cells is re-directed by monensin and nigericin. *Protoplasma* 122, 91-7.

Cram, W.J. (1980). Pinocytosis in plants. *New Phytol.* 84, 1-17.

Cunninghame, M.E. and Hall, J.L. (1985). A quantitative stereological analysis of the effect of indoleacetic acid on the dictyosomes in pea stem epidermal cells. *Protoplasma* 125, 230-4.

Dahl, G., Ekerdt, R., and Gratzl, M. (1979). Models for exocytotic membrane fusion. *Soc. exp. Biol. Symp.* 33, 349-368.

Dauwalder, M. and Whaley, W.G. (1982). Membrane assembly and secretion in higher plants. *J. Ultrastruct. Res.* 78, 302-320.

Ericson, M.C., Gafford, J.T. and Elbein, A.D. (1977). Tunicamycin inhibits GlcNAc-lipid formation in plants. *J. Biol. Chem.* 257, 3105-9.

Farquar, M.G. and Palade, G.E. (1981). The Golgi apparatus (complex) -(1954-1981)- from artifact to center stage. *J. Cell Biol.* 91, 77S-103S.

Fraley, R., Wilschut, J., Düzgünes, N., Smith, C. and Papahadjopoulos, D. (1980). Studies on the mechanism of membrane fusion: role of phosphate in promoting calcium ion induced fusion of phospholipid vesicles. *Biochemistry* 19, 6021-9.

Gardiner, M. and Chrispeels, M.J. (1975). Involvement of the Golgi apparatus in the synthesis and secretion of hydroxyproline-rich cell wall glycoproteins. *Plant Physiol.* 55, 536-41.

Harris, N. and Oparka, K.J. (1983). Connections between dictyosomes, ER and GERL in cotyledons of mung bean (*Vigna radiata L.*). *Protoplasma* 114, 93-102.

Hawes, C.R. (1985). Conventional and high voltage electron microscopy of the cytoskeleton and cytoplasmic matrix of carrot (*Daucus carota* L.) cells grown in suspension culture. *Eur. J. Cell Biol.* (in press).

Hawkins, E.K. (1974). Growth and differentiation of the Golgi apparatus in the red alga, *Callithamnion roseum*. *J. Cell Sci.* 14, 633-55.

Heath, I.B. and Seagull, R.W. (1982). Orientated cellulose fibrils and the cytoskeleton: a critical comparison of models. In: *The cytoskeleton in plant growth and development*. (ed. C.W. Lloyd). pp. 163-182. Academic Press, London.

Henry, Y. and Steer, M.W. (1985). Acid phosphatase localisation in the digestive glands of *Dionaea muscipula* Ellis flytraps. *J. Histochem. Cytochem.* 33, 339-44.

Herman, E.M. and Shannon, L.M. (1984). Immunocytochemical evidence for the involvement of Golgi apparatus in the deposition of seed lectin of *Bauhinia purpurea* (Leguminosae). *Protoplasma* 121, 163-70.

Herzog, V. (1981). Endocytosis in secretory cells. *Phil. Trans. R. Soc. Lond. B* 296, 67-72.

Heslop-Harrison, J. and Heslop-Harrison, Y. (1982). The growth of the grass pollen tube: 1. Characteristics of the polysaccharide particles ("P-particles") associated with apical growth. *Protoplasma* 112, 71-80.

Hickman, S., Kulczycki, A. Jr., Lynch, R.G. and Kornfeld, S. (1977). Studies on the mechanism of tunicamycin inhibition of IgA and IgE secretion by plasma cells. *J. Biol. Chem.* 252, 4402-8.

Howard, R.J. (1981). Ultrastructural analysis of hyphal tip cell growth in fungi: spitzenkorper, cytoskeleton and endomembranes after freeze-substitution. *J. Cell Sci.* 48, 89-103.

Howard, R.J. and Aist, J.R. (1979). Hyphal tip ultrastructure of the fungus *Fusarium*: improved preservation by freeze-substitution. *J. Ultrastruct. Res.* 66, 224-34.

Hübner, R., Depta, H. and Robinson, D.G. (1986). Evidence for endocytosis in maize root cap cells using heavy metal salt solutions. (in preparation).

Isenberg, G., Leonard, K. and Jockusch, B.M. (1982). Structural aspects of vinculin-actin interactions. *J. molec. Biol.* 158, 231-49.

Joachim, S. and Robinson, D.G. (1984). Endocytosis of cationic ferritin by bean leaf protoplasts. *Eur. J. Cell Biol.* 34, 212-6.

Juniper, B.E., Hawes, C.R. and Horne, J.C. (1982). The relationships between the dictyosomes and the forms of endoplasmic reticulum in plant cells with different export programs. *Bot. Gaz.* 143, 135-45.

Kristen, U. and Lockhausen, J. (1983). Estimation of Golgi membrane flow rates in ovary glands of *Aptenia cordifolia* using cytochalasin B. *Eur. J. Cell Biol.* 29, 262-7.

Kwiatkowska, M. and Maszewski, J. (1979). Changes in the activity of the Golgi apparatus during the cell cycle in antheridial filaments of *Chara vulgaris L. Protoplasma* 99, 31-8.

Lawson, D. and Raff, M.C. (1979). Some membrane events occurring during fusion and exocytosis in rat peritoneal mast cells. *Soc. exp. Biol. Symp.* 33, 337-347.

Mollenhauer, H.H. and Morré, D.J. (1976). Cytochalasin B, but not colchicine, inhibits migration of secretory vesicles in root tips of maize. *Protoplasma* 87, 39-48.

Mollenhauer, H.H., Morré, D.J. and Droleskey, R. (1983). Monensin affects the trans-half of *Euglena* dictyosomes. *Protoplasma* 114, 119-24.

Mollenhauer, H.H., Morré, D.J. and Norman, J.O. (1982). Ultrastructural observation on maize root tips following exposure to monensin. *Protoplasma* 112, 117-26.

Morré, D.J., Boss, W.F., Grimes, H. and Mollenhauer, H.H. (1983). Kinetics of Golgi apparatus membrane flux following monensin treatment of embryogenic carrot cells. *Eur. J. Cell Biol.* 30, 25-32.

Morré, D.J., Kartenbeck, J. and Franke, W.W. (1979). Membrane flow and interconversions among endomembranes. *Biochim. Biophys. Acta* 559, 71-152.

Morré, D.J. and Mollenhauer, H.H. (1974). The endomembrane concept: a functional integration of endoplasmic reticulum and Golgi apparatus. In: *Dynamic aspects of plant ultrastructure.* (ed. A.W. Robards). pp. 84-137. McGraw-Hill, London.

Morré, D.J. and VanDerWoude, W.J. (1974). Origin and growth of cell surface components. In: *Macromolecules regulating growth and development.* (ed. E.D. Hay, T.J. King, J. Papaconstantinou). pp. 81-111. Academic Press, New York.

Morré, D.J., Mollenhauer, H.H. and Bracker, C.E. (1971). Origin and continuity of Golgi apparatus. In: *Results and problems in cell differentiation. II. Origin and continuity of cell organelles.* (ed. J. Reinert and H. Ursprung). pp. 82-126. Springer-Verlag, Berlin.

Nieden, U., Manteuffel, R., Weber, E. and Neumann, D. (1984). Dictyosomes participate in the intracellular pathway of storage proteins in developing *Vicia faba* cotyledons. *Eur. J. Cell Biol.* 34, 9-17.

Orias, E., Flacks, M. and Satir, B.H. (1983). Isolation and ultrastructural characterisation of secretory mutants of *Tetrahymena thermophilia*. *J. Cell Sci.* 64, 49-67.

Parker, M.L. and Hawes, C.R. (1982). The Golgi apparatus in developing endosperm of wheat (*Triticum aestivum L.*). *Planta* 154, 277-83.

Pellegrini, L. (1976). Variations ultrastructurales de l'appareil de Golgi chez la *Cystoseira stricta* Dauvageau (Pheophycee, Fucale). *Protoplasma* 90, 205-28.

Picton, J.M. (1981). Quantitative ultrastructural studies of vesicle production and utilisation in pollen tubes. *Ph.D. thesis. Queen's University, Belfast.*

Picton, J.M. and Steer, M.W. (1981). Determination of secretory vesicle production rates by dictyosomes in pollen tubes of *Tradescantia* using cytochalasin D. *J. Cell Sci.* 49, 261-272.

Picton, J.M. and Steer, M.W. (1983a). Membrane recycling and the control of secretory activity in pollen tubes. *J. Cell Sci.* 63, 303-10.

Picton, J.M. and Steer, M.W. (1983b). The effect of cycloheximide on dictyosome activity in *Tradescantia* pollen tubes determined using cytochalasin D. *Eur. J. Cell Biol.* 29, 133-8.

Picton, J.M. and Steer, M.W. (1983c). Evidence for the role of Ca^{2+} ions in tip extension in pollen tubes. *Protoplasma* 115, 11-7.

Picton, J.M. and Steer, M.W. (1985). The effects of ruthenium red, lanthanum, fluorescein isothiocyanate and trifluoperazine on vesicle transport, vesicle fusion and tip extension in pollen tubes. *Planta* 163, 20-6.

Picton, J.M., Steer, M.W. and Earnshaw, J.C. (1983). Vesicle dynamics in pollen tubes. In: *The application of laser light scattering to the study of biological motion.* (ed. J.C. Earnshaw, M.W. Steer). pp. 383-388. Plenum Press, New York and London.

Pope, D.G., Thorpe, J.R., Al-Azzawi, M.J. and Hall, J.L. (1979). The effect of cytochalasin B on the rate of growth and ultrastructure of wheat coleoptiles and maize roots. *Planta* 144, 373-83.

Quaite, E., Parker, R.E. and Steer, M.W. (1983). Plant cell extension: structural implications for the origin of the plasma membrane. *Plant, Cell and Env.* 6, 429-32.

Reiss, H.-D. and Herth, W. (1979). Calcium ionophore A 23187 affects localised wall secretion in the tip region of pollen tubes of *Lilium longiflorum*. *Planta* 145, 225-32.

Robinson, D.G. (1981). The ionic sensitivity of secretion-associated organelles in root cap cells of maize. *Eur. J. Cell Biol.* 23, 267-72.

Robinson, D.G. (1985). Plant membranes. Endo- and plasma membranes of plant cells. pp. 331. John Wiley and Sons, New York.

Robinson, D. G., Eisinger, W.R. and Ray, P.M. (1976). Dynamics of the Golgi system in wall matrix polysaccharide synthesis and secretion by pea cells. *Ber. Deutsch. Bot. Ges.* 89, 147-61.

Robinson, D.G. and Kristen, U. (1982). Membrane flow via the Golgi apparatus of higher plant cells. *Int. Rev. Cytol.* 77, 89-127.

Rose, P.E. (1980). Improved tables for the evaluation of sphere size distribution including the effect of section thickness. *J. Microsc.* 118, 135-41.

Rothman, J.E. (1981). The Golgi apparatus: two organelles in tandem. *Science* 213, 1212-9.

Sabatini, D.D., Kreibich, G., Morimoto, T. and Adesnik, M. (1982). Mechanisms for the incorporation of proteins in membranes and organelles. *J. Cell Biol.* 92, 1-22.

Saermark, T., Jones, P.M. and Robinson, I.C.A.F. (1984). Membrane retrieval in the guinea-pig neurohypophysis. *Biochem. J.* 218, 591-9.

Schnepf, E. (1961). Quantitative Zusammenhange zwischen der Sekretion des Fangschleimes und den Golgi-Strukturen bei *Drosophyllum lusitanicum*. *Z. Naturforschung* 16b, 605-10.

Schnepf, E. (1974). Gland cells. In: *Dynamic aspects of plant ultrastructure*. (ed. A.W. Robards). pp. 331-357. McGraw-Hill, London.

Schnepf, E. and Busch, J. (1976). Morphology and kinetics of slime secretion in glands of *Mimulus tilingii*. *Zeit. Pflanzenphysiol.* 79, 62-71.

Schnepf, E., Witte, O., Rudolph, U., Deichgräber, G. and Reiss, H.-D. (1985). Tip cell growth and the frequency and distribution of particle rosettes in the plasmalemma: experimental studies in *Funaria* protonema cells. *Protoplasma*. (in press).

Schwab, D.W., Simmons, E. and Scala, J. (1969). Fine structure changes during function of the digestive gland of Venus's Fly trap. *Am. J. Bot.* 56, 88-100.

Shannon, T.M. (1983). Qualitative and quantitative ultrastructural investigations into the slime production by the outer root cap cells of *Zea mays L.* *Ph.D. thesis. Queen's Univeristy, Belfast.*

Shannon, T.M., Henry, Y., Picton, J.M. and Steer, M W. (1982). Polarity in higher plant dictyosomes. *Protoplasma* 112, 189-95.

Shannon, T.M., Picton, J.M. and Steer, M.W. (1984). The inhibition of dictyosome vesicle formation in higher plant cells by cytochalasin D. *Eur. J. Cell Biol.* 33, 144-7.

Shannon, T.M. and Steer, M.W. (1984a). The root cap as a test system for the evaluaiton of Golgi inhibitors. I. *J. Exptl. Bot.* 35, 1697-707.

Shannon, T.M. and Steer, M.W. (1985b). The root cap as a test system for the evaluation of Golgi inhibitors. II. *J. Exptl. Bot.* 35, 1708-14.

Steer, M.W. (1977). Differentiation of the tapetum in *Avena*. II. The endoplasmic reticulum and Golgi apparatus. *J. Cell Sci.* 28, 71-86.

Steer, M.W. (1981). Understanding Cell Structure. pp. 126. Cambridge University Press, Cambridge.

Steer, M.W. and Picton, J.M. (1984). Control of cell wall formation in pollen tubes: the interaction of dictyosome activity with the rate of tip extension. In: *Structure, function and biosynthesis of plant cell walls*. (ed. W.M. Dugger, S. Bartnicki-Garcia). American Society of Plant Physiology, Maryland, pp. 483-9.

Steer, M.W., Picton, J.M. and Earnshaw, J.C. (1984). Diffusive motions in living cytoplasm probed by laser Doppler microscopy. *J. Microsc.* 134, 143-9.

Steer, M.W., Picton, J.M. and Earnshaw, J.C. (1985). Laser light scattering in biological research. *Adv. Bot. Res.* 11, 1-69.

Steinman, R.M., Mellman, I.S., Muller, W.A. and Cohn, Z.A. (1983). Endocytosis and recycling of plasma membrane. *J. Cell Biol.* 96, 1-27.

Stossel, T.P. (1982). The structure of cortical cytoplasm. *Phil. Trans. R. Soc. Lond. B.* 299, 275-289.

Tanchak, M.A., Griffing, L.R., Mersey, B.G. and Fowke, L.C. (1984). Endocytosis of cationised ferritin by coated vesicles of soybean protoplasts. *Planta* 162, 481-6.

Tartakoff, A. and Vassalli, P. (1977). Plasma cell immunoglobin secretion: arrest is accompanied by alterations in the Golgi complex. *J. Exptl. Med.* 146, 1332-45.

Traas, J.A. (1984). Visualisation of the membrane bound cytoskeleton and coated pits of plant cells by means of dry cleaving. *Protoplasma* 115, 81-7.

Tsekos, I. (1981). Growth and differentiation of the Golgi apparatus and wall formation during carposporogenesis in the red alga, *Gigartina teedii* (Roth) Lamour. *J. Cell Sci.* 52, 71-84.

Tsukita, S., Tsukita, S., Ishikawa, H., Sato, S. and Nakao, M. (1981). Electron microscopic study of reassociation of spectrin and actin with the human erythrocyte membrane. *J. Cell Biol.* 90, 70-7.

Ueda, K. and Noguchi, T. (1976). Transformation of the Golgi apparatus in the cell cycle of a green alga, *Micrasterias americana*. *Protoplasma* 87, 145-162.

VanDerWoude, W.J. and Morré, D.J. (1968). Endoplasmic reticulum-dictyosome-secretory vesicle associations in pollen tubes of *Lilium longiflorum* Thunb. *Proc. Indiana Acad. Sci.* 77, 164-70.

VanDerWoude, W.J., Morré, D.J. and Bracker, C.E. (1971). Isolation and characterization of secretory vesicles in germinated pollen of *Lilium longiflorum*. *J. Cell Sci.* 8, 331-51.

Wilkins, J.A. and Lin, S. (1981). Association of actin with chromaffin granule membranes and the effect of cytochalasin B on the polarity of actin filament elongation. Biochim. Biophys. Acta 642, 55-66.

Wilschut, J., Duzgunes, N., Fraley, R. and Papahadjopoulos, D.
(1980). Studies on the mechanism of membrane fusion: kinetics
of calcium ion induced fusion of phosphatidylserine vesicles
followed by a new assay for mixing of aqueous vesicle contents.
Biochemistry 19, 6011-21.

7

Laser Doppler microscopy of plant cells

Richard P. C. Johnson and Graeme R. A. Dunbar

7.1 INTRODUCTION

During the past thirty years, the ultrastructure and the biochemistry of cells have been mapped in great detail. However, much of our knowledge of biochemical pathways has come from studying the average behavior of fractions of smashed cytoplasm and most of our views of ultrastructure have been obtained from specimens which were pickled or frozen rigid for electron microscopy. These methods destroy and leave little impression of the intensely complicated patterns of flow and diffusion which are essential to life in the structure of cytoplasm. During the past ten years there has been a surge of new methods for the light microscopy of living cells. Laser Doppler microscopy provides one which can measure motion. If developed further, it may contribute to a better understanding of factors which affect diffusion, viscosity and flow in the fluid regions between the more static, cytoskeletal, components of cytoplasm.

Motion in cytoplasm, or of whole microorganisms, can be measured in various ways. The simplest method, measuring the time taken for an object to move between marks under the microscope, can provide the mean velocity of a particle between the marks, but not the maximum or minimum velocities it attained. Another disadvantage is that it gives the mean velocity of a general flow from the mean velocity of individual particles, which may not behave typically; particles may pause at obstructions or their velocity may depend upon their size. Another method is to superimpose an image of a rotating ground-glass disc on the image of the specimen in the microscope. The calibrated velocity of the disc may then be adjusted, subjectively, to match the average velocity of particles in the specimen, but this method does not show spread

in velocity. Other methods include the laborious process of
measuring the displacements of particles from frame-to-frame in
cine films (Barclay and Johnson, 1982) or video-recordings. A
tracking microscope (Berg,1983) has been used to follow the
motion of microorganisms, but this ingenious instrument seems
unsuitable for following the motions of complex mixtures of
particles inside cells.

In laser Doppler microscopy, the distribution in the
velocity of particles and thus their mean velocity and diffusion
constants are measured by analysing the Doppler shifts given to
the frequency of laser light which has been scattered from them.
The parent method from which laser Doppler microscopy is
developing, laser Doppler spectroscopy (see Drain, 1980;
Degiorgio, 1983), was first demonstrated by Yeh and Cummins in
1964. Its development depended on the advent of efficient
electronics and on the laser as an intense source of coherent
light. This parent method has a variety of names; heterodyne or
optical beating, laser Doppler spectroscopy, photon correlation
spectroscopy and velocimetry, laser Doppler anemometry. It is
now well established in chemistry, biochemistry, physics and
engineering to measure, for example, the diffusion constants and
hence the molecular weights of proteins, or to measure flow,
diffusion and turbulence in combustion chambers or fluidised
beds.   In biology it has become established mainly as a method
for studying the diffusion of macromolecules or flow in blood
vessels, but it is also proving useful in work which ranges from
studies of phase transitions in membranes (Crawford and
Earnshaw, 1984) to the swimming of microorganisms (see Earnshaw
and Steer,eds., 1983) and the vibration of tympanic membranes
(Vlaming, Aertson and Epping, 1984). Because laser Doppler
spectroscopy uses the frequency of scattered light, and not
dimensions in an image, it can measure the movement, or sizes,
of particles with diameters less than the diffraction limit to
resolution in optical images.

In most applications of laser Doppler spectroscopy the
scattered light is collected from a volume of the specimen
which is as wide, or wider, than the diameter of the
illuminating beam where it emerges from the laser. This
diameter is generally greater than 100 μm and is thus more than
the diameter of a typical living cell. If laser Doppler
spectroscopy is to be used to study motion in identified regions
of single cells then the laser light must be applied to
specimens and collected from them via some form of microscope.
The first laser Doppler microscope was described by Maeda &
Fujime (1972). Since then various kinds of laser Doppler
microscopes have been constructed, mainly in departments of
physics or engineering. So far, most of the work with them has
been on the flow of blood in capillaries. There have been only

two publications describing results from the use of laser
Doppler microscopes to measure the motion of specified
structures in living cells (Johnson, 1983; Steer, Picton and
Earnshaw, 1984). There is little background of experience to
interpret the signals they produce.

7.2 BASIC PRINCIPLES OF LASER DOPPLER SPECTROSCOPY

7.2.1 The Doppler shift

The intensity of light scattered from a particle varies with
its radius, the wavelength and polarisation of the illuminating
light and the direction of the detector (Steer, Picton and
Earnshaw, 1985, figures 3 and 4). If the particle is moving, the
frequencies in the light scattered from it are increased or
decreased depending on whether the particle is moving towards or
away from the source of illumination. This shift in frequency,
the Doppler effect, may be explained if we consider a beam of
light to consist of regular peaks and troughs in electromagnetic
intensity travelling at the speed of light and if we imagine
that every time a particle encounters a peak it scatters it.
When the particle is moving towards the light source it will
encounter more peaks per second than when it moves away from
it. The light scattered by the particle will appear to have a
higher or a lower frequency than the beam which illuminates it,
according to the speed and direction of the particle relative to
the beam and the detector.

     Light from ordinary lamps contains a mixture of frequencies
which propagate as short independent trains of waves with phases
at random to each other. Thus a particle moving in a beam from
an ordinary lamp will scatter light containing a very complex
and changing mixture of intensities and phases - inconveniently
complicated to detect and analyse into Doppler shifts. However,
lasers emit light at one or a few constant frequencies. They can
provide monochromatic light which, at least over a useful length
of the beam, is coherent; ie all the wave trains in the beam are
in phase with each other. For these reasons, the relation
between the frequency of light in a laser beam and that detected
from a moving particle which scatters it will be simpler and
more constant than in light from an ordinary lamp; almost as
simple and constant as assumed in the previous paragraph.

7.2.2. How big a Doppler shift will particles in cells give to scattered light and how can it be measured?

The maximum Doppler shift given to the frequency of light
scattered from an object moving at 30 miles per hour is about

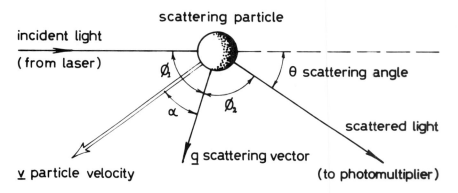

Fig.7.1. Diagram of the optical arrangement of a typical
laser Doppler spectrometer.

4.10^7 Hz. Resolvable particles in flow or in Brownian motion
in cytoplasm move at speeds of the order of 1 μm per second. The
maximum Doppler shift they would produce is only a few thousand
cycles per second. The light from a helium-neon laser has a
frequency of about 4.7×10^{14} Hz (ie a wavelength of 632.8 nm)
and thus we would need to measure changes in its frequency of
less than about one part in 10^8, or, possibly, much less.
Unfortunately, even the most selective optical filters are too
coarse to do this. Alternatively, the oscillations of light
cannot be followed directly by photoelectronic detectors; their
frequency response is not high enough. Furthermore, at
frequencies higher than radio frequencies (above about 10^{11}
Hz) the energy of electromagnetic waves is not detected
continuously, a quantum of energy must be absorbed before an
electron will jump between the orbits of an atom in a detector.

In spite of these obstacles, it is possible to measure
Doppler shifts in frequency even as low as one cycle per second
if the Doppler-shifted light is allowed to beat with unshifted
light on the surface of a detector, usually a photomultiplier.
The beat frequencies then detected represent the sums and
differences between the frequency of the beam used to illuminate
a specimen and the frequency of light scattered from it. In
practice, the Doppler-shifted light scattered from moving
particles is mixed to beat, either with a beam of reference
light at constant frequency direct from the laser ("heterodyne"
beating), or with other light which has also been Doppler
shifted by scattering from other moving particles nearby
("homodyne" or "self-" beating). Sometimes a mixture of
heterodyne and self-beating frequencies is detected. Generally,
the beat frequencies caused by summing are filtered out,
optically, to leave only the difference frequencies to be

detected and recorded for subsequent mathematical analysis.

     Fig.7.1. shows the optical arrangement of a typical laser
Doppler spectrometer for detecting laser light scattered from a
specimen. If the light is scattered from a single particle
moving at constant velocity, or from many particles which are
all moving with the same velocity, then a single beat frequency
will occur in the intensity of light detected from them. This
beat frequency will be directly proportional to their velocity.
However, if the particles are moving with random velocities and
directions, ie diffusing as in Brownian motion, then the
scattered light from them will contain a spectrum of beat
frequencies which can be detected and analysed to give the
translational diffusion constant (D) of the particles.

$$D = \frac{kT}{6\pi\eta a} \qquad\qquad (1)$$

Where k = Boltzmann's constant (ie. 1.3806 X 10^{23} J K^{-1}, T is
the temperature in Kelvin degrees, a is the radius of the
particles (metres) and η is the viscosity in pascal seconds (one
poise = 10^{-1} pascal seconds). Eq.1 shows that if D and two of
the other variables are known then the third may be calculated.
The derivation of D was described by Berg (1983).

7.2.3. The calculation of diffusion constants and velocities

Since the oscillations of light are detected as an intensity
(the square of their amplitude) the spectrum of frequencies
detected will be a power spectrum. If the particles are all of
the same size then this spectrum, as displayed in a spectrum
analyser, will be in the form of a Lorentzian, a curve which
declines on each side of a central peak. This peak represents
the mean velocity of the particles which, for a cloud of
particles in random thermal motion in the absence of a
concentration gradient, is zero. However, because all
frequencies are detected as positive, only the right-hand half
of the curve will be displayed (Fig.7.2a). If the particles are
flowing at a constant mean velocity in one direction as well as
diffusing, then the peak will be displaced positively by a
distance proportional to the velocity of this flow (Fig.7.2b).
The half-width of the Lorentzian at half its maximum height
(Figs.7.2a & 7.2b) is q^2D where D is the diffusion constant of
the particles, as shown in Eq.1. and q is the scattering
wavenumber (sometimes given the symbol k). For Doppler shifts as
small as those of interest to cell biologists, q is a convenient
constant used to describe a given arrangement of the instrument

and is described by Eq.2:

$$q = \frac{4\pi n}{\lambda} \sin\left(\frac{\Theta}{2}\right)$$

(2)

Where λ is the wavelength of the incident light (metres), n the refractive index of the medium in which the particles are suspended and Θ is the scattering angle (degrees), the angle between the illuminating beam and a line drawn between the scatterers and the detector (Fig.7.1).

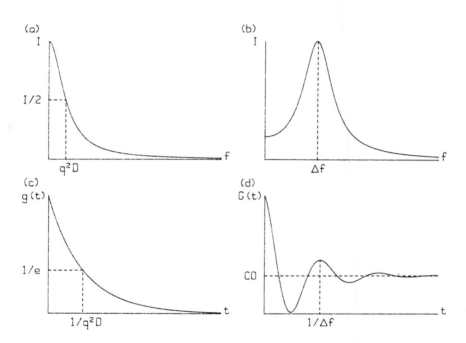

Fig.7.2. Theoretical curves: (a) Power spectrum (Lorentzian) of laser light scattered from diffusing particles. (b) Power spectrum from particles diffusing and flowing. (c) Normalised autocorrelation function (exponential) from diffusing particles. (d) Autocorrelation function from particles diffusing and flowing.

When light which is Doppler shifted by scattering from a particle or particles moving with uniform velocity (V) in relation to the laser and the detector (Fig.7.1) is allowed to beat with unshifted reference light, then the single Doppler frequency (Δf) which can be detected is proportional to the speed of the particles. But, it is also proportional to the

angle (α) between the direction of the motion and the
scattering (or Bragg) wave vector $\underline{q}$, which lies along the
bisector of the angle between laser, particle and detector, as
shown in Fig.7.1. Therefore the Doppler shift (Δf) is given by
Eq. 3:

$$\Delta f = fs - fd \ = \frac{2n}{\lambda} \sin(\frac{\Theta}{2})V \cos(\alpha) \tag{3}$$

where fs is the frequency of the light from the source (Hz) and
fd is the frequency of the light at the detector (Hz) or, if
angular frequencies are used, $\Delta\omega = 2\pi\Delta f$ radians/second.
Therefore, from Eqs.2 and 3:

$$\Delta f = \frac{qV \cos(\alpha)}{2\pi} \tag{4}$$

Or

$$V = \frac{2\pi\,\Delta f}{q \cos(\alpha)} \tag{5}$$

     In practice it has become usual to use, not a spectrum
analyser, but a signal correlator.   A signal correlator, when
fed with pulses from a photomultiplier, calculates and displays
the autocorrelation function of the detected intensity of
light as a series of correlation coefficients (Figs. 7.7 and
7.8). An autocorrelation function shows how the intensity of the
detected light changed with time, ie how intensity at a given
time was correlated with intensity at a series of later times.
The autocorrelation function of the intensity of light scattered
from diffusing particles may be taken, intuitively, to describe
the decay of a temporary structure made of particles which
proceed to diffuse in relation to each other. The autocorrelation function of signal intensity is a Fourier transform
of the power spectrum. In other words, it bears the same
relation to this frequency spectrum as an image does to a
diffraction pattern. Thus, a single frequency appears in a
correlation function as a cosine wave (Fig.7.2d); in a frequency
spectrum it would appear as a delta function, as a single spike.
Descriptions of signal correlators may be found in Oliver
(1974) and in Drain (1980). A correlator accumulates the
autocorrelation function of the signal obtained from the
detector and displays it, conveniently, above a baseline of
noise (C_0 in Eq.6) which can be subtracted easily.

     For diffusing particles, all of one size, the
autocorrelation function (Gt) of the intensity of light
scattered from them has the form of an exponential with a time
constant of $1/q^2 D$ (Fig.7.2c). It is a Fourier transform of the
corresponding Lorentzian which would have appeared if a spectrum
analyser had been used. The time constant is defined as the

point on the time axis (the x axis in Figs. 7.2c,d, 7.7 and 7.8) above which the exponential has fallen to 1/e of its initial value. Thus, after a time t, the autocorrelation will be:

$$G(t) = C_0 + C_1 \exp(-q^2 Dt) + C_2 \exp(-2q^2 Dt) \qquad (6)$$

The first term (C_0) represents a subtractable horizontal baseline due to the uncorrelated intensity of light scattered from stationary objects. It represents noise, the random, ie uncorrelated, parts of the signal which have not been Doppler shifted. The second term, with coefficient C_1, represents the scattered field autocorrelation of the heterodyne part of the signal, that is, of light scattered from moving particles beating with light scattered from stationary objects, for example, from lenses, cuvettes, coverslips, microscope slides and cell walls. If all the detected light came from moving particles then the coefficient C_1 would be zero, ie, this second term would be absent. The third term, with coefficient C_2, represents self-beating. It represents the intensity autocorrelation of light from moving particles which was beating with light from other moving particles. It has half the time constant of the heterodyne term because the reference light for it comes, not from stationary scatterers, but from other diffusing particles which, on average at any instant, move relative to each other twice as fast as to stationary scatterers.

Of course, for every size of particle present in a population of moving scatterers there would be a corresponding pair of self-beating and heterodyne terms; the number of exponential terms in Eq.6 increases with the diversity (polydispersity) of particle sizes. However, in practice, if the heterodyne signal is made large enough to swamp the self-beating signal, then the self beating terms may be neglected. Oliver (1974, pp.216) suggests that the self-beating signals might be negligible if unshifted reference light from stationary scatters is made to be more than 30 times greater. Conversely, the heterodyne terms might be made negligible by screening the detector from light scattered from stationary objects.

Eq.6 represents the autocorrelation of the intensity of light scattered from an illuminating beam with a Gaussian cross-section of intensity, as is usually emitted by lasers. If the scattering volume has some other kind of profile of intensity then other terms may be also be necessary (see Earnshaw, 1983, equation 6).

The first correlation coefficient, at t = 0, shows how the

signal intensity at t = 0 is correlated with itself. Since this correlation is exact it is convenient to give it a value of one. Therefore the autocorrelation functions obtained from experiments are usually normalised to a maximum value of one by dividing all the correlation coefficients by the value of the first, which is usually the greatest. The correlation at time t is then represented by $g(t) = G(t)/G(t=0)$.

If particles are flowing uniformly as well as diffusing then an extra term, a cosine, appears in the heterodyne part of the equation:

$$G(t) = C_0 + C_1\exp(-q^2 Dt)\cos(2\pi\Delta ft) + C_2\exp(-2q^2 Dt) \quad (7)$$

In Eq. 7, the cosine representing flow multiplies an exponential representing diffusion to produce an oscillating autocorrelation function as shown in Fig.7.2d. The cosine to describe flow does not appear in the self beating term(s) of the equation because in self beating the reference light is derived from scatterers which also flow; flow can only be detected when a heterodyne signal from a stationary source of scattered light is present. If there is no stationary source of reference light then only self-beating signals can appear to represent diffusion. One may see that, if the frequency of the cosine and thus the velocity of flow is to be measured, then the exponential multiplying the cosine must be long enough to allow at least one cycle of it to be expressed. This condition can be achieved, either by choosing particles of a size to match the flow, or by by changing the scattering angle to alter q. Alternatively, the magnitude of the component of flow parallel to the scattering wave vector (q) may be altered by adjusting the angle (α) between the direction of flow and the apparatus (Fig.7.1 and Eq.4).

7.2.4. Crossed beam instruments to measure flow

In the foregoing descriptions the reference light required to provide a heterodyne signal and thus to detect flow (Eq.7) has been assumed to be scattered from stationary parts of the specimen or apparatus. However, if the main intention is to measure velocities of flow, rather than diffusive motions, then it may be an advantage to divide the beam emerging from the laser and to redirect the two parts of it to intersect in the flow so that they produce a pattern of parallel interference fringes where they cross. Particles moving through these fringes scatter flashes of light. The rate at which they flash is directly proportional to the component of their velocity at right angles to the plane of the fringes. The rate of flashing

is, in fact, the beat frequency between the Doppler shifts
given to each beam by the moving particles. The speed of flow
can thus be measured directly from the rate of flashing if the
angle between flow and fringes is known.

    This, crossed beam, interference fringe, kind of laser
Doppler velocimetry has the advantage that the rate of flashing
is independent of the direction from which the detector views
the specimen. Thus a detector of wide numerical aperture may be
used to collect a maximum amount of the light scattered from the
specimen. However, the fringe method has the disadvantage that
errors arise when the diameters of flowing particles become
similar to the distance between the fringes. Also, the shape of
the fringes, and thus the rate of flashing may be distorted by
variation of refractive index in a specimen (see Johnson, 1983
for references to publications on these problems in Doppler
microscopes). A further disadvantage of the fringe method is
that diffusion appears as a spread in velocity (see, eg Mishina,
Ushizaka and Asakura, 1976, figures 8 and 10) and cannot be
separated from it by change of scattering angle or by using only
a self-beating signal.

                7.3 LASER DOPPLER MICROSCOPES

    Now that we have described the basic principles of Laser
Doppler spectroscopy we can proceed to outline, briefly, the
various kinds of laser Doppler microscopes which have been
described in the literature. They provide two main facilities
which are not available in conventional laser Doppler
spectrometers. They enable laser light to be directed into and
collected from a defined volume in the specimen with a much
smaller diameter than that of an unmodified laser beam, and
they allow the condition of the specimen and the position of
the scattering volume in it to be viewed microscopically. Laser
Doppler microscopes and their use were reviewed in greater
detail by Cochrane & Earnshaw (1978), Earnshaw and Steer
(1979), Johnson (1983),   Johnson & Ross (1984) and Steer, Picton
& Earnshaw (1985).

    There are two main kinds of laser Doppler microscopes. We
shall discuss these, in turn, under the headings 7.3.1. and
7.3.2. Under 7.3.3. we refer, briefly, to other kinds of
instruments which use laser light to measure motion inside
cells. Under 7.3.4. we describe our own instrument.

7.3.1. Crossed-beam laser Doppler microscopes

Most of the work published on the use of laser Doppler

microscopes to measure flow in biological specimens has been
done using instruments with crossed beams and most of it has
been done on the flow of blood through capillaries, as, for
example, by Koyama, Horimoto, Mishina and Asakura, (1982). The
greatest number of publications has come from their group at the
Research Institute of Applied Electricity at Hokkaido University
(see Johnson, 1983, for references). Crossed beam microscopes
have proved able to measure velocities of blood flow, and
distributions of its velocity, down to about 0.5 mm/s at
intervals of 5 μm from the wall in, for example, a venule in
the web of a frog's foot (Mishina, Ushizaka and Asakura, 1976).

In crossed-beam microscopes, the beam from a laser is
divided into two by prisms in a beam-splitter. The two beams are
then arranged to intersect in the plane of focus of the
microscope and to produce interference fringes there. Generally,
a single lens is used to direct the two beams into the
specimen, to collect the light scattered back from the specimen
and to view it. In some instruments the two beams have been
directed into the specimen via the condenser lens used to
illuminate the specimen for viewing. Horimoto, Koyama, Mishina,
Asakura and Murao (1979) described an instrument in which two
objective lenses were used side by side, one to direct two beams
to the specimen and the other to collect the scattered light
from it. This arrangement has the advantage of providing a long
working distance to enable flow in dissected specimens to be
investigated and flare from their surfaces to be avoided.
Generally, in crossed-beam microscopes, only about ten fringes
have been reported to be produced in the specimen. These produce
short trains of pulses which may not be adequately registered by
a spectrum analyser or signal correlator; a frequency tracker
may be necessary.

Crossed-beam microscopes have been favoured for the laser
Doppler microscopy of flow because of the advantages of the
fringe method, discussed in section 7.2.4. Also, they can
measure flow at right angles to the optical axis of the
microscope objective, ie parallel to the specimen stage; most
flows of interest in microscopic specimens are seen in this
orientation. Crossed-beam microscopes avoid a disadvantage of
single beam instruments in which backscattered light is not
Doppler shifted by flow at right angles to the beam emerging
from the objective lens. However, the use of interference
fringes may not be ideal for measuring the diffusion of
particles; the fringes lie in only one direction and, also, as
mentioned in section 7.2.4., errors may arise if the sizes of
particles are similar to the spacings between fringes. Therefore
it is probably better to use a single beam microscope if
diffusion, Brownian motion, is of more interest than flow.

7.3.2. Single- and reference-beam laser Doppler microscopes

In these, a single beam of laser light is directed into the
specimen, usually via the objective lens used to view it. Light
scattered from moving parts of the specimen is collected and
passed back to the detector by the same lens. In a single-beam
microscope, reference light, needed to produce a heterodyne
signal, may arise by scattering from stationary parts of the
microscope and specimen, eg from the objective lens, coverslip,
microscope slide or from cell walls. In a reference beam
microscope, a proportion of unshifted laser light is diverted to
the detector directly. A single beam microscope does not differ
much from a reference-beam microscope and may be converted to
one by adding an arrangement above the objective lens to reflect
part of the beam to the detector before it enters the specimen.

In single-beam and in reference-beam microscopes, the
scattering angle, the angle at which the scattered light is
detected (Fig.7.1) must be defined as well as the angle between
the scattering wave vector ($\underline{q}$) and any flow. The scattering
angle can be defined by placing the detector behind an aperture
in the back focal plane (the diffraction plane) of an extra lens
above the objective (Fig.7.3).

7.3.3. Other kinds of instruments

The fibre optic Doppler anemometer can detect Doppler shifts in
light scattered from microscopic volumes in cells (less than
0.06 mm^3) or from within opaque suspensions, eg of
spermatozoa, via an optical fibre (see Johnson and Ross,1984).

Microscopes have been built which measure motion by
detecting the fluctuations in intensity of light produced when
images of particles pass over arrays of photodiodes (Baker and
Wayland, 1974) or across Ronchi gratings (Reuter and Kratzer,
1979; Ushizaka and Asakura, 1983). These have the advantage of
simplicity but may record the passage of particles outside the
plane of focus. A differential Doppler laser microanemometer
which uses two detectors and a prism grating to measure flow
down to 1 µm/s with a spatial resolution of 0.5 µm in
capillaries was described by Reuter and Talukder (1980).

Fluctuations in intensity of light from fluorescent
molecules under the microscope can be analysed to measure their
motions. Three basic methods and apparatus for doing this were
reviewed by Koppel (1983); time-resolved emission, fluorescence
photobleaching and fluorescence correlation spectroscopy. They
enable the lateral or rotational diffusion of labelled
molecules or particles to be investigated in the presence of

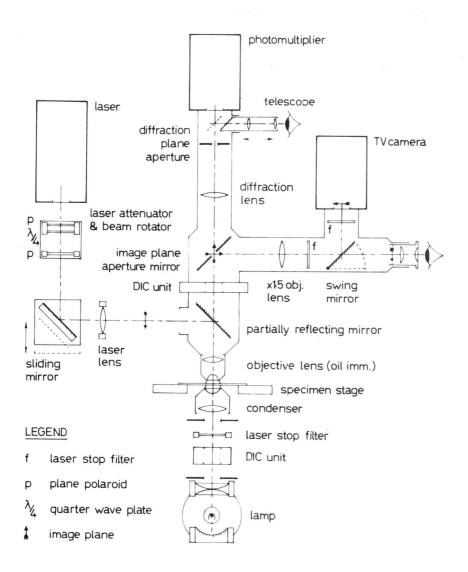

Fig.7.3. Optical arrangement of our single-beam laser
Doppler microscope with video-enhanced contrast.

unlabelled particles.

7.3.4 Our laser Doppler microscope

Our laser Doppler microscope uses a single beam from a 5 mW
helium-neon laser. The optical arrangement of it is shown in
Fig.7.3. The electronic apparatus for processing the laser
Doppler signals is outlined in Fig.7.4. The microscope is
assembled on a lathe bed for rigidity and on air bags to reduce
vibration. It is, incidentally, equipped with differential
interference contrast optics and Allen video enhanced contrast
("AVEC"; Allen and Allen, 1983) with a television frame-store.
Special features of our instrument are:

7.3.4a The diffraction-plane aperture. This can be moved in
the back focal (diffraction) plane of an extra lens to select
the angle at which scattered light from the specimen is passed
to the photomultiplier. The diffraction plane can be viewed
against a graticule in a retractable telescope. Positions on
this graticule correspond to angles at which light is scattered
from the specimen. They are calibrated against the positions of
spots from a diffraction grating placed instead of a specimen at
the focus of the objective lens.

7.3.4.b The image-plane aperture. This defines the
scattering volume; the volume of specimen from which scattered
light is to be passed to the photomultiplier. Single beam
microscopes without an image-plane aperture have the
disadvantage that light is detected from a volume which extends
over the full length of the illuminating beam where it passes
through the thickness of the specimen. The scattering volume may
thus be long and diabolo-shaped. Particles moving within this
extended scattering volume can contribute to the Doppler signal
even if they are out of focus in the image and not recognised.
This problem is avoided by our image-plane aperture which
consists of a small eliptical hole (least diameter 0.5 mm) in a
surface-aluminised mirror placed at 45 degrees in an image plane
(Figs. 7.3 and 7.5). It enables scattered light to be selected
from a volume which does not extend through the full thickness
of the specimen and which is centred at the plane of focus of
the objective lens. The image-plane aperture is arranged to do
this when the illuminating beam is tilted to emerge obliquely
from the objective lens, as shown in Fig.7.5. The beam can be
tilted by moving the mirror below the laser up or down
(Fig.7.3). The aperture then selects the light scattered from a
defined length of the illuminating beam (Fig.7.5). It reduces
stray light from outside the scattering volume even if the
illuminating beam is not tilted. Light passing through the
aperture goes to the diffraction lens and then to the photo-

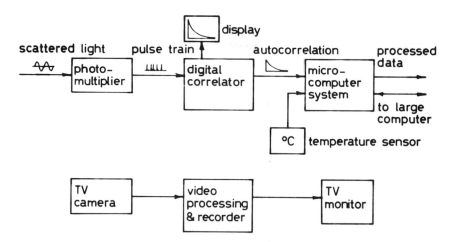

Fig. 7.4. The arrangement of electronics in our laser Doppler microscope.

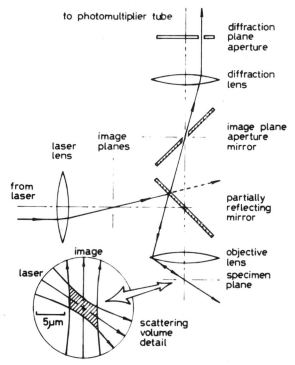

Fig.7.5. The optical arrangement for illuminating the specimen and for collecting backscattered light from a volume of it defined by a hole in an image-plane mirror.

multiplier at an angle selected by the diffraction-plane
aperture. Light reflected from the rest of the mirror passes to
eyepieces or to a television camera to form an image for
viewing. The image-plane aperture appears clearly and round in
the image, though slightly eliptical on television screens
(Fig.7.6), to show where it limits the scattering volume in the
specimen. It enables high contrast video images to be seen while
laser Doppler signals (Figs. 7.7 and 7.8) are being detected.

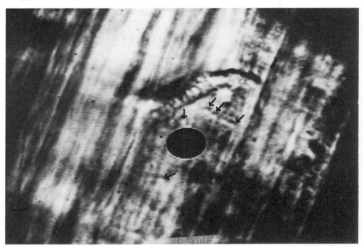

Fig.7.6. A single frame from a recording of a video-
enhanced   differential-interference contrast image of a
sieve plate between sieve elements in a dissected length
of fresh phloem from *Heracleum mantegazzianum*. Starch
grains (arrows) appear to one side of the sieve plate.
The image-plane aperture appears 6 µm wide and 6 µm high.

7.3.4c Measurements of Brownian motion and flow. We devised
our microscope to use a single beam, rather than crossed beams
and interference fringes, because our initial aim has been to
measure the Brownian motion of starch grains released from burst
plastids into the lumen of sieve elements, the pipe-like cells
through which food substances are translocated from source to
sink in higher plants (Fig.7.6). The background to this work was
described by Johnson (1983). We wish to determine whether the
root mean square (r.m.s.) displacement of these starch grains is
greater or less than that that to be expected for particles of
their size in Brownian motion in solution of the concentration,
and hence viscosity, generally assumed for sieve elements
(around 10-20% sucrose; 0.0015-0.0019 Pa s). A greater r.m.s.
displacement might indicate a random conversion of chemical to
mechanical energy; a lesser displacement might indicate a higher
viscosity than has previously been assumed in calculations of
the pressure required to drive solutes through sieve elements.

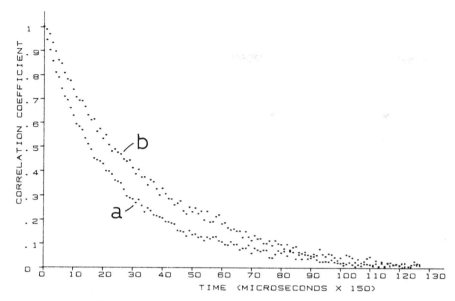

Fig. 7.7. Autocorrelation functions of the intensity of
light scattered from starch grains in the lumen (a) near
the centre and (b) 12 µm lower, just above the cell wall,
near a sieve plate like that shown in Fig. 7.6. In water
the mean diffusion constants would have been 1.2×10^{-13}
and 4.3×10^{-14} cm^2/s. If the proximity of the walls had
no effect and all the grains are assumed to have had the
same radii as the largest (about 0.5 um), the data give
viscosities of at least (a) 0.0018 and (b) 0.005 Pa s.

Although our microscope uses a single beam it can measure
flow parallel to the specimen stage if the beam is tilted to
emerge at an angle to the axis of the objective lens as
described above. Variations in refractive index in the specimen
might then cause undefined variations in the angle between beam
and flow (Fig.7.5). However, the effects of variation in
refractive index can be arranged to cancel out if backscattered
light from the specimen is collected, ie at the angle where
light returns along the path of the illuminating beam. A
procedure for adjusting microscopes to do this was described by
Johnson (1982).

The oscillatory parts of the curves that our microscope
produces from flow in the cytoplasm of Characean algal cells
(Fig.7.8) are much less regular than those produced with simple
photon correlation spectrometers (Sattelle, Green and Langley,
1979). The irregularities appear to arise because the scattering
volume in our microscope is about a thousandth of that used in
previous studies of these cells; it detects small, and

transient, variations in velocity which are averaged and remain
undetected in instruments with wider beams. Whether these small
flows can be resolved and analysed usefully remains to be seen.

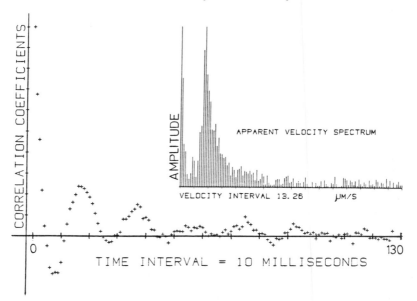

Fig. 7.8. Autocorrelation function of the intensity of
light scattered from cytoplasmic particles flowing near
stationary chloroplasts in a cell of *Chara australis*. The
Fourier transform of it (inset) indicates a mean velocity
of 24.9 um/s. That measured by timing large particles in
the image was 51.2 um/s, S.D. 17.5 (n=18).

7.4. DISCUSSION

7.4.1. Limits to accuracy

7.4.1a The scattering volume. Our microscope is designed to
have a smaller scattering volume than other single beam
instruments previously described because we seek to detect
scattering from small defined volumes in cells within specimens
which are multicellular. It will indicate differences in the
diffusion constants of latex microspheres suspended in liquid or
of starch grains in sieve elements. But, so far, it has not
enabled us to make absolute measurements of diffusion constants.
This is because the smallness and perhaps the shape of the
scattering volume we are able to attain affects the signals in
ways which we have yet to define and account for.

    We limit the extent of the scattering volume, not by the
length of the illuminating beam between the top and bottom of

the specimen, but by the short length of beam seen sideways
through the image-plane aperture. However, we have found that
the mean diffusion constant for a sample of microspheres appears
to increase with increase in size of this aperture, up to a
limit. We think that this may be because, as the area of the
aperture is decreased, the proportion of signal due to
fluctuations in the intensity of light caused by particles
moving back and forth across the edge of the aperture will
increase. Less of the signal will be due to Doppler shifts alone
as described by Eqs. 6 and 7. We have not yet obtained a
mathematical model to include these fluctuations.

It is necessary to calibrate the microscope against
particles with a known diffusion constant. We have used latex
microspheres. These are available with a variety of specified
mean and standard deviations of diameter. Those we have obtained
have given inconsistent results. Some samples have proved, by
light and electron microscopy, to contain particles which,
although few in number, are very much larger than the mean size
given by the suppliers. The amount of light scattered by
particles increases rapidly with their radii (see Steer, Picton
and Earnshaw, 1985). Therefore, the presence of one or a few
unexpectedly large particles for a short or for a longer time
in the small scattering volume attained in our microscope may
cause erratic results during an experiment. We have yet to
overcome this problem of calibration.

The only other published work on the laser Doppler
microscopy of particles in small volumes of cytoplasm appears to
be the elegant study of the motion of vesicles in pollen tubes
by Steer, Picton and Earnshaw (1984). They also calibrated their
microscope using latex microspheres and appear to have obtained
more predictable results with them than we have done. However,
the length of the scattering volume during their calibration
with microspheres was probably much greater than ours and
greater than during their work on pollen tubes. This is because
in their microscope the illuminating beam passed along the
optical axis of the objective lens so that their (backscattered) signal arose from the full length of the beam where
it passed through their specimens. The scattering volume would
have extended between the upper and lower surfaces containing
the latex beads. However, for pollen tubes, their scattering
volume was limited to a depth of about 5 μm by the thickness of
the tubes. It was thus about as small as the volume defined by
the image-plane aperture in our microscope (much technology may
be avoided by happy choice of specimen, especially if it is not
multicellular) and may have caused similar unquantified effects.
Also, the walls of tubes as small as pollen tubes are likely to
constrain the diffusion of particles within a few micrometres of
them (Hurd, Mockler and O'Sullivan, 1980); a scattering volume

extending across the width of a tube would show the effect of
this constraint as a spread in their diffusion constants.

7.4.1b Effects of radiation. A focused spot of laser light,
might disturb a cell in a variety of ways. Of these, heating is
probably the main one governing the balance between radiation
damage to a specimen and the shortest time for which it is
possible to accumulate a signal from it in laser Doppler
microscopy. In our microscope the spot of light (or waist when
the beam is seen sideways) is in the plane of the specimen
(Fig.7.5). It has a Gaussian radial distribution of intensity,
with a spot size (ie the diameter through which all but e^{-2} of
the energy passes) of about 3 μm. The laser has an output power
of about 5 mW. At least half of this is lost at the partially
reflecting mirror (Fig.7.3) and a further unknown amount is lost
by reflection and absorption in the other optical components. If
20% of the light reaches the specimen plane then the average
power density in it is about 1.10^8 Wm^{-2}. Most of this light
passes through the focus, but some is scattered and some is
absorbed to cause local heating. The amount of light absorbed
depends upon its wavelength and on properties of the specimen.
As best and worst cases the absorption coefficients at 633 nm of
water (2.10^{-6}; Driscoll and Vaughan, 1978) and of chloroplasts
(0.2; Wolken, 1975) give absorbed power densities of 2C0 Wm^{-2}
and 2.10^7 Wm^{-2} respectively over the depth of the
scattering volume. The damage done will depend also on the
dissipation of the resulting heat. For example, the dissipation
in freely flowing cytoplasm is probably quite rapid, but within
a stationary part of a cell, a chloroplast for example, it could
much greater. We have found that, in cells of *Chara australis,*
the chloroplast directly under the strongest beam we can pass to
the objective lens detaches from the stationary ectoplasm after
5-10 minutes and is carried downstream by the flowing endoplasm.
Similarly, we find that the cytoplasmic streaming in a cell of
Elodea canadensis may be halted.

7.4.1c Curve fitting. The distribution of velocity in flow
can be obtained directly from a signal analyser (Fig. 7.2b) or
frequency tracker, or if the autocorrelation is available then a
cosine Fourier transform can be used to give the power spectrum
and hence the velocity distribution. Requirements for the
preparation of data for flow are well established.

However, the measurement of diffusion, Brownian motion,
particle-size or viscosity poses harder problems. For these the
decay constants must be measured. Ideally one would like to
obtain a spectrum of time constants. Unfortunately, there is no
simple equivalent to the Fourier transform to convert between
time (or frequency) space and decay constant space. Furthermore,

the process of separating exponentials is extraordinarily
sensitive to noise. A variety of methods has been developed to
fit data to sums of exponentials and other curves (Chu, 1983), a
common one being the method of non-linear least squares. Several
comprehensive computer programmes are available (Provencher,
1982). Because sums of exponentials are so difficult to analyse,
care must be taken not to analyse them beyond the point where
curve-fits, good in the least-squares sense, give unreal
answers (Lanczos, 1956). At present we can extract two decay
constants with confidence using a computer program developed to
our own needs. This severe limit to our laser Doppler microscopy
is set, at present, by the quality of our data. It limits the
complexity of questions we can answer about diffusion in
cytoplasm.

7.4.2. Prospects for the use of laser Doppler microscopy to study the dynamics of living cytoplasm

So far, there have been few studies of cytoplasm in living
cells by laser Doppler spectroscopy, even without the aid of
microscopes. Most of them have been of cytoplasmic streaming in
giant algal cells (Sattelle *et al.* 1979; Piddington and Ross,
1985) or of contraction in muscle. Laser Doppler spectroscopy
without the aid of microscopes may produce signals which are
less noisy than ours. However, it then measures the average
motion in volumes of cytoplasm which are large compared to the
microscopic scale of many cytoplasmic movements which are of
interest. If the laser Doppler microscope is to reduce this
limit to the resolution of laser Doppler spectroscopy and to
produce new information about motion on a fine scale in living
cells then, as during the development of other kinds of
microscopes, better optical arrangements must be devised to
improve its signal-to-noise ratio and, when necessary, to reduce
the damage caused by radiation. Also, the effects of small
scattering volumes must be accomodated in the analysis of
signals. As with any new method seeking respectability,
results from ideal and familiar specimens must be used to
demonstrate accord with those already obtained in other ways.
Above all, if laser Doppler microscopes are to do more than
merely to produce measurements which appear consistent with what
is already known from other methods, ways must be devised to
process and display the data so that they stimulate the
development and use of new model equations rather than merely
being forced to fit equations already preconceived.

7. 5 ACKNOWLEDGEMENT

We are grateful to Dr. D.A. Ross and Dr. R.J. Hobbs for help and
encouragement and to the Science and Engineering Research
Council for financial support.

7.6 REFERENCES

Allen, R.D. and Allen, N.S. (1983). Video-enhanced microscopy with a computer frame memory. *Journal of Microscopy* 129, 3-17.

Baker,M. and Wayland, H. (1974). On-line volume flow rate and velocity profile measurement for blood in microvessels. *Microvascular Res.* 7 131-143.

Barclay, G.F. and Johnson, R.P.C. (1982). Analysis of particle motion in sieve tubes of *Heracleum*. *Plant Cell and Environment* 5, 173-178.

Berg, H.C. (1983). *Random walks in biology*. Princeton University Press, New Jersey.

Chu, B. (1983). Correlation function profile analysis in laser light scattering. I General review on methods of data analysis. In: Earnshaw, J.C. and Steer, M.W. (1983), pp.53-76.

Cochrane,T, and Earnshaw, J.C. (1978). Practical laser Doppler microscopes. *J. Phys. E: Sci. Instrum.* 11, 196-198.

Crawford, G.E., and Earnshaw, J.C. (1984). Photon correlation spectroscopy as a probe of planar lipid bilayer phase transitions. *European Biophysics J.* 11 25-34.

Degiorgio, V. (1983). Physical principles of light scattering. In: Earnshaw, J.C. and Steer, M.W. (1983), pp. 9-30.

Drain, L.E. (1980). *The laser Doppler technique*. Wiley, Chichester, New York.

Driscoll, W.G. and Vaughan, W. (1978). *Handbook of optics* McGraw-Hill, New York.

Earnshaw, J.C. Laser Doppler velocimetry in a biological context. (1983). In: Earnshaw, J.C. and Steer, M.W. (1983), pp. 147-163.

Earnshaw, J.C. and Steer, M.W. (1979). Laser Doppler microscopy. *Proc. R. micr. Soc.* 14, 108-110.

Earnshaw, J.C. and Steer, M.W. (1979). Studies of cellular dynamics by laser Doppler microscopy. *Pestic. Sci.* 10,358-368.

Earnshaw, J.C. and Steer, M.W. eds. (1983). *The application of laser light scattering to the study of biological motion*. NATO Advanced Study Institute Series A, 23. Plenum Press, New York.

Horimoto, M., Koyama, T., Mishina, H., Asakura, T. and Murao, M. (1979). Blood flow velocity in pulmonary microvessels in bullfrog. *Respiration Physiology* 37, 54-59.

Hurd, A.J., Mockler,R.C. and O'Sullivan, W.J. (1980). A light scattering study of constrained Brownian motion. In: *Proceedings of the Fourth International Conference on Photon Correlation Techniques in Fluid Mechanics,* (eds. W.T. Mayo and A.E. Smart). pp. 24-27, Joint Institute for Aeronautics and Acoustics, Stanford University.

Johnson, R.P.C.(1982). A laser Doppler microscope for biological studies. In: *Proc. Conference on Biomedical Applications of Laser Light Scattering.* (eds. B.R. Ware, W.L.Lee and D.B. Sattelle). pp. 391-402. Elsevier, North Holland.

Johnson, R.P.C. (1983). Laser Doppler microscopy: Especially as a method for studying Brownian motion and flow in the sieve tubes of plants. In: Earnshaw, J.C. and Steer, M.W. (1983), pp. 147-163.

Johnson, R.P.C. (1984). Laser Doppler microscopy. *Trans. Biochemical Society* 12, 635-637.

Johnson, R.P.C. and Ross, D.A. (1984). Laser Doppler microscopy and fibre optic Doppler anemometry. In: *The analysis of organic surfaces,* (ed. P. Echlin). Chapter 20, pp. 507-527, John Wiley and Sons, Inc., New York.

Koppel, D.E. (1983). Fluorescence techniques for the study of biological motion. In: Earnshaw, J.C. and Steer, M.W. (1983), pp. 245-273.

Koyama, T., Horimoto, M., Mishina, H. and Asakura, T. (1982). Measurement of blood flow velocity by means of a laser Doppler microscope. *Optik* 61, 411-426.

Lanczos, C. (1956). *Applied analysis.* Prentice Hall, Englewood Cliffs, New York.

Maeda, T. and Fujime, S. (1972). Quasielastic light scattering under optical microscope. *Rev. scient. Instrum.* 42, 56-567.

Mishina, H., Ushizaka, T. and Asakura, T. (1976). A laser Doppler microscope. *Opt. laser Tech.* 8, 121-127.

Oliver, C.J. (1974). Correlation techniques. In: *Photon correlation and light beating spectroscopy.* (eds. H.Z. Cummins and E.R. Pike). NATO Advanced Study Institute Series B 3. pp. 151-223, Plenum Press, New York and London.

Piddington, R.W. and Ross, D.A. (1985). Cross-bridge action in Characean cytoplasmic streaming? *Aust. J. Plant Physiol.* 12, In press.

Provencher, S.W. (1982). CONTIN: a general purpose constrained regularisation program for inverting noisy linear algebraic and integral equations. *Comp. Phys. Commun.* 27, 229-242.

Reuter, B. and Kratzer, M. (1979). Microanalysis of the velocity profile in blood vessel models. In: *Laser 79 Opto-electronic* (ed. W. Wallach). IPC Science and Technology Press. pp. 209-215.

Reuter,B. and Talukder,N. (1981). New differential laser microanemometer. *Proceedings of the Society of Photo-Optical Instrumentation Engineers* 236. 1980 European Conference on Optical Systems and Applications (Utrecht). pp. 226-230.

Sattelle, D.B., Green, D.J. and Langley, K.H. (1979). Subcellular motions in *Nitella flexilis* studied by Photon Correlation Spectroscopy. *Physica Scripta* 19, 471-475.

Steer, M.W., Picton, J.M. and Earnshaw, J.C. (1984). Diffusive motions in living cytoplasm probed by laser Doppler microscopy. *J. Microscopy* 143, 143-149.

Steer, M.W., Picton, J.M. and Earnshaw, J.C. (1985). Laser light scattering in biological research. In: *Advances in Botanical Research*. (eds. J.A. Callow and H.W. Woolhouse). Chapter 1. Academic Press, New York and London.

Ushizaka, T. and Asakura, T. (1983). Measurements of flow velocity in a microscopic region using a transmission grating. *Applied Optics* 22, 1870-1874.

Vlaming, M.S.M.G., Aertson, A.M.H.J. and Epping, W.J.M. (1984). Directional hearing in the grass frog *Rana temporaria* I. Mechanical vibrations of tympanic membrane. *Hear. Res.* 14, 191-202.

Wolken, J.J. (1975). *Photoprocesses, photoreceptors and evolution*. Academic Press, New York.

8

On the form and arrangement of cell wall microfibrils

J. H. M. Willison and R. M. Abeysekera

8.1 INTRODUCTION

Plant cells are surrounded by cell walls which are complex
assemblages, both chemically and architecturally (Frey-Wyssling
1976; Darvill *et al.* 1980; Cooper *et al.* 1984). As cells
develop, their walls undergo progressive transformation through
changing intracellular activity (Willison 1981), and cellular
differentiation is therefore often clearly expressed as
differences in the structure and composition of the cell walls
of different cells.

Cell walls are generally composed of at least two
structurally distinct phases: matrix and microfibrils (Preston
1974; Frey-Wyssling 1976; Darvill *et al.* 1980). Microfibrils
are long rods within which cellulose, the focus of the present
work, is located (Preston 1974). Models of cell wall
architecture (i.e. the detailed organization of microfibrils
within small portions of typical walls, and within complete
walls in a few cases) generally display cellulosic microfibrils
as straight, or slightly flexuous, rods having a highly
crystalline structure (see for examples: Preston 1974; Frey-
Wyssling 1976; Roland and Vian 1979). Although microfibrils
are sometimes found fairly randomly distributed within the plane
of the cell wall, they most usually display order, which in some
cases, as in *Glaucocystis* (Willison and Brown 1978) may be of
high degree. The present work is concerned with aspects of
detail in cell wall architecture.

While microfibrils are generally regarded as relatively
straight rods, helical twisting in cellulose has been described
at two principal levels: at the level of the microfibril and at
the level of the cell. In 1961, Colvin noted that *Acetobacter*
cellulose displayed "twisting of bundles of .... microfibrils".
These twists have since been illustrated by several authors,
including Brown and co-workers (1976), Zaar (1977) and Haigler

and Benziman (1982). Colvin (1972) considered that *Acetobacter* cellulose was generated at some distance from the cell surface by addition of glucose residues to growing microfibril tips from soluble, activated, intermediate carrier molecules generated by the cells. More recent evidence (Brown *et al.* 1976; Zaar 1977, 1979; Brown and Willison 1977; Haigler and Benziman 1982) has been interpreted to show that this cellulose, like that of higher plants, is generated at the cell surface. This fundamental conceptual dichotomy over the site of synthesis of bacterial cellulose has not, however, led to a dichotomy of views on the origin of the twists. Colvin (1961) considered that the twists "must reflect some sort of assymetric constraint internal to the microfibrils which is seeking a minimum position", and Haigler and Benziman (1982) state "the characteristic twisting of the ribbon ..... may result from relief of thermodynamic strain during cellulose I lattice formation (Stöckmann 1972)". If these authors are correct in supposing that bacterial cellulose twists inherently, then we might expect to find such twists in celluloses other than that generated by *Acetobacter*, and we report here the results of a search for them.

It is generally recognized that cell wall microfibrils are deposited at the cell surface and that cell walls are therefore layered, with secondary thickening always lying to the cyto-plasmic side of the primary wall, for example. But, does layering apply absolutely? If cellulosic microfibrils are deposited exclusively by synthetic complexes bound firmly to the plasma membrane, as is now widely supposed (see Herth, Chapter 12, this volume), then only layering should occur, weaving should not. In a comprehensive review, however, Frey-Wyssling (1976, p. 9) states unequivocally that micro-fibrils are "interwoven" and therefore that microfibril synthesis occurs effectively within three-dimensional rather than two-dimensional space. We reconsider this point.

8.2 MATERIALS

Acetobacter xylinum (wild-type strain no. NRC 17005) was obtained from J.R. Colvin (N.R.C., Ottawa, Canada) and was grown in a liquid culture medium (Hestrin and Schramm 1954). Within 4 days, the bacteria synthesized sufficient cellulose to make a mat, the "pellicle", which floated to the liquid-air interface.

The unicellular green alga *Oocystis apiculata* was grown on the surface of a semi-solid medium at the University of North Carolina by David Montezinos and Richard Santos (for details, see Montezinos and Brown 1979).

A lumpy suspension culture of alfalfa (*Medicago sativa* L.)

in logarithmic growth phase was kindly supplied by Dr. L.V. Gusta
(Crop Science Department, University of Saskatchewan, Canada).

Mature quince (*Cydonia vulgaris L.*) fruit were obtained
from bushes owned by Charles Payzant of Falmouth, Nova Scotia.
Seeds were placed in distilled water for 12-24 hours, whereupon
a mucus was released from which the seeds themselves were
removed.

Plants were raised in a variety of conditions. Radish
(*Raphanus sativus L.*) roots were grown over a period of 3-4
days from washed seeds on damp filter paper. New Zealand
spinach (*Tetragonia expansa*) and Pinto bean (*Phaseolus vulgaris
L.*) seedlings were grown in potting soil in a controlled
environment chamber. Cotton (*Gossypium hirsutum L.*) seeds hairs
were obtained from plants grown in a greenhouse at the
University of North Carolina (for details, see Willison and
Brown 1977).

8.3 METHODS

8.3.1 Negative staining

Quince slime was negatively stained directly, or after treatment
with either strong potassium hydroxide solution or an acetic-
nitric reagent (Updegraff 1969), using 2% uranyl acetate.
Optical diffractograms were prepared from positive images using
a Polaron Equipment Ltd. (Watford, U.K.) optical diffractometer.

8.3.2 Simple replicas

Replicas of cell wall fragments were prepared by freezing pieces
of tissue in liquid nitrogen (or sometimes, liquid Freon at
liquid nitrogen temperature), and grinding the frozen material
with pestle and mortar in the presence of liquid nitrogen. Root
hairs were scraped from the surfaces of frozen radish roots;
bean epidermis was excised from the hypocotyl before freezing.
After grinding, the material was thawed and rinsed 3 or 4 times
in distilled water before being spread as a thin smear on
freshly-cleaved mica. Quince slime was simply smeared on mica,
either in the presence or absence of latex spheres, and
air-dried. Replicas of the dried wall fragments were prepared
in Balzers 360M freeze-etch machines at room temperature, by
pre-shadowing with platinum/carbon and backing with a thin layer
of carbon. Replicas were cleaned with 60% chromic acid.
Details of replica preparative methods may be found in Willison
and Rowe (1980).

8.3.3 Freeze-fracturing

For freeze-fracture, materials were mounted on standard Balzers

gold specimen mounts, either in water or, in the case of tissue
cultures, in their culture medium. No specimens were
pretreated, other than by excision, and all were frozen by rapid
immersion (quenching) in liquid Freon 12 at its freezing point
(see Willison and Rowe 1980). Specimens were freeze-fractured
in the temperature range 168-173K, with etching times of 5-30
seconds, before platinum/carbon shadowing and conventional
replication (for detailed review, see Willison and Rowe 1980).

8.4   RESULTS AND DISCUSSION

8.4.1   The twisted ribbon

Acetobacter xylinum produces cellulose in the form of broad
thin ribbons, each composed of parallel subfibrils commonly
described as microfibrils (Brown et al. 1976; Zaar 1977;
Haigler and Benziman 1982). The ribbons tend to roll about
longitudinal axes (Brown *et al.* 1976). They also twist
helically about the central axis (Colvin 1961; Brown *et al.*
1976; Zaar 1977) throughout the length of the ribbon (Fig. 8.1),
including the first-formed tip (Fig. 8.2). The helical repeat
distance is about 1.5μm. Brown and Colpitts (1978) have
demonstrated microcinematographically that the whole bacterium
undergoes axial rotation in association with the generation of
the helical twisting of the ribbon. Within regions of the
developing pellicle having higher cellulose density, however,
the rotation may not occur (R.M. Brown Jr., personal
communication), and this is reflected in the many ribbons
lacking twists to be found in pellicles (Fig. 8.3).
    It is particularly striking that twisting extends to distal
ends of ribbons (Fig. 8.2, and see Brown *et al.* 1976; Zaar 1977).
If twisting were under biological control, via some mechanism
independent of cellulose crystallization which generates
bacterial rotation, distal twisting would not occur since the
unclamped ribbon would rotate with the bacterium. We must
assume, in agreement with previous conclusions (Colvin 1961;
Haigler and Benziman 1982), that the ribbon finds a lowest
free-energy equilibrium in the twisted state. The lack of
twists within the pellicle must therefore result from the
application of physical constraint to the complex of bacterium
and cellulose product, constraints which are readily imagined
within the confines of the pellicle. This notion may well find
correspondence with that of the metastability of cellulose,
discussed at length by Stöckmann (1972).
    If cellulose, unconstrained at the time of crystallization,
adopts a twisted form, then we might expect to find examples in
cellulose from other sources. We looked first to the highly
crystalline microfibrils of algae such as *Oocystis* and
Glaucocystis, which have in the past proved to be excellent

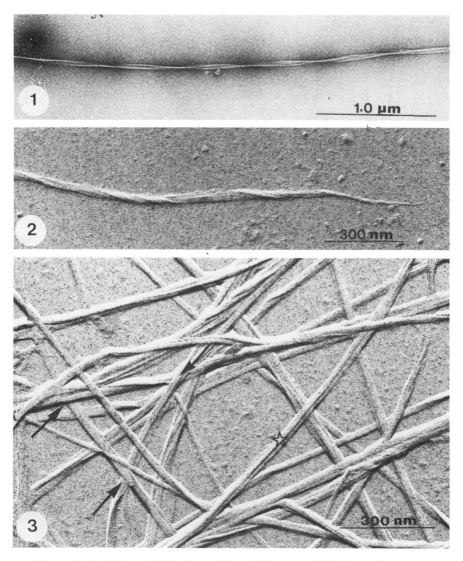

Figs. 8.1-8.3.   Cellulose from flimsy pellicles of
Acetobacter xylinum. 8.1) Regular twisting in a single
ribbon, negatively stained. 8.2) The tapered distal
end of a shadowed ribbon, showing that it is twisted as
it originates. 8.3) Twisting of the ribbons of this
thin shadowed pellicle is relatively rare (cf. the
ribbon indicated by a star and those in Figs. 8.1 and
8.2). The ribbons indicated by arrows are woven
together, as shown by close examination of the apices
of the triangle made by the three ribbons.

experimental systems in studies of cellulose (Preston 1974;
Brown and Willison 1977; Quader and Robinson 1981), since we
considered that twisting should be visible if present in these
unusual microfibrils which have cross-sectional dimensions of
about 10x20 nm (Harada and Goto 1982).

   Twists were found only very rarely in frozen-ground wall
fragments of *Oocystis* (Fig. 8.4), and a brief survey of
microfibrils from comparable sources has similarly revealed few
examples. There are two possible explanations: (1) *Acetobacter*
cellulose has special crystallographic properties, or (2) a
constraint mechanism like that within the *Acetobacter* pellicle
operates during cellulose crystallization in *Oocystis* and
similar algae. It seems reasonable to propose that the second
option is more probable. Such constraint would operate at the
cell surface if neither the extended microfibrils nor the
linear "terminal complex" of *Oocystis*, which synthesizes the
microfibrils (Brown and Montezinos 1976), were permitted to
rotate axially. It is worth noting that the walls of these
algae are strikingly elegant in design (Preston 1974; Willison
and Brown 1978; Willison 1982). The broad microfibrils describe
spirals about the cells, with ascending and descending spirals
having different signs, leading locally to the classic crossed
polylamellate wall form. These spirals are open helices,

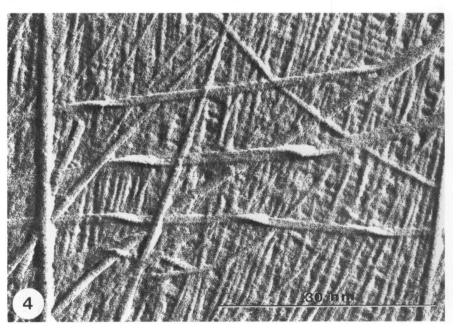

Fig. 8.4.   A fragment of cell wall of *Oocystis* shows
twists in a few superficial microfibrils. Such twists are
rare and possibly arise during specimen preparation.

morphologically distinct from the tight axial twisting exhibited
by bacterial cellulose, but a common physical basis seems
possible. Perhaps in these cells wall integrity is maintained
in part by tension arising from the residual expression, as open
spirals, of the intrinsic desire for twisting in cellulose.

8.4.2 Quince slime microfibrils

Seeds of the edible quince (*Cydonia vulgaris*) secrete a mucus
when placed in water (Schmidt 1844; Kirchner and Tollens 1874).
This mucus contains cell wall microfibrils which originate from
integumentary hair cells that disrupt during mucus release.
The microfibrils have been of special interest in the past
because of their presence in a mucilage and their small
dimensions (Mühlethaler 1950; Franke and Ermen 1969). They are
said to be cellulosic (Renfrew and Cretcher 1932; Mühlethaler
1950) and to be "ribbon-like", sometimes having a "beaded"
appearance (Franke and Ermen 1969).

In shadowed preparations, we found microfibril ends but no
evidence of weaving (Fig. 8.5). After negative staining, a
ribbon-shape (i.e. unequal cross-sectional dimensions) was often
evident (Fig. 8.6). The mean cross-sectional dimension
measured directly from negative-stained preparations was 4.3 nm,
and measured indirectly using the linear regression method of
Ohad *et al.* (1963) from bidirectionally-shadowed microfibrils
was 5.6 nm.

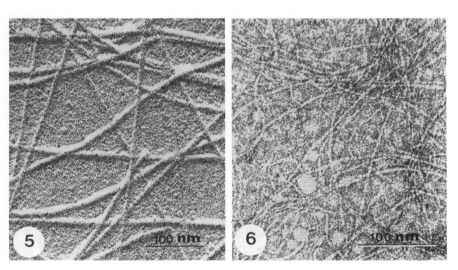

Figs. 8.5 and 8.6. Microfibrils from quince seed slime.
8.5) Shadowed unidirectionally. 8.6) Negatively stained.

At high resolution, in the infrequent instances where we judge that stain penetration had been optimal, a double helical structure with a helical repeat distance of about 7 nm appeared (Fig. 8.7). Similar results were obtained from untreated slime, strong KOH-treated slime (Fig. 8.7), and slime subjected to the acetic-nitric reagent of Updegraff (1969) which should have dissolved anything other than cellulose. From positive images of the apparent helices, optical diffraction indicated anastigmatism and lack of strong phase contrast halos (Fig. 8.8), but patterns corresponding with a double helix were not discerned. Given the small amount of repeated information available for integration, this is perhaps not surprising.

While we presently remain cautious in our interpretation of these high resolution micrographs due to uncertainty arising from possible beam damage to specimens and to phase effects (for discussion of the latter, see Hanna 1971), we consider that this double helical form is probably real. We do not know whether it is present rarely, or is generally present but rarely resolved. Similarly, we cannot tell whether it arises at the time of synthesis of the microfibrils, or secondarily during mucus release.

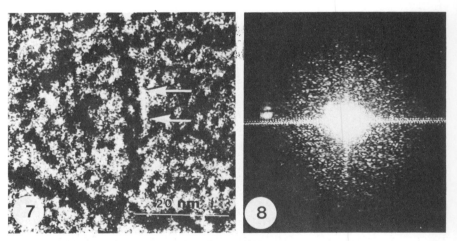

Fig. 8.7. A negative image of negatively stained quince slime microfibrils, KOH-treated. Note the double helical structure (arrows).
Fig. 8.8. Optical diffraction pattern from a masked region in the centre of Fig. 8.7, including the apparently double helical microfibril.

Very recently, Ruben (1985) has claimed evidence for helical strcuture in microfibrils from platinum/carbon replicas. In our studies of shadowed quince microfibrils (Figs. 8.5 and 8,9), we believe we have not resolved microfibril substructure. In this

regard, attention should be drawn to problems associated with
the interpretation of crystallite patterns on shadowed fibres,
previously discussed at length by Willison and Rowe (1980,
pp. 255-260). During shadowing, metal crystallites grow from
initial nuclei, and the growing crystallites may migrate and
fuse with neighbours. Nucleation and migration are physico-
chemical interactive processes between the substrate and the
crystallites involved, and ultimate crystallite location is not
determined solely by geometric parameters (see Willison and Rowe
1980, pp 8-11, 135-138 and 248-254). To complicate the process,

Fig. 8.9.   Quince seed slime microfibrils, shadowed
unidirectionally with platinum/carbon (shadowing direction
indicated by encircled arrow). Cross striations appear
on the microfibrils in many places (arrows), but their
orientation relative to the microfibril long axis is
variable and the periodicity is related to metal
crystallite size.

growing crystallites themselves capture incoming metal adatoms
and create shadows. The latter process is described as "self-
shadowing", and appears to be a particularly acute problem at
the edges of fibres. On shadowed quince seed slime microfibrils,
for example, repetitive striations may be found which sometimes
give the appearance of helical substructure (Fig. 8.9). We

regard these as preparative artefacts and reject then as
evidence of substructure, however, because the appearance of
the repetitive striations is dependent in part upon fibre
orientation relative to shadowing direction (Fig. 8.9), and the
spacing of the striations is dependent upon crystallite size
(full evidence not shown).

8.4.3 Spiral microfibrils

In primary walls of cotton fibres, observed after rapid
freezing, fragmentation by grinding, and replication, microfibrils commonly display flat wave forms within loose bundles
(Willison and Brown 1977). A survey of immature cells from
several sources suggests that such wave forms are commonplace
at the cytoplasmic face of primary walls (Figs. 8.10 - 8.13),
and perhaps elsewhere. Such waves have been found in several
regions of elongating bean hypocotyl (both pith and epidermal
regions), bromegrass suspension culture, the filamentous alga
Zygnema and New Zealand spinach leaf. In the last case, such
waves were visible on both the cytoplasmic face and the external
face of epidermal cell walls, the latter having been observed
as air-dried peels, the torn side replicated.
     We presently doubt that these waves are wholly artefacts of
preparative processes, for the following reasons. Firstly, they
are present after rapid (Freon-quenching) freezing, which can be
a very effective procedure for structural preservation.
Secondly, they may be seen sometimes in freeze-fractured
material as impressions in plasma membranes (Fig. 8.13). Thirdly,
they may be seen in leaf epidermal peels, which are extensive in
area and have not been subjected to freezing, grinding and
thawing. Finally, the wave form is not affected as microfibrils
undergo changes in direction, such as in diverging around pit-
fields (Fig. 8.10), nor are the waves always in phase (Fig. 8.11)
or limited to a single layer (Fig. 8.12). It remains possible
that some of the examples seen arise, or are exaggerated by, the
release of tension in the wall lamella adjacent to the plasma
membrane, notably during thawing in fragments obtained by
grinding. If so, this is reason to regard the phenomenon as
significant, because it indicates that this lamella may operate
dynamically in the maintenance of structural integrity. Just as
a sweater made of twisted yarn better follows the contours of a
body, so might the cell wall retain contact with the cell
surface under conditions which might tend to separate a more
rigid wall, such as limited turgor or local growth strain.
     Not only is it striking that these wavy microfibrils are so
widely distributed, but also that they sometimes give the
impression of forming spirals rather than flat waves (see Figs.
6-8 in Willison and Brown 1977; Fig. 10 in Willison 1982; and
Figs. 8.10-8.13 here). Close examination of many of the wavy
microfibrils seen both in frozen-ground wall fragments and

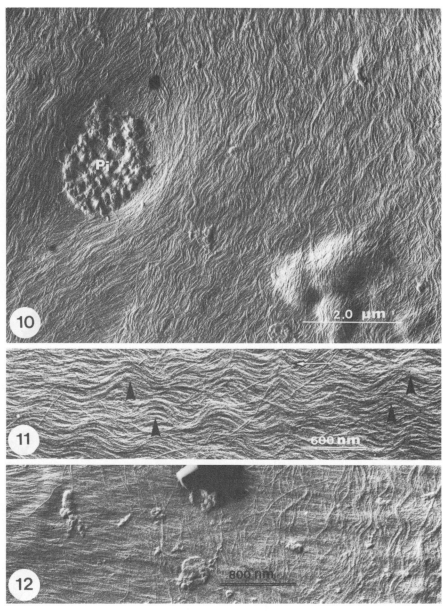

Figs. 8.10-8.12. Cytoplasmic faces of cell walls in frozen-
ground bean hypocotyls. 8.10) Wave forms maintained as
microfibrils deviate around pit-field (Pi). 8.11) Phase
differences between adjacent in-phase microfibril bundles
(arrow heads); kinks (arrow). 8.12) Waves in cross-oriented
lamellae. Wave amplitude is greater in the younger lamella.

after freeze-fracturing (Fig. 8.13) and freeze-etching (Fig.
8.15) shows kinks (sharp bends) in places. Intuitively, kinking
of microfibrils is likely to be artefactual. It occurs for
example in cellulose when spherical cells are squashed during
drying (Fig. 8.14). We suggest that the kinks which arise in
the wavy microfibrils are good evidence that these arrangements

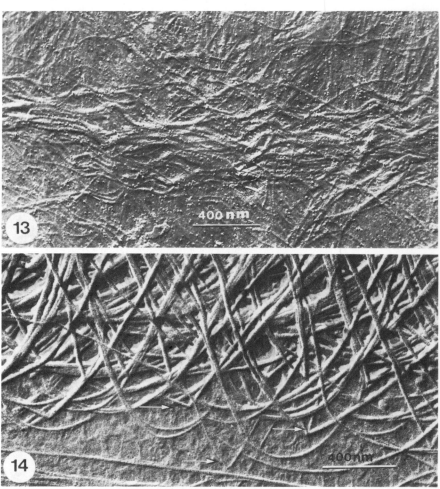

Fig. 8.13. Plasma membrane E fracture face of radish
root cortical cell in the root hair zone. A pronounced
band of wavy microfibrils, with many kinks, left impressions in the membrane fracture face.
Fig. 8.14. Replica of air-dried mother cell wall of
Glaucocystis (for detail, see Willison and Brown 1978),
showing kinking of some of these large microfibrils at
the point of greatest compressive stress (arrows).

are indeed actually spirals that have become flattened during
processing for electron microscopy. If this is so, then a rough
estimate of the helical diameter ($d°$) may be obtained from the
image of the flattened spiral by simple geometry ($d° \sim 4A/\Pi$,
where A = wave amplitude, i.e. mid-line to peak). In all cases
which we have measured, the helical diameter is less than 75 nm.
    Possible support for this spiral model comes from the
experimental work of Haigler (see Haigler and Benziman 1982) on
the effects of agents which bind to cellulose when applied to
Acetobacter cultures. Carboxymethylcellulose (CMC),
particularly, results in the transformation of the *Acetobacter*
twisted ribbon into a series of finer open spirals (see for
examples, Figs. 24 and 25 in Haigler and Benziman 1982). These
authors conclude that CMC binds to cellulose between the sites
of polymerization and crystallization (which are presumably in
close tandem at the bacterial surface), preventing the later
steps of heirarchical crystallization. They also point to the
chemophysical similarities between hemicelluloses and CMC,
suggesting that the former may "regulate the extent of fibril
aggregation and crystallization" (Haigler and Benziman 1982,
p. 294). Perhaps we should now extend this concept to include
regulation of helical form in higher plant primary cell walls.
    We have suggested that in *Oocystis* the linear terminal
synthetic complex is restricted from rotating axially. Since
higher plant microfibrils may be synthesized either by globular
entities associated with the cell surface and/or by
intramembranous rosettes (see Herth, Chapter 12, this volume),
it must be presumed that rotation of these entities occurs as
synthesis proceeds if spiral microfibrils are to arise.

8.4.4 The periplasmic zone

We have argued that some, if not all, of the wavy microfibril
forms seen in many primary walls are spirals which have been
flattened by preparative processes for electron microscopy. If
this is so, then a zone large enough to accomodate such spirals
should be present between the more compact older regions of the
primary cell wall, upon which cell extension has acted to
straighten spiral microfibrils, and the plasma membrane.
    Electron micrographs of thin sections, well prepared by
conventional means, commonly show a small space between plasma
membrane and cell wall (for examples, see Robards 1969; Olesen
1980). In the transfer apparatus of transfer cells, this zone
is reduced in well-frozen freeze-substituted material, but a
narrow periplasmic zone persists (Browning and Gunning 1977).
Since the appearance of this periplasmic zone varies with
preparative technique, might it be wholly artefactual?
    When plant material is frozen and freeze-fractured, the
plasma membrane is commonly so tightly pressed to the cell wall
that microfibril impressions in the plasma membrane are clearly

visible (Fig. 8.13, and see Willison 1975, 1976). Under these
conditions, there is clearly negligible room for a periplasmic
zone which could accomodate spiral microfibrils. However,
within the same tissue that exhibits tightly wall-adpressed
plasma membrane, periplasmic ice may also be found, but because
it distorts the plasma membrane it is rarely illustrated (for
examples, see Willison 1976; Pearce and Willison 1985).
Although little water will move across the plasma membrane to
contribute to the growth of extracellular ice during rapid
freezing (Pearce and Willison 1985), water redistribution within
the wall occurs. In Fig. 8.15, a zone of ice lies between the
plasma membrane of one cell and the adjacent face of the cell
wall, but in the neighbouring cell this cell wall has been
tightly pressed against the plasma membrane, leaving microfibril
impressions. We must presume that this ice has mostly formed
from water which was originally distributed throughout the
intercellular zone, and therefore that the tight apposition of
cell wall and plasma membrane is as much an artefact of
preparation as the large ice-filled zone which lies across the
cell wall in this figure.

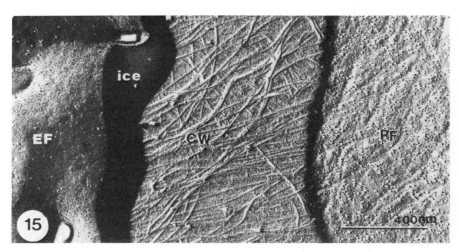

Fig. 8.15.   Part of radish root cortex in root hair zone.
At left, plasma membrane (EF) has been lifted from the
cell wall (cw) by growth of periplasmic ice (ice). At
right, the plasma membrane of the neighbouring cell (PF)
displays microfibril impressions, indicating tight
adpression to the cell wall.

     When material is cryo-fixed by rapid freezing for electron
microscopy, the heat transfer properties of biological material
make it inevitable that only a thin peripheral zone of the
sample will be frozen such that cytoplasmic damage by ice-

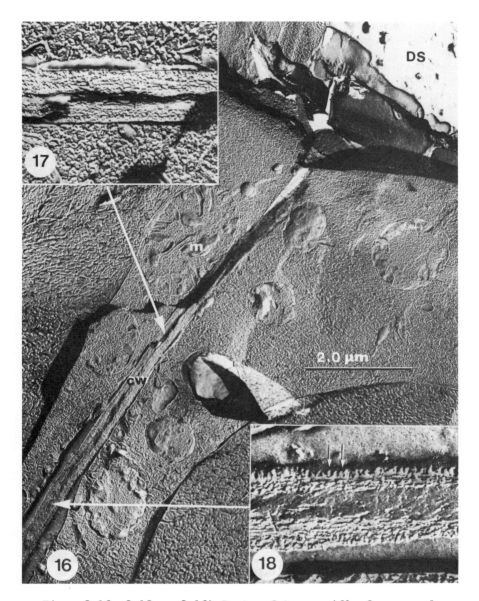

Figs. 8.16 - 8.18.   8.16) Parts of two rapidly frozen and
freeze-fractured alfalfa tissue culture cells mounted in
their culture medium, situated very close to the edge of
the sample droplet (droplet surface [DS] visible in top
right hand corner).   m - mitochondrion; cw - cell wall.
Insets, Figs. 8.17 & 8.18) Two regions of cell wall from
Fig. 8.16, at higher magnification, showing the
periplasmic zone and ice development within it (arrows, 8.18).

crystal growth is negligible. Using Freon-quenching, this zone
is about 10 μm thick (Elder *et al.* 1982). In practice,
suspension cultures are ideal materials for optimal cryoquenching of growing higher plant cells. If peripheral zones
of such specimens are scanned, optimally preserved material
reveals that microfibril impressions are not found on plasma
membranes, despite being plentiful in deeper regions of the same
specimens. Furthermore, cross fracture of the cell wall/plasma
membrane interface reveals a periplasmic zone of variable
thickness, generally in the range 30-60 nm (Figs. 8.16 - 8.18).
This periplasmic zone is distinguishable because of distinctive
fracture properties and distinctive freezing pattern relative to
the bulk of the cell wall. The fact that ice nucleation appears
to occur preferentially in this zone (Figs. 8.17 and 8.18),
suggests that it may be relatively hydrated.
    We conclude that a periplasmic zone containing, and perhaps
maintained in part by, recently synthesized spiral microfibrils
lies between the plasma membrane and primary wall of some
growing plant cells. We suggest that as these microfibrils age,
cell extension may progressively stretch and eliminate them,
leading to compaction of the resultant straight microfibrils
into the developing wall. We can only guess at the functional
significance of such a system, but in addition to the
conformational dynamism inherent in the seemingly spring-like
microfibrils, water movement in the periplasmic zone may be
facilitated.

8.4.5 Weaving versus layering

If microfibrils are synthesized by strictly membrane-bound
microfibril-terminal complexes, then the synthetic locus is
planar (two-dimensional) and microfibrils will be deposited as
layers. If the synthetic complexes operate in bulk (three-
dimensional) space, such as within a periplasmic zone, then
microfibril weaving becomes feasible. The question of whether
microfibrils are woven as opposed to layered is therefore an
important one to resolving the question of the site of
microfibril synthesis. Distinguishing between weaving and
layering is not easy, however. Frey-Wyssling (1976) provides a
simple diagram, which is drawn in slightly modified form below
(Fig. 8.19), to illustrate the distinction. Even this
triangular "woven" form, however, could arise through the agency
of membrane-bound synthetic complexes if, by chance, they were
suitably arranged spatially and temporally (Fig. 8.19c).
Nevertheless, the probability of suitable conditions arising to
create this triangular weave is clearly greater in bulk space
than planar space, and other more complex weaves are permitted
in the former space, while denied in the latter.
    In compliance with prediction implicit in the membrane-bound
synthetic-site hypothesis (Herth, Chapter 12, this volume), we

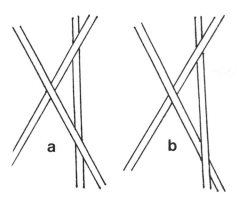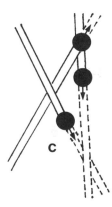

Fig. 8.19. a) Three microfibrils in layers. b) Three
microfibrils woven. (After Frey-Wyssling 1976).
c) An arrangement of membrane-bound terminal complexes
(black dots) which could produce the weave in "b" above.
The solid lines indicate completed microfibrils; dashed
lines, the incomplete. The diagram views the microfibrils as from the interior of the cell with which the
terminal complexes are associated.

have not yet found a clear example of the triangular weave (Fig.
8.19b) in cell walls (for example, Fig. 8.20) or their
derivatives (for example, Fig. 8.5), while such have been
relatively easily identified within the *Acetobacter* pellicle
(Fig. 8.3)in which synthesis occurs effectively in bulk space
(i.e., individual bacteria are synthetic terminal complexes for
each ribbon).

     The problem of distinguishing between microfibril terminal
complexes operating in planar versus bulk space is made easier
if these complexes can be identified on freeze-fractured
membranes. If we assume that the rate of synthesis of microfibrils is constant over short time periods and between
synthetic termini, then if terminal complexes operate in planar
space, where microfibril impressions cross, that complex closest
to the cross-over should always be asscciated with the microfibril closest to the membrane at the cross-over point (see Fig.
8.19c). A clear example complying with this rule can be seen
in Figure 4 in Willison and Grout (1978), and in the few clear
cases we have seen so far, there have been no exceptions to this
rule.

     The observations above tend to suggest that microfibril
synthesis occurs at a plasma-membrane bound site. It would be
false, however, to conclude that these results prove the case.
Three microfibrils will more probably be arranged as layers than
as a weave even if synthesized in bulk space, particularly if
this space is thin. We have not yet exhausted the need to
search. Furthermore, we have found patterns of microfibril

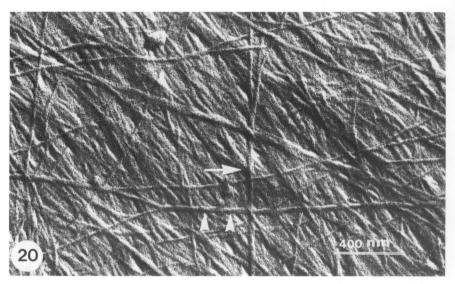

Fig. 8.20. Radish root hair cell wall inner (cytoplasmic)
surface revealed by freeze-grind replication. One pair
of microfibrils are twisted together (arrow-heads indicate
twists), as may be clearly seen by tilting the page and
looking along the fore-shortened microfibrils. A second
microfibril pair (large arrow) may also be twisted together.

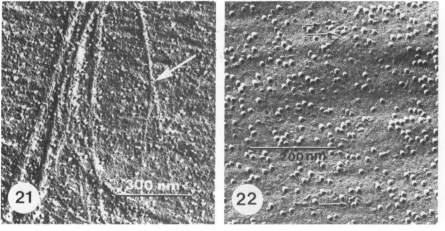

Fig. 8.21. Freeze-fractured plasma membrane (EF) from
radish root cortex in the root hair zone, showing a
possible twisted microfibril pair as impressions (cf. Fig.
8.20).   Fig. 8.22.   Freeze-fractured cotton seed hair
("fibre") plasma membrane (PF) showing rosettes both
singly (arrow) and in binary form (double arrow).

arrangement which can be accounted for only with difficulty on
the planar-site hypothesis. Microfibrils twisted as spirals,
for example, might arise if rosettes (Fig. 8.22, and see Herth,
Chapter 12, this volume) turned about their centres as they
proceeded in making microfibrils. But this microfibril form can
be more clearly envisaged to arise if a synthetic terminus were
to follow a helical path within the periplasmic zone. Rarely,
we have found two microfibrils twisted together (Fig. 8.20).
While this arrangement might have arisen post-synthetically,
during cell extension or as a preparative artefact, twisting in
conjunction with synthesis seems more probable for two reasons.
Firstly, at least two twisted pairs are present in Fig. 8.20,
with their long axes at a right angle to each other; post-
synthetic forces would presumably operate unidirectionally.
Secondly, comparable Y-shaped arrangements may be seen as
impressions in freeze-fractured membranes (Fig. 8.21), indicating
that preparative artefact is improbable. Such twisted
arrangements may be readily conceived to arise through the
agency of two synthetic sites moving in bulk space, but would
require that a pair of linked membrane-bound synthetic complexes
rolled about each other in the membrane plane, if a planar site
is envisaged. In compliance with the latter proposition,
rosettes may be found in pairs (Fig. 8.22); if synthetically
active in producing a twisted double strand, then they are a
binary system which spins slowly in the membrane plane about
their point of junction.

    We conclude that present evidence remains equally compatible
with both the membrane-bound cellulose-synthetic site hypothesis
and the periplasmic-zone cellulose-synthetic site hypothesis.
Supporting arguments for this equivocal proposal can be found in
Willison (1983).

8.5 SUMMARY

Twisting of *Acetobacter* cellulose is demonstrated and it is
concluded, considering all the available evidence, that this
cellulose twists inherently. It is argued that other helically
twisted celluloses should exist. Rare twists were found in
broad *Oocystis* (*Valonia*-type) microfibrils, but these are
considered preparative artefacts, and it is argued that
constraints in the biosynthetic mechanism may eliminate twisting
in this case. A heuristic example of possible tight double
helical twisting in microfibrils from quince seed mucilage, led
us to recognition that apparent flat waves in many primary cell
wall microfibrils were more probably spirals (open helices),
with a helical diameter of less than 75 nm. Freeze-etch
evidence of a periplasmic zone large enough to accomodate such
spirals, has lent credence to the proposal. It is suggested
that spiral microfibrils may help to maintain a relatively

hydrated periplasmic zone and may enhance cell wall flexibility
in the maintenance and development of form. Consideration has
also been given to biosynthetic requirements in permitting
twisting and spiralling. It is argued that present evidence is
compatible both with a membrane-bound biosynthetic site and with
a periplasmic zone biosynthetic site. If membrane-bound
structures, such as rosettes, are involed, these must rotate
(variously) in the membrane plane; if terminal complexes located
periplasmically are involved, these may spiral within this
limited zone. Thus, recognition that composite cellulosic
microfibrils inherently twist has led to new conceptions both
of detail in cell wall architecture and of biosynthetic
mechanisms.

8.6   ACKNOWLEDGEMENTS

We are grateful to: R. Malcolm Brown Jr. for provision of
facilities in the preparation of materials in Figs. 8.4, 8.13–
15, and 8.20–22, and for his support and generosity when the
senior author worked in his laboratory; K.B. Easterbrook for
access to, and instruction in the use of, the optical
diffractometer; J.R. Colvin and L.V. Gusta for provision of
biological materials; Charles Payzant for an unlimited supply
of quince fruit; Christine Mason and Gary Faulkner for
maintenance of, and assistance with the use of, electron
microscopical equipment; and Cong-Nghiem Nguyen for excellent
assistance with photography and plate preparation. This work
was suppported by grants from the Natural Sciences and
Engineering Research Council of Canada to J.H.M.W.

8.7   REFERENCES

Brown, R.M. Jr. and Colpitts, T.J. (1978). Direct visualization
of cellulose synthesis by high resolution darkfield microscopy
and time-lapse cinematography. *J. Cell Biol.* **79**, 157a.

Brown, R.M. Jr. and Montezinos, D. (1976). Cellulose
microfibrils: visualization of biosynthetic and orienting
complexes in association with the plasma membrane. *Proc. Natl.
Acad. Sci. U.S.A.* **73**, 143-147.

Brown, R.M. Jr. and Willison, J.H.M. (1977). Golgi apparatus
and plasma membrane involvement in secretion and cell surface
deposition, with special emphasis on cellulose biogenesis. In:
International Cell Biology 1976-1977. (ed. B.R. Brinkley, K.R.
Porter). Rockefeller University Press, New York, pp. 267-283.

Brown, R.M. Jr., Willison, J.H.M. and Richardson, C.L. (1976).
Cellulose biosynthesis in *Acetobacter xylinum*: visualization of
the site of synthesis and direct measurement of the *in vivo*
process. *Proc. Natl. Acad. Sci. U.S.A.* **73**, 4565-4569.

Browning, A.J. and Gunning, B.E.S. (1977). An ultrastructural and cytochemical study of the wall-membrane apparatus of transfer cells using freeze-substitution. *Protoplasma* 93, 7-26.

Colvin, J.R. (1961). Twisting of bundles of bacterial cellulose microfibrils. *J. Polymer Sci.* 49, 473-477.

Colvin, J.R. (1972). The structure and biosynthesis of cellulose. *CRC Crit. Rev. Macromol. Sci.* 1, 47-81.

Cooper, J.B., Chen, J.A. and Varner, J.E. (1984). The glyco-protein component of plant cell walls. In: *Structure, function and biosynthesis of plant cell walls*. (ed. W.M. Dugger and S. Bartnicki-Garcia). American Society of Plant Physiologists, Rockville, pp. 75-88.

Darvill, A., McNeil, M., Albersheim, P.A. and Delmer, D.P. (1980). The primary cell walls of flowering plants. In: *The biochemistry of plants, Vol. 1.* (ed. P.K. Stumpf, E.E. Conn). Academic Press, New York, pp. 91-162.

Elder, H.Y., Gray, C.C., Jardine, A.G., Chapman J.N. and Biddlecombe, W.H. (1982). Optimum conditions for cryoquenching of small tissue blocks in liquid coolants. *J. Microscopy* 126, 45-61.

Franke, W.W. and Ermen, B. (1969). Negative staining of plant slime cellulose: an examination of the elementary fibril concept. *Z. Naturforsch.* 246, 918-922.

Frey-Wyssling, A. (1976). *The plant cell wall. Handbuch der Pflanzenanatomie, III4*, Borntraeger, Berlin.

Haigler, C.H. and Benziman, M. (1982). Biogenesis of cellulose I microfibrils occurs by cell-directed self-assembly in *Acetobacter xylinum*. In: *Cellulose and other natural polymer systems, biogenesis, structure and degradation* (ed. R.M. Brown Jr.), Plenum Press, New York, pp 273-297.

Hanna, R.B. (1971). The interpretation of high resolution electron micrographs of the cellulose elementary fibril. *J. Polymer Sci., pt. C.* 36, 409-413.

Harada, H. and Goto, T. (1982). The structure of cellulose microfibrils in *Valonia*. In: *Cellulose and other natural polymer systems, biogenesis, structure and degradation* (ed. R.M. Brown Jr.), Plenum Press, New York, pp. 383-401.

Hestrin, S. and Schramm, M. (1954). Synthesis of cellulose by *Acetobacter xylinum*, 2. Preparation of freeze-dried cells capable of polymerizing glucose to cellulose. *Biochem. J.* <u>58</u>, 345-352.

Kirchner, W. and Tollens, B. (1875). Über den Pflanzenchleim. *Liebig's Annalen* <u>175</u>, 205-226.

Montezinos, D. and Brown Jr., R.M. (1979). Cell wall biogenesis in *Oocystis:* experimental alteration of microfibril assembly and orientation. *Cytobios* <u>23</u>, 119-139.

Mühlethaler, K. (1950). The structure of plant slimes. *Exp. Cell Res.* <u>1</u>, 341-350.

Ohad, I., Danon, D. and Hestrin, S. (1963). The use of shadow-casting technique for measurement of the width of elongated particles. *J. Cell Biol.* <u>17</u>, 321-334.

Olesen, P. (1980). The visualization of wall-associated granules in thin sections of higher plant cells: occurrence, distribution, morphology and possible role in cell wall bio-genesis. *Z. Pflanzenphysiol.* <u>96</u>, 35-48.

Pearce, R.S. and Willison, J.H.M. (1985). Wheat tissues freeze-etched during exposure to extracellular freezing: distribution of ice. *Planta* <u>163</u>, 295-303.

Preston, R.D. (1974). *The physical biology of plant cell walls.* Chapman and Hall, London.

Quader, H. and Robinson, D.G. (1981). *Oocystis solitaria*: a model organism for understanding the organization of cellulose synthesis. *Ber. Deutsch. Bot. Ges.* <u>95</u>, 75-84.

Renfrew, A.G. and Cretcher, L.H. (1932). Quince seed mucilage. *J. Biol. Chem.* <u>97</u>, 503-510.

Robards, A.W. (1969). Particles associated with developing plant cell walls. *Planta* <u>88</u>, 376-379.

Roland, J.C. and Vian, B. (1979). The wall of the growing plant cell: its three-dimensional organization. *Int. Rev. Cytol.* <u>61</u>, 129-166.

Ruben, G.C. (1985). Are cellulose microfibrils organized in a helical structure? *Proc. R. Microsc. Soc.* <u>20/2</u>, 10LT.

Schmidt, C. (1844). Über Pflanzenschleim und Bassorin. *Liebig's Annalen* <u>51</u>, 29-62.

Stöckmann, V.E. (1972). Developing a hypothesis: native

cellulose elementary fibrils are formed with metastable structure. *Biopolymers* <u>11</u>, 251-270.

Updegraff, D.M. (1969). Semimicrodetermination of cellulose in biological material. *Analyt. Biochem.* <u>32</u>, 420-424.

Willison, J.H.M. (1975). Plant cell-wall microfibril disposition revealed by freeze-fractured plasmalemma not treated with glycerol. *Planta* <u>126</u>, 93-96.

Willison, J.H.M. (1976). An examination of the relationship between freeze-fractured plasmalemma and cell-wall microfibrils. *Protoplasma* <u>88</u>, 187-200.

Willison, J.H.M. (1981). Secretion of cell wall material in higher plants. In: *Encyclopedia of plant physiology, vol. 13B.* (ed. W. Tanner and F.A. Loewus), Springer Verlag, Berlin, pp. 513-541.

Willison, J.H.M. (1982). Microfibril-tip growth and the development of pattern in cell walls. In: *Cellulose and other natural polymer systems, biogenesis, structure and degradation.* (ed. R.M. Brown Jr.), Plenum Press, New York, pp. 105-125.

Willison, J.H.M. (1983). The morphology of supposed cellulose-synthesizing structures in higher plants. *J. Appl. Polymer Sci.: Appl. Polymer Symp.* <u>37</u>, 91-105.

Willison, J.H.M. and Brown Jr., R.M. (1977). An examination of the developing cotton fiber: wall and plasmalemma. *Protoplasma* <u>92</u>, 21-41.

Willison, J.H.M. and Brown Jr., R.M. (1978). A model for the pattern of deposition of microfibrils in the cell wall of *Glaucocystis.* Planta <u>141</u>, 51-58.

Willison, J.H.M. and Grout, B.W.W. (1978). Further observations on cell-wall formation around isolated protoplasts of tobacco and tomato. *Planta* <u>140</u>, 53-58.

Willison, J.H.M. and Rowe, A.J. (1980). *Replica, shadowing and freeze-etching techniques.* North-Holland, Amsterdam.

Zaar, K. (1977). The biogenesis of cellulose by *Acetobacter xylinum. Cytobiologie* <u>16</u>, 1-15.

Zaar, K. (1979). The Visualization of pores (export sites) correlated with cellulose production in the envelope of the Gram-negative bacterium *Acetobacter xylinum. J. Cell Biol.* <u>80</u>, 773-777.

9

Immunofluorescence applications in plant cells

R. B. Knox and M. B. Singh

9.1 INTRODUCTION

It is now fifteen years since the first successful applications of immunofluorescence in plant tissues were published (Knox et al., 1970; Graham and Gunning 1970). Since then there have been several reviews documenting progress made in this rapidly developing field, the first originating at the last Botanical Microscopy meeting at York in 1979 (Knox et al., 1980; and see Jeffree et al., 1982; Knox 1982). All these developments are dependent on the fidelity of recognition by fluorescent-labelled antibodies of the original antigen against which they were made.

With the development of immuno-gold and other probes utilizing the higher resolution of electron microscopy, there might seem little need for immunofluorescence, which was the first immunocytological technique developed for animal cells by Coons in the early 1940s. Fluorescence technology, however, has some important benefits:
- maximum sensitivity and resolution of light microscopy especially with video-enhancement;
- quantitation of isolated cells by flow cytometry and fluorescence-activated cell sorting;
- rapid screening for specific antigens and hybridoma cell lines in monoclonal antibody (mab) technology.

In this review, we have focused on two areas in which real advances have occurred during the past five years:
(1) new methods for purification of antigens, including Western blots, and use of crude extracts for mab production;
(2) selection of mabs or polyclonal antibodies (pabs), and new techniques for testing their specificity.

9.2   ANTIGEN PURIFICATION

Most macromolecules are antigenic, and antigen preparation is a most important first step in raising antibodies for immuno- fluorescence studies. The antigen can be used in the form of crude mixtures, or highly purified proteins or carbohydrates. However, it is desirable to purify the antigen as much as possible, limiting impurities to no more than 1-2%. Antibodies to contaminants may still occur because the impurity can be highly immunogenic while the major antigen is only weakly so (see section 9.2.3). Molecules of very low M_r, < 1000 Kd, i.e. haptens, are generally very poorly immunogenic but may evoke strong responses when coupled to macromolecules that act as carriers. Suitable carriers include bovine serum albumin (BSA), ovalbumin, haemocyanin or fowl immunoglobulin. The chemistry of coupling depends on the availability of suitable groups on the hapten (see Bauminger and Wilchek 1980). In this way anti- bodies have been prepared to plant hormones and to low M_r secondary products.

    Antigens at a high level of purity can be obtained from tissue homogenates by high resolution methods such as isoelectric focussing, chromatofucusing, high performance liquid chromato- graphy, affinity chromatography or polyacrylamide gel electrophoresis (see Scopes 1982). A problem in most plant studies is the need for large masses of raw material. In the initial stages of plant cell or tissue homogenisation for isolation of a particular antigen it is important to inhibit proteolytic activity in crude extracts by addition of inhibitors such as PMSF and leupeptin. Phenolic constituents originating in plant cell vacuoles may bind to proteins making them possibly inaccessible as antigens.

9.2.1 Polyacrylamide gel electrophoresis

In a number of studies, the extreme resolving power of SDS- polyacrylamide gel electrophoresis (PAGE) has been used to purify protein antigens to produce highly specific antisera (see Goding 1983). Proteins treated with SDS can act as strong immunogens (Stumph et al., 1974), and the antibodies produced can complex with native protein. In the majority of cases, the Coomassie blue-stained protein band is excised, and the gel crushed and emulsified in adjuvant. The limitation of this approach is that antibodies may also be formed against the support or stain (Granger and Lazarides 1980)

    These problems can be avoided by electrophoretically eluting protein from the desired band before use for antibody production. A simple procedure utilizes a dialysis sac, and a standard tube gel apparatus (Stephen 1975; Goding 1983).

9.2.2 Western blotting

A rapid and efficient alternative to electrophoretic elution of
protein into dialysis tubing is the use of Western blots.
Protein which has been electrophoretically transferred from a
polyacrylamide gel to nitrocellulose paper is used as the
antigen (Burnette 1981). The transferred protein is detected
by staining a strip of the antigen-containing nitrocellulose
paper with a protein stain or other probe. The band on the
paper is cut out, dried thoroughly, and dissolved in the least
amount of dimethylsulfoxide (e.g. 0.5 ml) required to completely
solubilize the paper, and the extract used for antibody
production.

9.3 ANTIBODY PREPARATION AND SPECIFICITY

Whether mabs or pabs are more suitable for immunofluorescence
analysis is a question often raised. There is no specific
answer. Although by definition, mabs are more specific, their
use is not always warranted. Provided the antigen is available
in highly purified form, is immunogenic in rabbits (or other
animals) polyclonal antibodies can be satisfactory.   One
advantage of mab technology is in the preparation of highly
specific antibodies from crude plant cell extracts or even
intact cells. In this section, we will review methods for
preparation of antibodies, and their subsequent purification.

9.3.1 Polyclonal antibody production

When dealing with highly purified antigen, doses of the order
of 10 to 100 μg are often sufficient for pab production. In
some instances, even lower doses may be used. If the antigen
is readily available and extremely pure, higher dose (1-2 mg)
may result in quicker and stronger responses. However, the
risk of production of antibodies to impurities increases with
the dose. Rabbits, goats and sheep are perhaps the most suitable and widely used recipient animals for production of
specific antisera. These animals will all respond vigorously
to antigen emulsified in Freund's adjuvant, especially when
given repeatedly. For details of immunization protocols,
bleeding of animals and collection of serum refer to Knox and
Clarke (1978) and Jeffree *et al*. (1982).

9.3.2 Checking the specificity of antibodies

Immunological specificity of the antiserum against its original
antigen must be tested before use, and a wide range of methods
has been used for plant antigens. While antibody-binding is to
specific antigenic determinants, there is always the possibility
that the antibodies may recognise similar determinants in other
macromolecules in the tissue. Alternatively the antiserum may
contain antibodies against impurities in the antigen preparation.
It needs to be recognised that once the primary antibody
incubation step has been performed, the subsequent probes will
detect all bound antibodies including those bound to contaminants. It is therefore essential to incorporate a series of
specificity assays in the protocol.

     In the 1970s, antibody specificity was tested using the
double diffusion-in-gel method of Ouchterlony as a test of
antibody specificity, on account of its simplicity (see Wang
1982). Gel immunodiffusion and its modifications, like radial
immunodiffusion, and immunoelectrophoresis, have proved to be
crude tests of specificity and are mainly useful as a first
test for presence of antibody. If an antiserum is to be used
in immunofluorescence analysis then these gel diffusion methods
may be misleading because of their low sensitivity. The
specificity of the antiserum must be tested in a system which
is at least as sensitive as the immunofluorescence technology
which is to be used.

     The enzyme-linked-immunosorbent assay (ELISA) has proved
to be far more sensitive than any gel diffusion method. The
precondition in this case is that the antigen must be
available in purified form. This antigen is immobilized by
coating in wells of PVC microtitre plates. The remaining
binding sites are blocked by addition of an inert protein like
BSA, and the wells are incubated with specific antibody and
then with enzyme-conjugated secondary antibody. The most
commonly used enzyme labels are peroxidase, alkaline phosphatase,
β-galactosidase and urease. After washing and removing unbound
enzyme-conjugated secondary antibody, the enzyme reaction
mixture is added to the wells, and the level of reaction
product formed is read colorimetrically. The ELISA test can
detect antibody-binding to even nanograms of antigen and is
thus accurate and economical.

     A modification of the ELISA test is the dot immunobinding
assay developing by Hawkes *et al.* (1982). In this test,
antigenic protein is applied as dots on nitrocellulose paper
which has been marked into grids. The nitrocellulose sheet

is then incubated in the antibody to be tested, followed by enzyme-labelled or ^{125}I-labelled Protein A or secondary antibody. The radioactivity is detected by autoradiography, or enzyme-labelled antibody viewed by using the appropriate chromophore substrate. The presence of bound antibody is indicated by staining of the dots. The advantage of this method is that a number of different antigens can be spotted on the same strip and screened for binding to one individual antiserum simultaneously. The total amounts of both antigen and antibody required are much less than the microtitre plate-based ELISA test. We have used this test to study specificity of antibodies to rye grass pollen allergen (Singh and Knox 1985a).

If the antigen is an enzymatic protein whose activity can be assayed readily, then immunoprecipitation can be applied. In this case a series of dilutions of the antiserum to be tested are incubated with suitably diluted enzyme to form an immunoprecipitate, after several hours of incubation. The precipitate can be removed by centrifugation. The activity of the enzyme in the supernatant is then assayed. If antibodies against the enzyme are present in the antiserum, then much less activity will be detected in the supernatant. Burmeister and Hosel (1981) tested the specificity of antiserum against chickpea β- glucosidase by this method, while Kehrel and Wiermann (1985) detected antibodies to tulip phenylalanine ammonia lyase.

Recently, immunoprecipitation methods have been developed with a solid-phase second step such as heat-killed and fixed staphylococci (Kessler 1981) or Protein A- Sepharose (Pharmacia). In our studies on rabbit antibodies against β- galactosidase, we added Protein A- Sepharose to precipitate β- galactosidase-antibody complexes (section 9.5). The advantage of this method is that precipitation is almost instantaneous and antigen-antibody complexes bound to Sepharose beads are easily removed from the solutions. Immunoprecipitation methods do have some fundamental limitations if used with crude cell extracts. It is seldom possible to know with absolute certainty whether precipitated antigen represents the polypeptide recognised by the antibody, or whether it is precipitated because it is physically attached to molecules possessing the antigenic determinant.

These problems can be overcome by probing the proteins with antibodies after electrophoretic separation. The procedure is often known as Western blotting. All the previously described methods, even ELISA dot blotting, depend upon the availability of purified homogeneous antigens. In Western blotting even that need is obviated (see section 9.3.4). The heterogeneous mixture of proteins is separated by electrophoresis in a polyacrylamide gel and the proteins eluted electro-

phoretically in a transverse electric field on to a membrane
which binds proteins tightly. Nitrocellulose is the most
convenient membrane to use. The membrane is probed with
antiserum in the same way as described for the dot blotting
assay, and the protein band to which antibodies have bound
'lights up', and so is easily identified.

Western blotting is one of the most useful immunological
techniques to emerge in the 1980's. The method is not only
valuable in checking antibody specificity, but even purification
of antigens for antibody production and the highly specific
antibodies required for immunofluorescence (see section 9.5).

9.3.3 Monospecific polyclonal antibodies

In immunofluorescence, it is desirable to reduce the amounts of
non-reactive components in antisera to a minimum by means of
IgG fractionation. In view of its simplicity, Protein A-
Sepharose chromatography is the method of choice. This depends
on the binding of the Fc portion of IgG to staphylococcal Protein
A. Rabbit antisera have the distinct advantage that the sub-
classes of IgG bind to protein A at pH 7.4, and all can be
eluted at pH 4.0 (see review by Goding 1983). IgG cuts of
antiserum can also be obtained by a combination of ammonium
sulphate precipitation and ion-exchange chromatography on DEAE
cullulose (see reviews by Knox and Clarke 1978; Jeffree
et al. 1982 for details). All these methods yield highly
purified IgG fractions from the antiserum.

For immunofluorescence applications, it is not sufficient
only to have pure immunoglobulin but to have highly specific
IgGs directed against the antigen to be localized, i.e.
monospecific antibody. The above procedures do not separate
specific from non-specific IgG. Purified antigen-specific IgG
can be obtained by affinity chromatography.   The antiserum is
applied to a column containing antigen coupled to Sepharose
beads;   the column is then washed thoroughly with buffered
saline;   and the specific antibodies are eluted with
chaotropic agents   like sodium thiocyanate or a buffer of low
pH.   Such antigen-affinity-purified antibodies are monospecific
because they bind a specific antigen. Burmeister and Hosel
(1981) used purified β-glucosidase bound to Sepharose beads to
produce monospecific antibodies to the enzyme. The column-bound
antibodies were eluted with glycine-HCl buffer at pH 2.5, and
immediately neutralized by adding suitable amounts of solid Tris
base. Other examples of antigen-affinity purification, include
the preparation of monospecific antibodies to *Phaseolus vulgaris*
lectins (Manen and Pusztai 1982) and to soybean seed lipoxygenases
(Vernooy-Gerritsen *et al.* 1983).

Another important application of antigen affinity purification is not only preparation of monospecific antibodies to an antigen but to separate antibodies specific for different molecular forms of the antigen, for example, isoenzymes. Vernooy-Gerritsen et al., (1983) showed differential localization of two forms of lipoxygenases 1 and 2 in sections of germinating soybean seeds. Antibodies to lipoxygenase 1 and 2 were raised by injecting the individual isozymes into different rabbits. For purification, the antibodies to lipoxygenase 1 were passed through a lipoxygenase 2-conjugated Sepharose 4 B column, so that only antibodies specific to lipoxygenase 1 were obtained. Monospecific antibodies to lipoxygenase 2 were prepared by the reciprocal procedure. Earlier, Manen and Puztai (1982) had prepared monospecific antibodies to two different forms (E_4 or L_4) or Phaseolus vulgaris lectin.

We conclude that monospecific antibodies are useful for immunofluorescence analysis when the project involves highly purified antigen, that is available in sufficient quantities for affinity column preparation.

9.3.4 Purification of antibodies using western blots

In many cases, it is extremely difficult or even impossible to obtain purified antigens, as many antigenic proteins are minor constituents of cells. Olmsted (1981) published a procedure which eliminated most of the steps in purification of antibodies and antigens as a prerequisite for affinity-purification of specific IgG. In this procedure, the separation of the desired antigen from contaminants is achieved by SDS-PAGE and then transfer to diazophenyl thioester paper by the Western blotting procedure (Towbin et al. 1979). The blot is incubated in pabs, and the specific IgGs are then eluted from the blot.

Nitrocellulose blots have been used to prepare fluorochrome-conjugated affinity-purified antibodies. The regions of the blots to be eluted are located by probing parallel strips excised from the same blots with peroxidase-labelled antirabbit IgG. This strip is then realigned with the unprobed portion and the desired part identified, cut out, and antibodies eluted from it. This method is perfectly valid, except that it is difficult to distinguish with confidence two closely-migrating polypeptides. Smith and Fisher (1984) have refined the original procedure to resolve this problem. We have applied their procedure to isolate monospecific anti β - galactosidase antibodies of oilseed rape pollen proteins (see section 9. 5).

9.3.5 Monoclonal antibody technology

Although it is possible to isolate a single antibody fraction
directed against a specific antigen from the mixture of
polyclonal antiserum, the method is time consuming, and the
yield is low (Goding 1983). Therefore, the development of
monoclonal (single specificity) antibody technology has
attracted the attention of a large numbers of researchers.
This system has three advantages: 1) the yield is high, 2) the
antibody may be specific for a single antigenic determinant
and 3) the cell which produces the antibody (hybridoma) is
immortal.

The hybridoma is formed by the chemically-induced fusion
of two "parent" cells, the antibody-producing cell from the
spleen of an immunized mouse, and a myeloma (spleen cancer
type) cell. The mouse spleen cell contributes the antibody
producing information, and the myeloma contributes the
immortality. The potential of the hybridoma technology is
enormous and is likely to replace the conventional pab method
in several applications. There are some cases where the
extreme specificity of mabs can be limiting, especially in the
case of a mab recognising a 'conformational determinant'
(Goding 1983). The process of fixation and tissue processing
may alter the antigen conformation and this change can affect
the binding of the mab. The heterologous mixture of antibodies
in pabs ensures that some antigenic determinants will be
recognised in the processed antigens in tissue sections. It
has been argued that the ideal immunocytochemical reagents will
be a mixture of defined mabs which have specificity towards
different epitopes (specific determinants) on the same antigen
molecule (see Bosman 1983). This is an ideal situation that
may be seldom realised because of the cost of production of mabs.

The main limitation of monospecific pabs is the problem of
cross reactivity. The large numbers of heterologous antibodies
may crossreact with antigenic determinants with structural
similarities. The problem is more severe in the case of
glycoproteins, where some antibodies may be directed to the
carbohydrate moiety, especially to short sugar sequences. There
is a high probability of other non-related glycoproteins or
polysaccharides possessing similar sequences recognised by the
antibody. It may be appropriate in certain cases to
deglycosylate the glycoprotein prior to immunization.
Carbohydrates of plant origin evoke strong immunogenic responses
in animals, perhaps because of their foreignness (Anderson et
al. 1983).

Mabs may be produced to complex mixtures of cell components,
provided a suitable screening procedure is developed. An

example is the production of a library of mabs specific for allergenic compoents of perennial rye grass, *Lolium perenne* pollen by immunisation of mice with crude unfractionated pollen extract (Smart *et al.* 1983). In this case, hybridoma culture supernatants were screened for antibody specificity using western blots. Cultures were selected that secreted antibodies binding to allergenic proteins, as determined by IgE binding assay.

Screening for antibody specificity was initially carried out on nitrocellulose blots of SDS-PAGE-separated proteins present in denatured form. The implication of this observation for immunofluorescence is that the antigen-binding ability should not be adversely affected by conformational changes in the antigen during tissue processing. Rye grass pollen allergens are of low molecular weight, the major component, Group I allergen, is 32 Kd and is a highly soluble cytoplasmic and cell surface protein (see section 9.3.6). Immunofluorescent localization has unique difficulties (see Knox *et al.* 1980; Howlett *et al.* 1981).

Production of mabs by immunization with intact plant cells has been used in the study of Hardham *et al.* (1985). Glutaraldehyde-fixed whole cysts and zoospores of the fungal pathogen *Phytophthora cinnamomi* were used to immunize mice to obtain mabs specific for surface components. Screening of the mabs was done by ELISA test using intact cells immobilized in a 96 well microtitre plate as well as by immunofluorescence (see section 9.6). The authors detected mabs which bound only to zoospores or to cysts of *P. cinnamomi* but not to both. In this case, the nature of the antigen is not yet known. The existence of antibodies that are specific for a particular pathogen, at a particular stage in its life cycle, will be of real significance in understanding the molecular basis of the infection process. Morever, these antibodies could be immobilized on affinity supports and the antigens isolated and characterized.

This type of work in which whole cells are used for immunization is quite well established for animal cells (see Goding 1983). Immunization with cultured cells has allowed the preparation of mabs which in immunofluorescence studies recognised determinants present only in the central (Cohen and Selvendran 1981) or peripheral (Valliamy *et al.* 1981) mammalian nervous systems. Schachner *et al.* (1981) have shown that glial cells in the nervous systems can be classified convincingly by using libraries   of mabs against unknown surface determinants, followed by immunofluorescence techniques.

Such methodologies have obvious implications for plant
cell biology especially for plant pathology where immunofluorescence with mabs raised to unknown surface antigens can
help in distinguishing different pathogenic races.

9.3.6 Antibodies to enzymes and other markers: use of antibodies raised to antigens isolated from a different biological source

The structure of antigens with a defined biological function
can be highly conserved from one plant source to another, or
between plant, animal and microbial sources. This implies
that antibodies raised to an antigen isolated from one source
can recognise the same antigen from another biological source,
i.e. they show cross-reactivity. The structural homologies are
more likely to exist in taxonomically-related sources than in
distant ones. Madhavan and Smith (1982) used antibodies raised
to tobacco RuBP carboxylase to localize the enzyme in leaf
chloroplasts of 21 plant species. The same authors used
antisera raised to phosphoenolypyruvate (PEP) carboxylase from
Pennisetum glaucum (Madhavan and Smith 1984). When *E. coli*
PEP carboxylase antiserum was used for immunofluorescence,
specific fluorescence was observed in the guard cells of
Nephrolepis exaltata only and not in any of the other species
tested. It was claimed that these antibodies raised to an
E. coli enzyme localize the same enzyme in *Nephrolepis*.

This conclusion seems rather unlikely without supporting
biochemical or immunochemical evidence. It is surely not valid
to incubate plant tissue sections in antibodies raised to an
enzyme (or other antigen) from a totally unrelated source, and
on the basis of immunofluorescence observed, claim that the
same enzyme has been detected. In such cases, it is important
to show by other means such as immunoprecipitation, ELISA or
western blots (see section 9.3.4) that the antibody-binding is
really monospecific for a particular antigen. For example,
when we incubated *Brassica campestris* pollen sections in
antiserum to *E. coli* β-galactosidase followed by FITC - anti
rabbit IgG, some specific fluorescence was recorded in the
sections. However, when we developed SDS-PAGE nitrocellulose
blots of pollen proteins with the same antiserum, the β-
galactosidase protein bands were not stained at all (see section
9.5). The problem of cross-reactivity with unrelated antigens
is reduced with mabs. For example, mabs raised to a major rye
grass pollen allergen, Group I allergen, specifically
recognised the homologous antigen in related grass taxa (Fig. 9.1
and Singh and Knox 1985). Such fully characterized and specific
antibodies raised to *Lolium perenne* can thus be used for
immunofluorescence analysis.

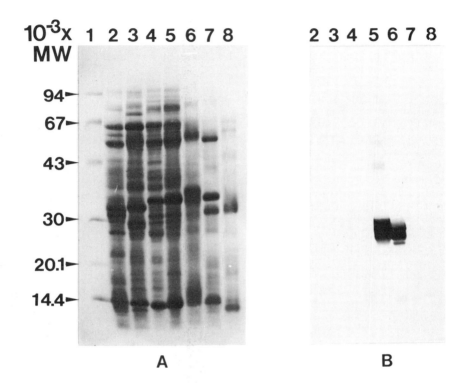

A   B

Fig. 9.1. Western blotting analysis of taxonomically related
antigens of grass pollens. A shows the SDS-PAGE patterns
stained with Coomassie blue of 7 different species of grass
pollen in lanes 2 to 8, with Mr Standards in lane 1.   B shows
a Western blot of this gel which has been incubated in mab-A7,
one of a library of mabs of *Lolium perenne* pollen screened for
binding to human IgE (i.e. anti-allergen) which recognises
specific bands in lanes 5 and 6 (i.e. *Lolium perenne* and
Festuca pratensis respectively).   From Singh and Knox (1985a).

9.4   IMMUNOFLUORESCENCE TECHNOLOGY

Immunofluorescence is based on the principle that when light is
absorbed by the fluorochrome to which the antibody is
covalently linked, the energy of the photons is transferred to
electrons. In the excited state, the electrons assume a higher
energy level, giving off heat, and a photon of lower energy
than the original photon. The efficiency, or quantum yield, is
a measure of fluorescence.

The excitation spectrum is the range of wavelengths
capable of causing the fluorochrome to fluoresce. The emission
spectrum is the range of wavelengths of the emitted fluorescence
and is shifted to a longer wavelength, as determined by Stoke's
Law. Spectral characteristics, and the intensity of
fluorescence are dependent on pH, and the physical environment.
Adjacent fluorochromes may interact, transferring energy to
each other, and therefore quenching each other. An excellent
review of the principles is given by Goding (1983).

9.4.1   Video-intensification immunofluorescence

Silicon-intensified target (SIT) video cameras are > 1000 times
more sensitive than conventional video cameras, and have been
used for analysis and recording of fluorescent images in
microscopy. The contrast of the fluorescent image can be
enhanced by the techniques of computerized image analysis.
Quantitation may be carried out by monitoring the brightness
or light intensity (in volts) along a vertical cursor that
intersects each of the horizontal TV lines of a particular
field, and the information digitized and analysed by microcomputer. Video-intensification has potential for the study of
fluorescent-labelled antibodies in living cells, e.g.
protoplasts.

9.4.2   Flow cytometry and fluorescence-activated cell sorting

The development of flow cytometry is probably the single most
important technological advance made in immunofluorescence
analysis (see Kruth 1982). By using a powerful argon laser beam
for excitation of fluorochrome-labelled cells passing in a
single file, these instruments can rapidly analyse the emitted
fluorescence of thousands of cells. A rate of 5000 cell/s is
achieved. Other parameters such as low-angle forward light
scattering or Coulter volume, can be analysed simultaneously
and recorded on a cell to cell basis, so that the size of a
given cell as well as its fluorescence emission for two
different fluorochromes (e.g. FITC and TRITC) is determined

simultaneously.

The fluorescence-activated cell sorter (FACS) is capable of separating a suspension of cells into individual fluid droplets, by means of electrostatic deflection into separate tubes. This separation can be on the basis of fluorescence, light scatter or any desired combination of parameters. In case of plant cells, FACS has been used to separate protoplasts and cell suspension (see Brown 1984). The application of FACS with plant cells is still in its early stages and it does not seem to have been applied in relation to immunological analysis. We have used FACS to determine the extent of quantitive variation in peptidases of rye grass pollen derived from a single parental genotype using a fluorescent probe (Singh and Knox 1985). Separation of rare mutants or variants of mammalian cells have been obtained with FACS (see Goding 1983). FACS should prove a valuable tool for these purposes in isolated plant cell systems.

9.4.3 Choice of fluorochromes

There are a number of different fluorochromes available (e.g. FITC, TRITC, Texas red and phycoerythrin) and the choice will be governed by colour, fluorescence intensity and cost. FITC gives a green fluorescence, has high quantum efficiency, its absorption and emission spectra are separated, and it is least expensive. Until recently the main problem in its use was photobleaching but this can now be overcome by use of non-fading mounting media (see section 9.4.7). TRITC gives higher levels of red fluorescence compared to rhodamine isothiocyanate. Also photobleaching (fading) of TRITC is much slower than FITC (Giloh and Sadat 1982). However, TRITC is more hydrophobic than FITC, and this can increase problems of non-specific adsorption of TRITC to proteins via hydrophobic interactions (see review by Goding 1983). Both FITC and TRITC have been used in two colour fluorescence experiments although the absorption and emission spectra overlap slightly (Goding 1983). Texas red a sulphonyl chloride derivative of rhodamine isothiocyanate, has proved to be a much superior fluorochrome (Titus et al. 1982) and is now commercially available. Because the emission spectra of FITC and Texas red are widely separated, these will be fluorochromes of choice for two-colour fluorescence. An excellent review of the filter combinations available for these fluorochromes is given by Goding (1983).

9.4.4 Direct and indirect immunofluorescence

In the direct immunofluorescence method, a one-step procedure, purified specific antibody is covalently bound with the selected fluorochrome, and incubated with the tissue, and after the unbound antibody is removed by washing, the specimen is examined by fluorescence microscopy. Considerable enhancement of fluorescence is achieved by use of the indirect method, a two-step procedure. The primary antibody is first bound to the antigen in the tissue section. The location of the primary antibody/antigen complex is visualized by fluorochrome – conjugated secondary antibody raised against the primary IgG antibody in another animal species. There are three advantages of the indirect method:
- it is not necessary to conjugate each individual primary antibody with fluorochrome;
- commercially available fluorochrome-conjugated secondary antibodies can be used;
- 6-8 fold amplication of fluorescence is achieved because many secondary antibodies can bind to one primary antibody molecules.

The indirect method is important where mabs are employed, because their specificity is directed to individual antigenic determinants and there will be less binding sites on each antigen molecule.

There are a number of refinements available, which may be useful in particular circumstances. In cases where whole cells or protoplasts are under investigation, enchanced permeability to Fab or F (ab')$_2$ fragments of antibodies, of lower M_r, may be found. These fragments are obtained by proteolytic digestion, and Fab fragments are univalent containing one antigen-combining site, while F (ab')$_2$ fragments contain two antibody-combining sites and the intact light chain but only the N-terminal portion of heavy chains. In the indirect method, if fragments are employed, they must be adopted for both steps (see review by Goding 1983).

9.4.5 The Biotin-Streptavidin system

An alternative option to the indirect method is the biotin-avidin system. Biotin is a small water soluble vitamin of M_r 244 d, to which avidin, a protein present in egg white, binds with extremely high affinity (approximately $10^{15}M^{-1}$). Avidin is a tetramer of identical subunits, each of M_r 15 Kd, with an isoelectric point of 10.5. The strength of the interaction between biotin and avidin has enabled it to be used in an immunological sandwich system (see review by Wilchek and Bayer 1984). Biotin can be readily conjugated to antibodies without damage to their antigen-binding ability or their physical characteristics. Several biotin molecules can be bound to each antibody molecule, with an obvious benefit to sensitivity. Labelling proceeds under mild conditions and yields a highly biotinylated antibody with essentially complete retention of biological activity.

Fluorescent-labelled avidin can be used as a stable high affinity probe, since only one conjugate needs to be used for different antibody or fluorochrome detection systems. Problems with the biotin- avidin system arise due to two specific disadvantages of avidin in immunological detection systems:
- Because of its high isoelectric point, avidin is positively charged at neutral pH. In consequence it binds electro- statically to acidic structures such as nucleic acids in chromosomes and phospholipids in cell membranes (Hoffman et al. 1980).
- Avidin is a glycoprotein and thus can bind to carbohydrate- reactive molecules such as lectins. This non-specific binding property of avidin has been the major obstacle in its use in immunofluorescence studies.

Both of these problems have been solved by using streptavidin, a biotin-binding protein from *Streptomyces avidini*. The molecular weight of streptavidin is the same as avidin and it has four high affinity sites for biotin. Since its isoelectric point is close to neutral pH, it shows correspondingly less non-specific binding. It is not a glycoprotein and does not bind to tissue lectins. Consequently, streptavidin has the desirable biotin- binding properties of avidin, without the undesirable physical characteristics. This protein is available commercially from BRL and Amersham. The biotin- streptavidin system has been used in localization of antigens in pollen grains (see section 9.5).

9.4.6 Use of fluorochrome-conjugated Protein A

Protein A is isolated from the cell wall of *Staphylococcus aureus* (Grov et al., 1964). This protein binds specifically to the Fc region of immunoglobulins of a number of mammalian species including human immunoglobulins. Protein A is a single polypeptide chain of M_r 41 Kd and has a very stable native structure. Fluorochromes can be conjugated to Protein A and this conjugation does not interfere with its binding properties. FITC - Protein A can be used as a fluorescent probe in place of fluorochrome-conjugated species-specific secondary antibody. Recently, Mau and Clarke (1983) have shown that *Staphylococcus aureus* Protein A binds to a plant glycoprotein. The implication

is that although FITC-labelled Protein A is a very effective
fluorescent probe, a range of controls are needed before
drawing conclusions regarding antigen localization in plant
cells.

9.4.7 Fading of fluorescence and its prevention

A major problem accompanying the use of fluorescent dyes
coupled to protein probes in microscopy is light-induced
bleaching, apparent as fading of the emitted fluorescence.
Photobleaching is very rapid with fluorescence microscopy in
which epifluorescent illumination is used under conditions of
high magnification and resolution because the excitation light
beam is intense and exacerbated by use of laser sources for
illumination. The problem is more severe for FITC, but TRITC
also shows photobleaching although at a slower rate. Texas red
is considered to be more resistant to photobleaching as compared
to TRITC (Titus et al., 1982). A solution has recently been
found by the addition to non-fluorescent mounting medium of
o or p-phenylene-diamine (1 to 10 µg. ml^{-1}), (Johnson et al.,
1982; Goding 1983) or n-propyl gallate (0.1 to 0.25 M in
glycerol, Giloh and Sadat 1982). These compounds are effective
in retarding the fading of the image. Apparently, molecular
oxygen or oxygen-induced free radicals may be involved in
photobleaching. As these chemicals apparently exert their
effect when they are in direct contact with fluorescent probes,
it is best to pre-mount the slides some time before observation.
An anti-bleaching mounting medium, "Citifluor" is available
commercially from the Department of Chemistry, The City
University, London, England.

9.4.8 Autofluorescence of plant tissues

Autofluorescence of plant tissues is a serious problem which
can undermine the specificity of immunofluorescence. Most
biological materials autofluoresce slightly in the yellow-green
when excited by blue or violet light, and this may be greatly
enhanced after glutaraldehyde fixation. The most intense
autofluorescence is observed in the case of lignified tissues
and tissues with vacuolar deposits of phenolic compounds. In
some situations use of specific barrier filters can reduce this
autofluorescence (see Jeffree *et al.* 1982).   Verbelen
et al.(1982) treated glutaraldehyde-fixed plant tissue with
sodium borohydride to eliminate background fluorescence. Non-
specific autofluorescence can be quenched if the tissue is
lightly stained with dyes, for example, toluidine blue (Knox
et al. 1980), Schiff's reagent (Carnegie *et al.*, 1980) or Evans

blue (Perrot-Rechenmann *et al.* 1983) after immunofluorescent staining.

9.4.9 Specificity of immunofluorescence in plant systems

Besides problems associated with antibody specificity and tissue autofluorescence there can be other difficulties. The tissue sections may have non-specific protein-binding sites which will fluoresce in test and control preparations. To enhance fluorescence specificity it is useful to add a carrier protein to all buffers, in order to saturate these sites. The most frequently used protein is bovine serum albumin (0.1 - 1%) or lactoglobulins from skim milk powder.

Inadequate washing can result in non-specific fluorescence due to the retention of unbound conjugate. The amount of washing required can be kept to a minimum if all the reagents are used at their optimal dilutions. The detergents Tween 20 and Triton X-100 severely reduce binding of antibody to antigen, so they should not be used in antibody dilution buffers. Once antigen-antibody complexes are formed these detergents will not cause their dissociation, so that they can be used at very dilute (< 0.1%) concentrations in washing buffers.

Naturally-occurring lectins, often present at high concentrations in certain plant tissues, may bind IgG molecules non-specifically. This kind of binding can be abolished by incorporating a monosaccharide inhibitor of lectin-binding in the antibody preparation (see review by Knox and Clarke 1978). Another possible problem is the 'sticky' nature of arabino-galactan proteins present in some tissues, and the protein-binding properties of polyphenols (see Knox 1982). Prevention of non-specific fluorescence depends upon recognitionof the fact that it is occurring, and the diagnosis of its cause. The importance of both negative and positive controls in immunofluorescence experiments cannot be over emphasized.

9.5 CASE HISTORY: IMMUNOFLUORESCENCE ANALYSIS OF POLLEN β-GALACTOSIDASE

In 1982, we discovered that in certain plants of *Brassica campestris*, the anthers shed pollen half of which is β-galactosidase deficient. The cytochemical test for enzyme activity involved incubation of the pollen in an artifical cytochemical substrate, 5-bromo- 4-chloro- 3-indoxyl β-galactoside   (Singh and Knox 1984). Genetic studies of inheritance of the deficiency indicated it is due to the presence of the single recessive gene *gal* in pollen. Strains were

222 R.B. Knox and M.B. Singh

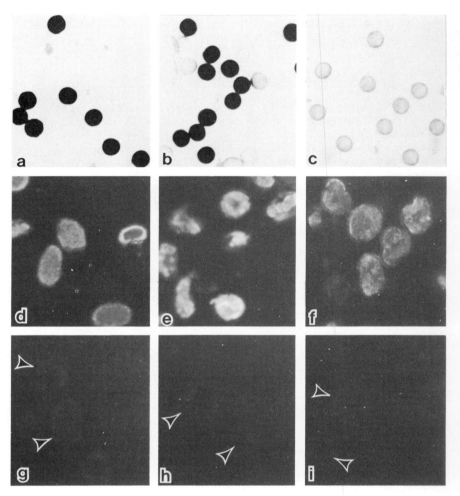

Fig. 9.2. Immunofluorescence analysis of β-galactosidase in pollen
of oilseed rape, *Brassica campestris*. (a)-(c) enzyme cytochemistry using the indoxyl method (described by Singh and Knox,
1984). (a) normal *Gal/Gal* pollen; (b) heterozygous *Gal/gal*
pollen and (c) *gal/gal* pollen. *Gal* pollen strains opaque blue-
green, while *gal* pollen is colourless indicating absence of
enzymic activity. (d)-(i) detection of the enzyme as an antigen
in frozen semithin sections of pollen using monospecific pab.
(d) *Gal/Gal* pollen, indirect method with FITC-secondary antibody,
showing positive fluorescence of grains; (e) *Gal/gal* pollen,
biotin-streptavidin method with Texas red, all grains show
positive fluorescence; (f) *gal/gal* pollen, indirect method with
FITC-secondary antibody, all grains show positive fluorescence;
(g)-(i) controls for (d)-(f) omitting primary antibody. Other
controls utilizing normal serum also showed no positive fluorescenc

available whose anthers showed the following pollen phenotypes: normal (*Gal/Gal*, Fig. 9.2a), heterozygous, i.e. 50% deficient (*Gal/gal*, Fig. 9.2b) and deficient (*gal/gal*, Fig. 9.2c) Singh and Knox (1984).

We have used immunofluorescence analysis to indicate whether in mutant *gal* pollen, the genetic defect is expressed at the transcriptional, translational or post-translational level. An answer might be obtained by probing with pabs for inactive enzyme molecules in *gal* pollen.

9.5.1 β-galactosidase as an antigen

Preliminary studies on the isolation and purification of β-galactosidase from *Brassica* pollen had indicated that on a protein to protein basis, this enzyme is present as a minor component of the total cellular proteins, and is quite unstable in a purified form, making monitoring of purification very difficult (Singh and Knox, unpublished results).

Initially we attempted to cross-react pabs prepared from *E. coli* β-galactosidase with pollen extracts by immuno-fluorescence. Faint fluorescence was observed in semithin frozen sections or whole pollen grains. Western blot analysis showed that this appears to be non-specific, due to cross-reactivity with non-related components. Expecting that β-galactosidase from one dicot plant source may have strong structural homologies to that from another source, we decided to prepare pabs to a commercially available enzyme preparation from cotyledons of jack bean, *Canavalia ensiformis*. Li et al. (1975) showed that this β-galactosidase is homogenous, giving a single band with PAGE after Coomassie blue staining. We carried out SDS-PAGE of the enzyme preparation and detected several bands indicating that the preparation is not 100% pure (Fig. 9.3).The presence of antibodies to β-galactosidase was demonstrated by a dot immunobinding assay (see Hawkes *et al.* 1982).

When the pabs were mixed with a diluted series of purified enzyme, and the antigen- antibody complex precipitated by addition of Protein A-Sepharose followed by assay of β-galactosidase in the supernatant, the absence of enzyme activity confirmed that the antiserum does, in fact, contain antibodies to the enzyme. Similar observations were made when jack bean cotyledon, *Brassica campestris* leaf and pollen extracts were used in place of diluted purified jack bean β-galactosidase.

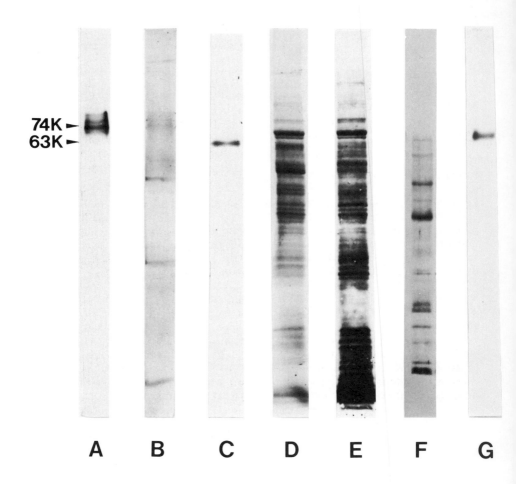

74K ►
63K ►

A   B   C   D   E   F   G

Fig. 9.3. Use of western blotting for isolation of monospecific
pab against β-galactosidase isolated from jack bean, *Canavalia
ensiformis*. A: SDS-PAGE of enzyme preparation stained with
enzyme reaction mixture (indoxyl substrate, Singh and Knox, 1984);
B: as A, but stained with Coomassie blue for proteins; C: SDS-
PAGE of *Brassica* pollen proteins, stained for enzyme as A; D: as
C, but stained with Coomassie blue; E: Western blot of gel
shown in D stained with india ink method (Hancock and Tsang,
1983); F: Western blot of *Brassica* pollen proteins probed with
pabs to jack bean β-galactosidase; G; Western blot of *Brassica*
pollen proteins probed with monospecific pab isolated by
Western blotting method (see text).

9.5.2 Monospecific pabs to β-galactosidase isolated from Western blots

The next step was to isolate pabs monospecific for the pollen β-galactosidase. A highly purified pollen antigen was not available, so we obtained monospecific pabs by using the Western blot method (see section 9.3.4). Recently, in several animal systems, Blank et al (1983) have shown that following SDS-PAGE of cell extracts, the detergent SDS can be removed from the gel by repeated washing of the gel with 20% isopropanol in 10 mM Tris buffer pH 7.4. This procedure results in renaturation of enzyme molecules in the gel with restored catalytic activity, detected in SDS-PAGE gels using artificial substrates.

Using this method, successful renaturation of β-galactosidase in SDS-PAGE gels was obtained. One lane of the gel was incubated in a fluorescent substrate for β-galactosidase for 10-15 min, revealing a single fluorescent band of enzyme activity which could be marked before the same gel was stained for enzyme using the artificial indoxyl substrate, and subsequently stained with Coomassie blue. The turquoise blue enzyme-stained band could be easily identified in such double-stained gels (Fig. 9.3).

A number of SDS-PAGE gels were run and, in each case, small strips were removed to carry out double enzyme/protein staining and the proteins from the rest of the gel were electro-blotted on to nitrocellulose paper. One strip each was cut from two sides of the blot and stained for proteins with amido black, or the india ink method (Hancock and Tsang 1983). The position of the β-galactosidase protein band was identified by comparison with the earlier stained acrylamide gel. After blocking the remaining protein binding sites, the nitrocellulose blot was incubated for 16 h in 1:50 diluted anti-β-galactosidase serum. Subsequently the β-galactosidase band was cut out, and the paper shredded into small pieces. The bound antibodies on the paper were eluted, giving a monospecific antibody preparation. An aliquot was tested by probing a strip of pollen proteins using Western blot analysis. The antibodies bound only to a single band corresponding to the β-galactosidase.

The monospecific antibodies were used for immunofluorescent analysis of β-galactosidase in cryostat sections (4 μm in thickness) of pollen (Fig. 9.2). Both *Gal* and *gal* pollen grains showed positive fluorescence for β-galactosidase. This indicated that *gal* pollen possesses antigenically active but enzymically inactive molecules. The presence of antigenic activity was also confirmed by probing the Western blot of proteins from *gal/gal* pollen. Thus, immunofluorescence analysis has provided some evidence suggesting that post-transcriptional processes are implicated in the genetic defect in pollen β-galactosidase expression.

9.6 PROTOCOL FOR IMMUNOFLUORESCENCE ANALYSIS OF PLANT CELLS

We conclude this chapter with a procedure for immunofluorescence analysis of plant cell surface antigens using mabs. There are usually several reasons for undertaking such projects:
- when no previous background information is available on the cellular location of antigens;
- distinguishing which particular cells in a multicellular tissues or suspension possess a particular antigen;
- when the cellular location of the antigen is already known but developmental or phase-specific changes in the amount of antigen or in the nature of the antigenic determinants are being investigated.

Fresh plant cells or tissue have been widely used for immunofluorescence (see Knox 1982) either as a suspension, or in frozen cryostat sections. Brief fixation is desirable to stabilize the antigen in its cellular sites, and Hardham et al. (1985) have successfully used low concentrations of paraformaldehyde/glutaraldehyde for immunofluorescent studies with fungal zoospores. This fixative has the advantage of not inducing autofluorescence which is a problem with glutaraldehyde alone. Such low concentrations probably cause little loss of antigenicity (Howlett et al. 1981). This technique is given in the protocol below.

The method of Hardham, Suzaki and Perkin (1985) for immunofluorescence analysis of mabs specific for surface antigens of fungal zoospores and cysts.

1. Fix cells in 0.2% glutaraldehyde and 4% paraformaldehyde in 50 mM Pipes buffer pH 7.0 for 30 min.
2. Wash 3 x in PBS (10 mM Naphosphate, 100 mM NaCl, pH 6.8).
3. Disperse clumped cells by passage through a 25 gauge hypodermic needle.
4. 10 µl cell suspension (c. 5 x 10^5 cells/ml) air-dried at room temperature on 10 mm diam. coverslips *or* air-dried at 37°C for 1 h on microscopic slides (multi-well pattern delineated by plastic coating, Carlson Scientific, Peotone, IL 60468).
5. Rinse in PBS, and add 5 µl hybridoma supernatant and incubate at 37°C for 45 min.
6. Remove supernatants individually by absorption with filter

paper and wash cells 2 x in PBS (1 min, 5 min).
7. Add 5 µl FITC - goat F (ab')$_2$ anti-mouse IgG (Tago Inc.,
    Burlingame, CA 94010), 1:40 dilution. Incubate at 37°C
    for 45 min.
8. Rinse 2 x in PBS, 1 x in distilled H$_2$0, mount in glycerol
    and examine by fluorescence microscopy.

9.6 CONCLUSIONS

Immunofluorescence analysis is a valuable cytochemical probe.
Its potential applications are, however, limited by the
specificity of the antibodies. The need to understand antibody
specificity is fundamental to an appreciation of the reliability
and precision of immunofluorescence. Both approaches must be
developed simultaneously, and can lead to the fuller utilization
of monoclonal antibody libraries that are being assembled in
several laboratories. It has two principal functions in plant
cell biology:

(a) monoclonal antibody (mab) technology

- rapid screening of the large numbers of hybridoma supernatants;
- rapid screening of developmental, spatial and taxonomically
   related antigens;
- potential of video-enhanced immunofluorescence for screening
   of living plant cell protoplasts or surfaces based on a
   library of mabs to surface antigens.

(b) flow cytofluorometry and cell sorting

- rapid quantitation of antigens in isolated plant cell systems;
- detection and sorting of isolated plant cells for mutants
   or surface markers.

9.7 ACKNOWLEDGEMENTS

We thank the Australian Department of Education (Special Research
Centre Program) for financial support, Dr A. Ashford, Prof. D.
Cass, Prof. J. Goding, Dr A. Hardham for helpful discussion,
Glenda Beresford for valued technical assistance and Debbie
Irvine for skilled typing.

9.8 REFERENCES

Anderson, M.A., Sandrin, M.S. and Clarke, A.E. (1984). A high proportion of hybridomas raised to a plant extract secrete antibody to arabinose or galactose. *Pl. Physiol.* 75, 1013-6.

Bauminger, S. and Wilchek, M. (1980). The use of carbodiimides in the preparation of immunizing conjugates. *Meth. Enzymol.* 70, 151-9.

Blank, A., Silber, J.R., Thelen, M.P. and Dekker, C.A. (1983). Detection of enzymatic activities in SDS-polyacrylamide gels: DNA polymerastras model enzymes. *Anal. Biochem.* 135, 423-430.

Bosman, F.T. (1983). Some recent developments in immuno-chemistry. *Histochem. J.* 15, 189-200.

Brown, S. (1984). Analysis and sorting of plant material by flow cytometry. *Physiol. Veg.* 22, 341-9.

Burmeister, G. and Hosel, W. (1984). Immunohistochemical localization of β-galactosidase in lignin and isoflavone metabolism in *Cicer arietinum.* L. seedlings. *Planta* 152, 578-86.

Burnette, W.N. (1981). "Western Blotting": Electrophoretic transfer of proteins from sodium dodecyl-sulfate-polyacrylamide gels to unmodified nitrocellulose and radiographic detection with antibody and radioiadinated protein A. *Anal. Biochem.* 112, 195-203.

Carnegie, J.A., McCully, M.E. and Robertson, H.H. (1980). Embedment in glycol methacrylate at low temperature allows immunofluorescent localization of a labile tissue protein. *J. Histochem. Cytochem.* 28, 308-10.

Cohen, J. and Selvendran, S.Y. (1981). A neuronal Cell-surface antigen is found in the CNS but not in peripheral neurones. *Nature* 291, 421-3.

Giloh, H. and Sadat, J.W. (1982). Fluorescence microscopy: reduced photobleachin of rhodomine and fluorescein protein conjugates by n-propylgallate. *Science* 217, 1252-5.

Goding, J.W. (1983). Monoclonal Antibodies: Principles and Practice. *Academic Press, New York Inc.*

Graham, T.A. and Gunning, B.E.S. (1970). Location of legumin and vicilin in bean cotyledon cells using fluorescent antibodies. *Nature (Lond.)* 228, 81-2.

Granger, B.L. and Lazarides, E. (1980). Synemin: a new high molecular weight protein associated with desmin and vimentin filaments in muscle. *Cell* 22, 727-38.

Grov, A., Myklestad, B. and Oeding, P. (1964). Immunochemical studies on antigen preparations from *Staphylococcus aureus*. I. Isolation and chemical characterisation of antigen A. *Acta. Pathol. Microbiol. Scand.* 61, 588-96.

Hancock, K. and Tsang, V.C.W. (1983). India ink staining of proteins on nitrocellulose paper. *Anal. Biochem.* 133, 157-62.

Hardham, A.R., Suzaki, E. and Perkin, J.L. (1985). The detection of monoclonal antibodies specific for surface componenets on zoospores and cysts of *Phytophthora cinnamomi*. *Expt. Myeol.* (in press).

Hawkes, R., Niday, E. and Gordon, J. (1982). A dot-immunobinding assay for monoclonal and other antibodies. *Analyt. Biochem.* 119, 142-146.

Hofmann, K., Wood, S.W., Brinton, C.C., Montibeller, J.A. and Finn, F.M. (1980). Iminobiotin affinity columns and their application to retrieval of streptavidin. *Proc. Nat. Acad. Sci. U.S.A.* 77, 4666-8

Howlett, B.J., Vithanage, H.I.M.V. and Knox, R.B. (1981) Immunofluorescence localization of two water-soluble glycoproteins including the major allergen from pollen of rye grass *Lolium perenne*. *Histochem. J.* 13, 461-80.

Jeffree, C.E., Yeoman, M.M. and Kilpatrick, D.C. (1982). Immunofluorescent studies on plant cells. *Int. Rev. Cytol.* 80, 231-65.

Johnson, G.D., Davidson, R.S., McNamee, K.C., Russell, G., Goodwin, D. and Holborow, E.J. (1982). Fading of immuno-fluorescence during microscopy: a study of the phenomen and its remedy. *J. Immunol. Methods* 11, 265-72.

Kehrel, B. and Wiermann, R. (1985). Immunochemical localization of phenylalanine ammonia-lyase and chalcone synthase in anthers. *Planta* 163, 183-90.

Kessler, S.W. (1981). Use of protein A-bearing staphylococci for immunprecipitation and isolation of antigens form cells. *Meth. Enzymol.* 73, 442-59.

Knox, R.B. (1982). Methods of locating and identifying antigens in plant tissues. In: G. Bullock and P. Petrusz (eds).

Immunocytochemistry, Vol. 1. 205-238, Academic Press, London.

Knox, R.B. and Clarke, A.E. (1978). Localization of proteins
and glycoproteins by binding to labelled antibodies and lectins
in Hall, J.L. (ed). *Electron Microscopy and Cytochemistry of
Plant Cells* Elsevier/North Holland, Biomedical Press Amsterdam,
150-83.

Knox, R.B., Heslop-Harrison, and Reed C. (1970). Localization
of antigens associated with the pollen grain wall by immunofluorescence. *Nature (Lond.)* 225, 1066-8.

Knox, R.B., Vithanage, H.I.M.V. and Howlett, B.J. (1980).
Botanical immunocytochemistry - a review with special reference
to pollen antigens and allergens. *Histochem. J.*   247-72.

Kruth, H.S. (1982). Flow cytometry: rapid biochemical analysis
of single cells. Anal. Biochem. 125, 225-42.

Li, S.-C., Mazzotta, M.Y., Chien, S.-F., Y.-T. (1975).
Isolation and Characterization of Jack Bean β-galactosidase.
J. Biol. Chem. 250, 6786-91.

Madhavan, S. and Smith, B.N. (1982). Localization of ribulose
bisphosphate carboxylase in the guard cells by an indirect,
immunofluorescence technique. *Plant Physiol.* 69, 273-7.

Madhavan, S., and Smith, B.N. (1984). Phosphoenolpyruvate
carboxylase in guard cells of several species as determined by
an indirect, immunofluorescent technique. *Protoplasma* 122,
157-61.

Manen, J.F. and Pusztai, A. (1982). Imunocytochemical
localisation of lectins in cells of *Phaseolus vulgaris* L.
seeds. *Planta* 155, 328-34.

Mau, S-L, and Clarke, A.E. (1983). Binding of a plant
glycoprotein to Staphylococcus Protein A. *Phytochemistry*
22, 91-5.

Olmsted, J.B. (1981). Affinity purification of antibodies
from diazotized paper blots of hetrogenous protein samples.
J. Biol. Chem. 256, 11955-7.

Perrot-Rechenmann, C., Jacquot, J.P., Gadal, P., Weeden, N.F.,
Cseke, C. and Buchanan, B.B. (1983). Localization of NADP-
malate dehydrogenase of corn leaves by immunological methods.
Plant Sci. Lett. 30, 219-26.

Perrot-Rechenmann, C., Chollet, R. and Gadal, P. (1984). In-situ immunofluorescent localization of phosphoenolpryuvate and 1,5-bisphosphate carboxylases in leaves of C_3, C_4 and C_3 - C_4 intermediate *Panicum* species. *Planta* 161, 266-71.

Schachner, M., Kim, S.K. and Zehnle, R. (1981). Developmental expression in central and peripheral nervous system of oligodendrocyte cell surface antigens (O antigens) recognised by monoclonal antibodies. *Devel. Biol.* 83, 328-38,

Scopes, R.K. (1982). Protein Purification: Principles and Practice, Springer-Verlag, New York.

Singh, M.B. and Knox, R.B. (1984). Quantitative cytochemistry of β-galactosidase in normal and enzyme deficient (gal) pollen of *Brassica campestris* : application of the indigogenic method. *Histochem. J.* 16, 1273-96.

Singh, M.B. and Knox, R.B. (1985a). Grass pollen allergens: antigenic relationships detected using monoclonal antibodies and dot blotting immunoassay. *Int. Archs Allergy Appl. Immun.* (in press).

Singh, M.B. and Knox, R.B. (1985b). Detection of peptidase activity in pollen of ryegrass *Lolium perenne* by flow cytometry. *Plant Science* (in press).

Smart, I.J., Heddle, R.J., Zola, H. and Bradley, J. (1983) Development of monoclonal mouse antibodies specific for allergenic components in Ryegrass *(Lolium perenne)* pollen. *Int. Archs. Allergy Appl. Immun.* 72, 243-8.

Smith, E.D. and Fisher, P.A. (1984). Identification, developmental regulations, and response to heat shock of two antigenically related forms of a major nuclear envelope protein in *Drosophila* embryos: application of an improved method for affinity purification of antibodies using polypeptides immobilized on nitrocellulose blots. *J. Cell Biol.* 99, 20-8.

Stumph, W.E., Elgin, S.C.R. and Hood, L.E. (1974). Antibodies to proteins dissolved in sodium dodecyl sulfate. *J. Immunol.* 113, 1752-6.

Titus, J.A., Hangland, R., Sharrow, S.D. and Segal, D.M. (1982) Texas red, a hydrophilic red-emitting fluorophore for use with fluorescein in dual parameter flow microfluorometric and fluorescence microscopic studies. *J. Immunol. Methods.* 50, 193-204.

Towbin, H., Staehlin, T. and Gordon, J. (1979). Electrophoretic transfer of proteins from polyacrylamide gels to nitrocellulose sheets: procedure and some applications. *Proc. Natl. Acad. Sci.* U.S.A. <u>76</u>, 4350-4.

Vergelen, J.-P., Pratt, L.H., Butler, W.L. and Tokuyasu, K. (1982). Localization of phytochrome in oats by electron microscopy. *Plant Physiol.* <u>70</u>, 867-71.

Vernooy-Gerritsen, M., Bos, A.L.M., Veldink, G.A. and Vliegenthart, J.F.G. (1983). Localization of lipoxygenases 1 and 2 in germinating soybean seeds by an indirect immuno-fluorescent technique. *Plant Physiol.* <u>73</u>, *262-7.*

Vulliamy, T., Rattray, S. and Mirsky, R. (1981). Cell surface antigen distinguishes sensory and autonomic peripheral neurones from central neurones. *Nature* <u>291</u>, 418-20.

Wang, A-C. (1982). Methods of Immune diffusion, immuno-electrophoresis, precipitation and agglutination. In: *'Antibody as a Tool.'* (ed. J.J. Marchalonis and G.W. Carr). John Wiley and Sons Ltd. 139-61.

Wilchek, M. and Bayer, E.A. (1984). The Avidin-biotin complex in immunology. *Immunology Today* <u>5</u>, 39-43.

10

The plant cytoskeleton

P. K. Hepler

10.1 INTRODUCTION

The presence of cytoplasmic fibers, especially in the mitotic
apparatus (MA) and phragmoplast, has been known for many
years from studies using polarized light microscopy (for
review, Inoue 1964). However, it was the electron microscopic
observations in the mid-sixties of microtubules (MTs) in the
cortical cytoplasm of plant cells (Ledbetter and Porter 1963;
Hepler and Newcomb 1964) that marked the discovery of the
cytoskeleton and sparked the interest in this topic that is
evident to this date. The importance of these elements to the
growth and development of plant cells became apparent from the
outset since it was evident that the cortical MTs were
oriented parallel to the cellulose microfibrils of the
surrounding cell wall. Here, then, was a candidate for the
cytoplasmic entity that might control cell wall formation and
thus the shape of the plant cell.

In the intervening years, numerous reports have appeared
revealing the ubiquity of the cortical MT system and its
mutual organization with that of the overlying cellulose
microfibrils. The concept of the cytoskeleton in higher plants
has also expanded such that we now recognize at least four
distinct but interrelated MT-containing structures: 1.) the
cortical MTs, mentioned above, 2.) the preprophase band (PPB),
3.) the mitotic apparatus (MA) and 4.) the phragmoplast. In a
typical growing, dividing plant cell, these microtubular
systems arise in sequence and appear to control the following
processes: 1.) the orientation of cellulose deposition, 2.)
the establishment of the plane of division, 3.) the separation
of chromosomes, and 4.) the formation of the cell plate. In
addition there are nuclear migrations and asymmetric shape
formations, as well as directed organelle movements and

alignments that owe their regulation to special manifestations
of the cytoskeleton. Within this framework, the importance of
the cytoskeleton to the most fundamental aspects of growth and
development cannot be overemphasized.

Virtually all aspects of the plant cell cytoskeleton have
been authoritatively reviewed; an entire book devoted to the
subject has recently appeared (Lloyd 1982) as well as review
articles to which the reader is directed for further
information (Hepler and Palevitz 1974; Gunning and Hardham
1982; Lloyd 1984). It is the purpose of this essay to present
an overview of the cytoskeleton of higher plants, to identify
emerging areas of endeavour, and to provide direction for
future experimentation.

10.2   THE CORTICAL CYTOSKELETON

Ultrastructural studies, with very few exceptions, indicate
that cortical MTs are aligned parallel to the cellulose
microfibrils of the wall (Hepler and Palevitz 1974; Robinson
and Quader 1982). Together with a variety of cytophysiological
investigations, they support the view that these MTs control
the orientation of the cellulose microfibrils that are
deposited during wall formation. Evidence against this view
has appeared over the years in studies in which it is reported
that cellulose microfibrils and MTs are not coaligned. The
bulk of these examples has been carefully analyzed by Robinson
and Quader (1982) who assert that none holds up under critical
examination. They conclude that "...the weight of evidence is
now so great in favour of microtubules that it is difficult to
deny them a role in microfibril orientation."

Since the review by Robinson and Quader (1982), the
reports by Emons (1982) and Emons and Wolters-Arts (1983)
again raise the possibility of nonalignment between cellulose
and MTs. Studies on root hairs of *Equisetum* show that the
wall forms layers in which the cellulose is first random, then
helicoidal, and finally longitudinal (Emons and Wolters-Arts
1983). The MTs, however, always show a predominately
longitudinal orientation and thus it is argued, for this
example, that they cannot control the orientation of
deposition of the cellulose. The initial study by Emons (1982)
suffers from the fact that cellulose microfibril orientation
is determined by one procedure while MT orientation by
another. Thus, there is no image showing both cellulose and
MTs in the same view where it can be unambiguously determined
that the two components in question possess different
orientations. Nevertheless, the matter deserves attention and
given the interest in this topic, further studies will be most

surely forthcoming.

Examples that provide compelling support for the
MT/microfibril hypothesis are numerous (Hepler and Palevitz
1974; Robinson and Quader 1982). Elongating cylindrical cells
that have transversely oriented cellulose microfibrils on
their inner layer invariably have transversely oriented MTs.
Two systems that show an especially high degree of congruence
between MTs and microfibrils are differentiating tracheary
elements and stomatal guard cells (Hepler 1981a; Palevitz
1982). In both examples, MTs are grouped specifically over
the wall thickenings at their inception and are aligned
strictly parallel to the cellulose. Guard cells of grasses
undergo major changes in orientation and depostion of their
wall during differentiation, but the microtubules appear to
anticipate these changes closely (Galatis 1980; Palevitz
1982). Additional support for a role for MTs is provided by
studies in which the cytoskeleton has been severely disrupted
by application of antimicrotubule agents. In both tracheary
elements and guard cells, the normal pattern of the wall as
well as that of the underlying cellulose microfibrils is lost
(Hepler 1981a; Palevitz 1982).

Among the recent developments in the study of the
cytoskeleton, the introduction of fluorescent antibodies and
stains to localize fibrous components at the light microscopic
level is helping us to obtain a more global view of MT
organization (Lloyd 1984; Wick et al. 1981). Bundles of MTs
are easily visualized and it has become clear that they are
quite long; lengths greater than the circumference of the cell
have been measured. In onion root hairs the images reveal
that the MTs are wound in a helical pattern, and even cover
the dome tip (Lloyd 1983). One of the most exciting
discoveries has been the realization that the MT cytoskeleton
is extremely resilient and does not collapse easily. MTs
bound to one another and to the plasma membrane appear to
account for the stability. Even after cells are made into
protoplasts and the plasma membrane solubilized by treatment
with up to 10% Triton X-100, the MT cytoskeleton remains on
those fragments that had adhered to the cover glass, causing
Lloyd et al. (1980) to argue that MTs are linked to
transmembrane proteins, which in this example have bound to
the glass substratum.

The ultrastructural view of the cytoskeleton is also
benefiting from improved and modified procedures for fixing
and preparing specimens for the eletron microscope. The
ability to dry cleave protoplasts stuck to glass slides has
permitted an ultrastructural examination of the cortical
cytoskeleton (Traas 1984; Traas et al. 1984). Long MTs can be

viewed by this procedure as well as 5-10 μm filaments of
unknown composition.

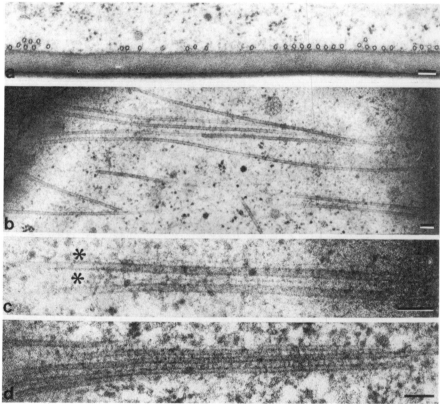

Fig. 10.1   The cortical cytoskeleton in elongating
staminal hair cells of *Tradescantia* (Fig. 10.1a-c) and
Gibasis (Fig. 10.1a) prepared by rapid freeze fixation
and substitution. Figure 10.1a shows the plasma membrane
tightly appressed to the cell wall. The cortical MTs,
usually in a monolayer, reside close to the plasma
membrane. When tangential sections are examined (Fig.
10.1b) the cortical MTs appear as gently curving arcs.
Observation of the pair of MTs in the lower right corner
of Fig. 10.1b at high magnification (Fig. 10.1c) reveals
fine filaments (*) alongside and between the MTs. Fig
10.1d shows an especially high degree of cross-bridging
between cortical MTs of *Gibasis*. Fig. 10.1a x55,000.
Fig.10.1b x41,650. Fig. 10.1c x108,800. Fig. 10.1d
x90,000. Bar = 0.1 μm.   (Fig. 10.1 from Lancelle,
Callaham and Hepler, unpublished results).

Another approach has been the application of rapid freeze-fixation followed by freeze substitution as a means for faithfully preserving the integrity of the cytoskeleton (Tiwari *et al.* 1984; Figs 10.1, 10.3). Even the best chemical fixation often takes minutes to stop vital processes such as cytoplasmic streaming and one can imagine therefore that the fixation process itself introduces significant artefact. In rapid freeze-fixation, the tissue is cooled at rates of $10,000^{\circ}C/sec$ and thus vital processes are almost instantaneously stopped. When the tissue is then substituted at low temperature, the subcellular structures are locked in place with a minimum of distortion.

Electron microscopic observation of the cortex of freeze-substituted tissues reveals a plasma membrane that is even and tightly appressed to the cell wall (Fig. 10.1a). Immediately adjacent to the plasma membrane are cortical MTs (Fig 10.1a, 10.1b), some of which virtually abut against the membrane. Cross-bridges between MTs (Fig. 10.1d) and between MTs and the plasma membrane (Fig. 10.3a) are commonly observed. Tangential grazing sections reveal additional substructures, notably fine filaments (MFs) (Fig. 10.1c) running parallel to the MTs. Similar filaments have been occasionally observed in chemically fixed tissues (Hardham *et al.* 1980; Seagull 1983; Seagull and Heath 1979). The identification of the root hair filaments as F-actin by virtue of their reaction with heavy meromyosin (Seagull and Heath 1979) lends support to the idea that those seen after freeze substitution might also be F-actin. These filaments also appear to be linked to the MTs, thus creating a complex of MTs, MFs and membrane (Fig. 10.1c).

Actin MFs, as a cytoskeletal component, have received much less attention than MTs in studies of plant cells; nevertheless bundles of MFs, often near the cortex, are observed in some cell types (Pesacreta and Parthasarathy 1984; Goosen de Roo et al. 1983). However, the relationship of F-actin to the cortical cytoskeleton per se has been less clear. Although the recent report by Clayton and Lloyd (1985) fails to denote by rhodamine-phalloidin staining a cortical organization of F-actin, it nevertheless seems possible that some filaments are commonly present. For example, the relatively small number of filaments observed after rapid freeze-fixation might be too few to give a positive phalloidin stain, but might still be crucial to the structure and function of the cytoskeleton. Further work is needed to confirm the identity of these MFs and to decipher their relationship to MTs and the plasma membrane.

Repeated mention has been made of the close relationship between the cortical MTs and the plasma membrane. There is, in

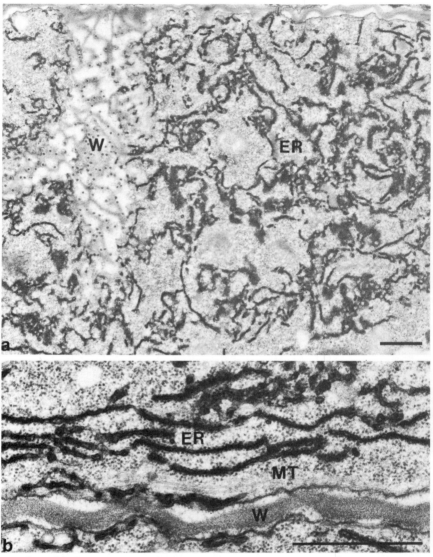

Fig. 10.2   The cortical cytoplasm of a root tip cell of
Lactuca postfixed in OsFeCN showing the ER. Examination
of tangential sections (Fig 10.2a) reveals a rich array
of interconnected tubular and lamellar ER.   Generally
this occurs interior to the MTs but occasionally
elements of membrane appear to slip in between the MTs
and the plasma membrane (Fig. 10.2b). Fig. 10.2a
x13,000. Fig. 10.2b x40,000.   Bar = 1 μm.   (Fig. 10.2a
from Hepler, 1981, Fig. 10.2b from Hepler, unpublished
results)

addition, an extensive network of ER that resides at the
perimeter of the cell and might justifiably qualify as a
component of the cortical cytoskeleton (Hepler 1981b; Fig.
10.2). When cells are post-fixed with osmium tetroxide-
potassium ferricyanide, the endoplasmic system is selectively
contrasted. Electron micrographs of cells stained by this
method invariably reveal an accumulation of reticulate tubular
and fenestrated lamellar ER in the cortical cytoplasm
subjacent to the cortical MT zone (Fig. 10.2). Some of this
ER is directly associated with the membrane strands that
traverse the plasmodesmata (Hepler 1982). No structural
connections with the cortical MTs have been observed but the
presence of this membrane system might be important in
creating the proper conditions (i.e. low [Ca^{2+}]) that permits
MTs to polymerize.

10.3   PREPROPHASE BAND OF MTS

In 1966 Pickett-Heaps and Northcote (1966) discovered a
specialized band of cortical MTs that appeared shortly before
the onset of mitosis, i.e. preprophase, in a spatial
localization that accurately predicted the site where the cell
plate would fuse with the parent wall during cytokinesis.
Called the preprophase band (PPB) of MTs, it constitutes an
extremely noteworthy but enigmatic specialization of the
cytoskeleton. The enigmatic feature is due to the fact that
the MTs of the PPB disappear early in mitosis, long before
cytokinesis. The PPB has been observed in numerous cell types
undergoing either symmetrical or even highly asymmetrical
division, and in each instance the band occurs at the site
where the cell plate joins the parent wall. Indeed the
precision is good enough to cause Gunning and co-workers
(Busby and Gunning 1980; Gunning 1982) to assert that the
fusing cell plate bisects the PPB site. Given the importance
to morphogenesis of the cell plate position, it should be
recognized that the PPB is at least a direct indication of
cytoplasmic polarity and it becomes important to know where
this structure comes from and what role it plays.

    The PPB forms through the accumulation of cortical MTs at
the presumptive division site. Using immunofluorescence
microscopy, Wick and Duniec (1983) are able to depict the
clumping of PPB MTs as the emergence of a bright fluorescent
equatorial band. At the same time the PPB arises fluorescence
also appears in association with the nucleus. The latter seems
to be tubulin that is not yet polymerized into discernible MTs
(Wick and Duniec 1983). The PPB may initially appear as a
double structure, especially in long cells, before joining to
form a single, more tightly defined band. At maturity the

Fig. 10.3   The preprophase band (PPB) of MTs in
staminal hair cells of *Gibasis* that have been fixed by
rapid freeze-fixation. Examination of cross sections
(Fig. 10.3a) reveals a high degree of cross bridging
between MTs and between MTs and the plasma membrane
(arrows). Observation of tangential sections (Fig.
10.3b) reveals that the MTs of the PPB are generally
more densely packed and closely aligned than the
cortical MTs of an elongating cell. Fig. 10.3a x96,000.
Fig. 10.3b x78,000. Bar = 0.1 µm.   (Fig. 10.3 from
Lancelle, Callaham and Hepler, unpublished results)

band may be composed of hundreds of MTs or it may contain a
relatively small number; the outermost are closely associated
with the plasmamembrane while those more interior may be
linked to one another (Fig. 10.3.a; 10.3.b). In recent
studies on staminal hair cells of *Gibasis* that have been
processed by freeze-substitution (Lancelle *et al.*, abstract
these meetings), a high degree of alignment and regularity of
spacing of the PPB MTs is observed (Fig. 10.3.b). They seem
much more tightly packed than normal cortical MTs, thus
supporting the contention that during formation of the PPB the

MTs become grouped together. Although not apparent in Fig.
10.3.b, recent studies on the PPB in staminal hair cells of
Tradescantia (Gunning and Hepler 1984) reveal the presence, in
some instances, of F-actin as indicated by positive rhodamine-
phalloidin staining.

As the cell progresses from preprophase to prophase the
PPB becomes more tightly constricted and some MTs become
associated with the nucleus. Schnepf (1984) is able to
identify an early transition in dividing leaflet cells of
Sphagnum in which MTs, apparently dissociated from the PPB,
are found within the interior cytoplasm but are still parallel
to the MTs of the PPB. These interior MTs soon disappear and
MTs become closely associated with the nucleus as part of the
forming MA; they are now oriented perpendicular to the PPB.
Immunofluorescence microscopy also reveals the decay of the
PPB concomitant with the emergence of nuclear-associated
spindle MTs (Wick and Duniec 1984).

The existence and location of the PPB is thus well
established; the primary question that remains is how it
functions (for review, Gunning 1982). Various ideas have been
put forth including ones that suggest that the band controls
nuclear migration or positioning, that it supplies tubules or
tubulin for the developing MA, that it represents a cortical
MTOC or region that attracts MTs, that it causes local wall
deposition (Galatis et al. 1982), and most recently, that it
blocks a "latent tendency" of the cell to form a cleavage
furrow (O'Brien 1983). Because the MTs of the band disappear
long before cytokinesis, we have long thought that the MTs
themselves did not determine PPB function (Hepler and Palevitz
1974; Palevitz and Hepler 1974a). It seems that the cortical
site may possess the ability to nucleate, attract or stabilize
MTs, and this property may explain why the centrifugally
expanding phragmoplast comes to fuse at this location.

There are a variety of studies indicating that the PPB
site has unique properties and further, that the site
expresses its function late in division, even towards the
terminal phases of cytokinesis. The uniqueness of the band
site as opposed to the phragmoplast, for example, as a
nucleating or stabilizing location is demonstrated by the
ability of CIPC, a potent disruptor of MT organization, to
markedly alter phragmoplast structure while having no effect
on the formation of the the PPB MTs (Clayton and Lloyd 1984).
That the site retains specialized properties long after the
MTs have disappeared is supported by recent studies in which
the PPB cortical location in stamen hair cells of *Tradescantia*
has been mechanically damaged with microneedles during mitosis
(Gunning and Hepler 1984). Disruption of the plasma membrane

and cortical cytoplasm at the PPB site itself either slows or
prevents the cell plate from fusing at the place of damage.
However, similar damage to the plasma membrane away from the
PPB does not alter cell plate growth or fusion. In these
studies and earlier ones by Ota (1961) the observation is made
that if the MA is physically displaced to one end of the cell
the growing cell plate curves towards the PPB site and
attempts, as it were, to divide in the correct plane. There
are limits however to the ability of the cell to perform
corrective realignment maneuvers. In guard cell complexes, for
example, Galatis *et al*. (1984) have shown that centrifuging
subsidiary cell nuclei long distances from their proper
division site can severely disrupt final placement of the
plate. There are also developmental anomalies that show up
from time to time in which, for example, the cell plate has
more than one PPB to respond to and again a variety of
abnormal cell plate fusions exist (Galatis, *et al*. 1983b).

Under normal circumstances though the PPB site appears to
effectively attract the growing edges of the phragmoplast and
cause the cell plate to be locked into correct alignment. The
studies by Palevitz and Hepler (1974a;b) on guard mother cell
divisions in *Allium* show that while the MA may be oriented
diagonally across the cell, during late anaphase or early
telophase the entire phragmoplast and associated nuclei rotate
into correct position to ensure longitudinal orientation of
the cell plate. Spindle-phragmoplast reorientation has been
observed in many different plant cell types and appears to be
a common process. That the PPB site or cortical cytoplasm is
involved is supported by the observation that nubs of wall
material may grow inward (centripetally) from this region and
further that the distal edges of the growing plate initially
curl towards the cortical region, only later pulling into
alignment the rest of the plate (Palevitz and Hepler 1974a).
The PPB site thus appears to have special properties related
to its apparent role in the attraction of and fusion with the
expanding cell plate. What gives the PPB site these
properties though remains a fascinating mystery.

10.4   THE MITOTIC APPARATUS (MA)

Whereas the MTs of the cell cortex and PPB were discovered
through electron microscopic observations, the fibrous nature
of the MA was firmly established much earlier by polarized
light microscopy. The pioneering work of Inoue and co-workers
(for review, Inoue 1964) showed in different living cells,
including notably two higher plants, pollen mother cells of
Lilium and the endosperm of *Haemanthus* that both the MA and
phragmoplast were composed of positively birefringent fibers.

Only later has it been shown that these birefringent fibers
are composed largely of MTs, indistinguishable in structure
from those of the cell cortex and PPB. Although there are
variations in detail, an MA composed of MTs is formed in all
eukaryotic cells, plant or animal, higher or lower and thus
constitutes, from an evolutionary view point, an extremely old
cytoskeleton. Despite its ubiquity and the attention it has
received we still do not know how the MTs work, although most
researchers would agree that they participate in chromosome
separation. The literature on this topic is vast and will not
be covered in this brief overview. There are many excellent
reviews including entire symposia devoted to the structure and
function of the MA (Little et al. 1977; Zimmerman and Forer
1981), and the reader is directed to these volumes for
additional information.

     In an effort to be brief, attention will be directed only
to certain aspects of the mitotic cytoskeleton in plants that
have emerged in recent years and provide fresh insight about
the workings of the MA. One recent noteworthy development has
been the determination of MT polarity in the MA of *Haemanthus*
Using to advantage the ability to decorate MTs with curved
sheets of tubulin protofilaments Euteneuer et al. (1982) have
been able to show that all MTs in a half spindle have the same
polarity; their plus (+) or growing ends reside at the mid
zone while the minus (−) or nongrowing ends are at the pole.
For kinetochore MTs this means that growing point is
presumably the point of attachment of the MT to the
chromosome. These results constitute important, basic factual
information about the structure of the MA and further they may
help us to sort out the function of MTs. These findings have
already provided evidence against one model of mitosis that
had been predicated on the concept of antipolarity between
kinetochore and non-kinetochore MTs (McIntosh *et al.* 1969).

     Another development in our quest to decipher the
structure of the mitotic cytoskeleton has been the realization
that endomembranes are common components of the MA (Hepler and
Wolniak 1984). These membranes, derived largely from the
endoplasmic reticulum, in some species form extensive and
specific associations with kinetochore MTs (Fig. 10.4).
Membranes also occur in abundance at the spindle poles and may
even surround the MA, creating for it a compartment that is
separate from the rest of the cytoplasm.

     Among the functions for these membranes, two in
particular seem likely: 1.) They may control the $[Ca^{2+}]$ in
the MA and thus regulate, for example, the
polymerization/depolymerization of spindle MTs, the activity
of a spindle dynein ATPase or the structure/function of a

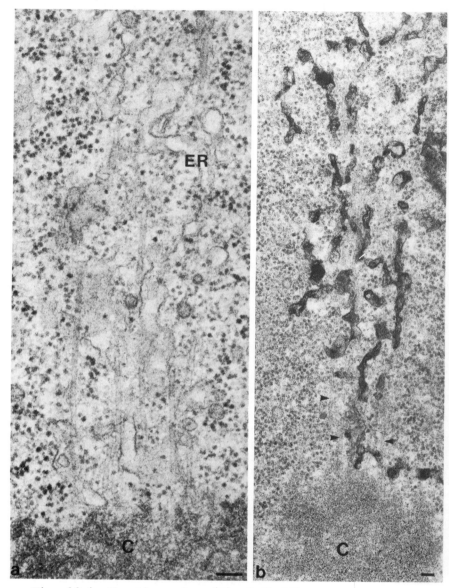

Fig. 10.4   The kinetochore(C) of dividing leaf epidermal
cells of *Hordeum*.   Gultaraldehyde-osmium fixation (Fig.
10.4a) depicts the kinetochore MTs and faintly shows the
ER. OsFeCN fixation (Fig. 10.4b), however, stains the ER
markedly and reveals the specificity of ER
interdigitation with kinetochore MTs (arrows). Fig 10.4a
x80,000. Fig. 10.4b x40,000.   Bar = 0.1μm.   (Fig. 10.4a
from Hepler, 1980, Fig. 10.4b from Hepler *et al*. 1981)

spindle-associated actomyosin system. 2.) Alternatively or in addition membranes might be structural components to which MTs can cross-bridge and develop shear force. For example, MTs might crawl along membranes and pull chromosomes to the poles (Hepler and Wolniak 1984) (Section 10.6).

10.5 THE PHRAGMOPLAST

The phragmoplast, the fibrous structure associated with cell plate formation, initially appears to be derived from remnants of the MA during late anaphase but soon grows to become a distinct birefringent structure (Inoue 1964). The birefringence is largely due to the aligned MTs that are oriented at right angles to the plane of the cell plate. Electron microscopic examination reveals that the phragmoplast is composed of two interdigitating sets of oppositely polarized MTs (Hepler and Jackson 1968) (Fig. 10.5). The (+) ends are embedded in a dense matrix of material and vesicles in the plane of the plate while (-) ends extend outward. (Euteneuer et al. 1982)

Like the MA, the phragmoplast has an extensive and closely associated membrane system (Hepler 1982) (Fig. 10.6). Some of these elements are discrete vesicles, derived from the dictyosomes, that flow in between the MTs to the mid plane and fuse to form the growing cell plate. There are also elements of the ER among the MTs and intertwined with the golgi vesicles (Fig. 10.6).

An exciting recent observation has been the demonstration by Clayton and Lloyd (1985) that the phragmoplast is enriched in F-actin. Using cells that had been doubly stained with fluorescein-labelled antitubulin and rhodamine-labelled phalloidin the authors show that only the phragmoplast (and not the cortical cytoskeleton, the PPB or the MA) becomes stained with rhodamine phalloidin, indicating the presence of actin MFs. On seeing these results one is reminded of the earlier studies on the control of cell plate orientation in guard mother cells of *Allium* (Palevitz and Hepler 1974b; Palevitz 1980), which show that inhibitors of actin MFs (cytochalasin-B and phalloidin) block the motile process of spindle-phragmoplast rotation. Cells treated with these inhibitors therefore do not align their cell plate correctly. Taken together these inhibitor and staining studies suggest that an actomyosin system drives phragmoplast rotation.

Because the phragmoplast arises from the MA and is, along with the MA, involved with cell division, there may be a tendency to consider these two cytoskeletal units as

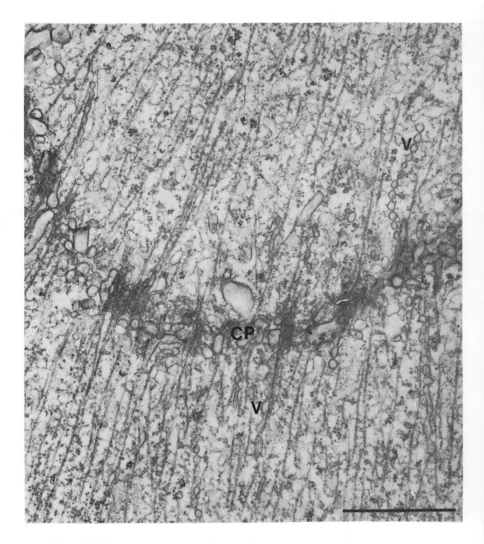

Fig. 10.5   The phragmoplast from an *Haemanthus*
endosperm cell prepared by glutaraldehyde-osmium
tetroxide fixation.   MTs accumulate in bundles to from
the phragmoplast.   Within individual bundles the MTs
overlap at the mid zone.   Chains of vesicles (V) can be
observed on either side of the plate (CP) and they
appear to be caught in the act of flowing into the mid
zone where they fuse to form the new membrane and cell
wall that separates daughter nuclei.   x35,000. Bar = 1
µm.   (From Hepler and Jackson, 1968)

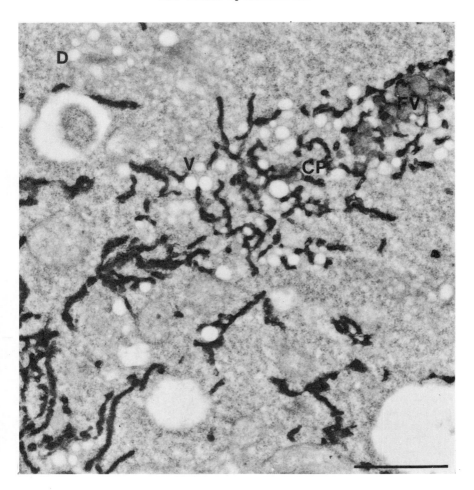

Fig. 10.6   The phragmoplast from a *Lactuca* root tip
cell that has been postfixed in OsFeCN. Once again the
OsFeCN staining causes the ER membranes to stand out
against other elements and reveals their intimate
association with the edges of the phragmoplast. MTs are
preserved by the fixation procedure but they are poorly
stained and cannot be revealed in this micrograph. From
comparative studies it is evident that the ER heavily
invades the region occupied by the phragmoplast MTs.
(From Hepler 1982)

manifestations of the same structure. The phragmoplast does
appear to arise from the MA and is composed of MTs, presumably
derived from the same tubulin subunits that made the MA.
Moreover there is a constant polarity in these MTs that is
maintained throughout mitosis and cytokinesis. In addition

both the MA and the phragmoplast are enriched for calmodulin (Wick et al. 1985).

However there are marked differences between these two cytoskeletons that deserve attention and emphasis. For example, the phragmoplast is unique in being composed of two interdigitating sets of oppositely polarized MTs (Euteneuer et al. 1982; Hepler and Jackson 1968). In addition it contains an abundance of F-actin (Clayton and Lloyd 1985), as just mentioned, and displays responses to drugs that have no effect on the structure or function of the MA (Palevitz 1982). Besides the sensitivity of phragmoplast rotation to cytochalasin and phalloidin, I draw attention to the well known inhibition of cytokinesis, but not mitosis, by methylxanthines, e.g. caffeine (Paul and Goff 1973). When treated with caffeine, dividing stamen hair cells of *Tradescantia* always initiate a cell plate and furthermore the plate grows normally until about 80% complete (Bonsignore and Hepler 1985). Thereafter, though, the plate gradually dissolves until no discernible structure remains and the cell is binucleate. In this instance the drug does not appear to be affecting F-actin since cell plate orientation and cytoplasmic streaming occur normally. Similarly MTs appear unaffected as evidenced by their normal appearance in the electron microscope and by the fact that chromosome motion proceeds without delay. In addition vesicle aggregation and fusion occur in the presence of the drug. Nevertheless the plate does not become fully stabilized or fused with the side walls and thus there may be some aspect of the cytoskeletal-membrane relationship that becomes altered by caffeine. It has long been thought that caffeine may impair the membrane-Ca^{2+} relationship and thus block fusion of vesicles (Paul and Goff 1973). However, our preliminary data show that fusion does occur (Bonsignore and Hepler 1985); much more structural work is needed to show which components, if any, are altered by caffeine.

A further special aspect of the phragmoplast that deserves attention is its ability, especially in vacuolate cells or cambial initials, to persist for many hours after the MA has decayed and the nuclei reverted to interphase (Venverloo et al. 1980). The edges of the phragmoplast retain their normal MT morphology and grow to their proper place of attachment along the parent wall. In this instance there is an additional cytoplasmic structure, the phragmosome, that also has MTs and may guide the phragmoplast to its destination (Goosen-de Roo *et al.* 1984). The ability of the phragmoplast to generate its own growth suggests a considerable degree of independence and autonomy from the MA. Further support for this idea may be derived from studies of certain cell types,

notably endosperm, that are able to form phragmoplasts and
cell plates in the absence of a preceding mitosis (Fineran et
al. 1982).

The phragmoplast is thus a separate and distinct
cytoskeletal structure. In certain respects it is more
tractable than the MA, and enterprising plant cell biologists
might find its study rewarding. The role of actin in the
other three general cytoskeletal structures is either unknown
or not suspected. In the phragmoplast evidence is now
beginning to show that F-actin is present and that, among a
possible variety of functions, it may participate in cell
plate orientation. Understanding this process in detail will
mark an important step forward in our attempts to elucidate
the structure and function of the cytoskeleton.

10.5.1 From Cell Division to Cell Growth

Upon completion of the cell plate the phragmoplast MTs break
down and cortical MTs reappear. There is considerable
evidence showing that the initiation of new cortical MTs
occurs in cell corners and along cell edges (Galatis et al.
1983a; Gunning et al. 1978; Palevitz 1982). Short segments of
MTs appear to radiate out from regions that contain clusters
of small vesicles or particles and an accumulation of finely
fibrillar matrix material. These regions are thought to be
MTOCs and it is interesting that the location which initially
gives rise to MTs is the cell edge formed by the fusion of the
cell plate with the parent wall (Gunning 1980; Gunning et al.
1978). Reflecting back to earlier stages of development, the
position along the parent wall where the cell plate fuses is
the PPB site. Here as in earlier discussions it is attractive
to imagine that this site, despite the lack of PPB MTs
themselves, retains the ability to attract or nucleate MTs.
In addition it should be pointed out that the centrifugally
growing cell plate also brings with it the capacity to
nucleate new MTs at its perimeter. Thus the edge formed by
the fusion of the cell plate and parent wall contains the
remnants of two MT organizing or stabilizing centers. Their
combined activity may control the initiation of new cortical
MTs.

10.6 MEMBRANES AND THE CYTOSKELETON

Throughout this essay repeated mention has been made of the
close structural relationship of MTs to membranes. Cortical
MTs and the outermost rank of PPB MTs are closely appressed to
the plasma membrane, and cross-bridges between the two are
evident (Fig. 10.3.a). MTs of the MA and phragmoplast also

have numerous associations with membranes; the interdigitation
of elements of smooth ER with kinetochore MTs in the MA is
especially dramatic (Hepler 1980). ER and golgi vesicles
likewise infiltrate among the phragmoplast MTs. While the
initial evidence concerning the structural interaction between
membranes and MTs has been gained from electron microscopic
studies, increasingly this fundamental idea is being supported
by studies using fluorescence microscopic localization of
tubulin antibodies (Lloyd 1984). Cortical MTs, for example,
appear to be so tightly linked to transmembrane proteins that
removal of the membrane lipids and soluble proteins with a
detergent (Triton X-100) does not destroy the cytoskeleton.
From theoretical grounds it is also attractive to postulate a
coupling of MTs to membranes. For example, if cortical MTs in
the cytoplasm control the oriented deposition of cellulose on
the outside surface of the plasma membrane then the MTs must
somehow extend their directional influence through the plasma
membrane.

    In an attempt to decipher the nature of membrane-MT
interactions it seems pertinent to consider some recent
exciting developments on this problem in studies of animal
cells. During the last few years there has been an explosive
interest in the membrane cytoskeleton with the realization
that certain key proteins, notably spectrin and ankyrin, which
heretofore had been found only in erythrocytes, occur widely
in virtually all cell types (Baines 1984; Lazarides and Nelson
1982; Mangeat and Burridge 1984). In the erythrocyte, ankyrin
forms a link between an integral membrane protein and
spectrin, while the latter binds to actin. Studies of
nonerythrocyte spectrin and ankyrin reveal in addition cross
reactivity between the high molecular weight MT-associated
protein MAP-2 and the α subunit of spectrin (Davis and
Bennett 1982), and between MAP-1 and ankyrin (Bennet and Davis
1981). These associations could provide the molecular basis
for a structural interaction between spectrin, ankyrin and
MTs. It may, however, be possible for MTs to join directly
with the membrane since evidence exists for the presence of
tubulin as an integral component of membranes (Stephens
1985). Together with the observation that actin MFs are also
associated with the spectrin/ankyrin complex as well as with
MTs one has ample evidence for a multiplicity of membrane-
cytoskeletal associations. The MT-associated proteins (MAPS)
would appear to play a central role in these linkage schemes.
It is not known whether spectrin and ankyrin occur in plants
but their general ocurrence in animal cells and the
fundamental nature of their associations with MTs and MFs,
elements which are present in plants, suggests that these
membrane proteins are ubiquitous.

If MTs interact with membranes, what is the significance
of this association to the function of the cytoskeleton in the
development of the plant cell? At the least the MT-membrane
associations would impart greater stability to the MTs. The
strength of the combined elements might be important in
permitting MTs to retain certain shapes during cellular
morphogenesis. Beyond this there may be an active, motile
component to the interaction that could participate in a
variety of processes from cellulose orientation to chromosome
movement. From their discovery it has been thought that MTs
are involved in generating motion. Their role in the beating
of cilia and flagella is undeniable; they are assumed to cause
motion in the MA. Recent studies on axonal transport reveal
that vesicles move along single MTs (Vale *et al*. 1985).
Perhaps the relative motion of membrane and MT is a common
factor underlying many basic aspects of MT function.

The idea that MAP cross-bridges might be mechanochemical
units that are capable of generating shearing forces in the
direction of the MT long axis seems attractive. The activity
of an oar-like cross-bridge could cause a MT to move relative
to a membrane surface, or alternatively the membrane, or
component therein, to move relative to the MT. How this
motile activity could account for the function of the
cytoskeleton in its different manifestations is given below.
In the cortical cytoplasm where the MTs appear to control
cellulose orientation it is suggested that the cross-bridge
action moves components in the membrane, i.e. integral
proteins, possibly even the cellulose synthetases themselves,
in the direction of the MT (for review, Heath and Seagull
1982). The shearing forces thus created by the directed flow
would align the growing cellulose microfibril. In the MA the
motion generated by the cross-bridges might cause a MT to
slide relative to the membrane (Hepler and Wolniak 1984). In
developing these ideas about the MA I have been influenced by
the observations from barley leaf cells and other cell types
of extensive arrays of ER that invade the spindle specifically
along kinetochore MTs (Hepler 1980). Here I suggest that the
membrane is anchored and the kinetochore MTs crawl along the
membrane, pulling their attached chromosome to the pole.
Kinetochore MT depolymerization would accompany this process,
but it is viewed as the rate limiting step rather than the
motor.

The final example considered is the phragmoplast, in
which it is known that vesicles are moved to the mid plane
where they fuse to form the cell plate. The phragmoplast
differs from the previous two cytoskeletons in consisting of
two overlapping sets of oppositely polarized MTs. The MTs are
anchored against themselves and through cross-bridge action

they could move vesicles to their mid plane. Note that the
types of motion in the MA and phragmoplast that have been
suggested are consistent with the known polarity of the MTs
(Euteneuer et al. 1982). In both examples it is implied that
the cross-bridge has a power stroke towards its (+) end. Thus
the kinetochore MTs, with oars that work towards the point of
chromosome attachment, would move the chromosome to the poles,
assuming that the membrane system against which the MTs work
is sufficiently anchored. In the phragmoplast it is assumed
that the MTs are anchored and thus the vesicles would be moved
towards the mid zone or (+) end. Understanding how membranes
and MTs interact may help explain a variety of cytoskeletal
functions that currently are unknown.

10.7 CONCLUSION: COMMENTS ON METHODS

The cytoskeleton occupies a central position in our thinking
about how cells control their division, growth and
differentiation. Whereas our attention initially has been
focussed largely on MTs, with the increasing realization that
the cytoskeleton is complex and includes, for example,
membranes and MFs besides MTs, we are forced to accommodate
this complexity in developing new models and experiments. The
importance of membranes to the structure and function of the
cytoskeleton, discussed in the previous section, is widely
acknowledged. Our understanding of the relationship of MFs to
the cytoskeleton, however, is much more fragmentary but
nonetheless deserves mention (Jackson 1982).

There is no question that MFs composed of actin are
common components of plant cells, and further it is
conclusively established that they participate in cytoplasmic
streaming. Here I draw attention to the possible association
of actin MFs with the four principal cytoskeletal structures.
Although there are relatively few reports, filaments which
react with heavy meromyosin have been observed in the cortical
cytoskeleton of radish root hair (Seagull and Heath 1979) and
in the MA of Haemanthus (Jackson 1982), and phalloidin-stained
material has been observed in the PPB (Gunning and Hepler
1984) and in the phragmoplast (Clayton and Lloyd 1985). While
the amounts of actin in the cortex may be small and
undetectable by some procedures the association nevertheless
may be crucial to the function of the cytoskeleton. It is
evident, for example, that the fine filaments (actin?) which
are observed to lie between MTs in the cortical cytoskeleton,
hold a precise and close structural relationship to the
neighboring MT (Fig. 10.1.c). Through MAPs, MTs and the
presumptive MFs may form stable linkages that are functionally
important to the operation of the cytoskeletal, MT-membrane

complex (Arakawa and Frieden 1984). It is imperative that more work be done on these filaments of the plant cytoskeleton. The cortical ones mentioned above require definitive identification as well as some means of probing their function, since the standard so called, anti-MF agents, such as cytochalasin B, have little or no effect on the operation of the cortical cytoskeleton or MA. There is, of course, the possibility that filaments other than actin are involved; if that is true then considerable ground work is needed to elucidate their identity, composition and function.

Studies on the structure of the cytoskeleton are benefiting from the introduction of new methods. Not only are we learning about filament systems which had heretofore escaped attention, e.g. actin in the phragmoplast, but we are also uncovering more information about the MTs themselves. The application of antibody localization procedures, in particular, has allowed us to obtain a global, pan-cellular, view of the cytoskeleton and to more readily decipher its changes and reorganization during development. For example the helical nature of the MT cytoskeleton (Lloyd 1983) would have been almost impossible to glean by conventional electron microscopy. Similarly the occurrence of double PPBs that eventually fuse into one (Wick and Duniec 1983) would have been extremely difficult to decipher by older methods. Given these impressive gains we can anticipate a surge in studies that will use these procedures to localize a variety of antigens besides tubulin.

Despite the elegance of these immunofluorescence localization studies a word of caution is in order regarding their interpretation. A quick perusal of the methods used to prepare cells reveals the application of relatively harsh procedures. Cells are commonly fixed in paraformaldehyde, then treated with a wall digesting enzyme together with detergents (Triton X-100). Thereafter the cells are stuck to a glass cover slip and treated with methanol, or air dried, before rehydration and staining with a fluorescent antibody (Wick et al. 1981; Clayton and Lloyd 1985). If the cells at this stage were embedded and examined in thin section by electron microscopy the images of the cytoplasm would be severely distorted and normal structures unrecognizasble or simply missing. It is perhaps a testimony to the strength of the cytoskeleton that at least some of it remains after these procedures. I fully recognize that these methods are useful, however, they are not good enough. If membranes are an important part of the cytoskeleton then any procedure that removes them, e.g. detergent treatment, is unacceptable. And although cytoskeletal MTs do remain after triton extraction the remnants may be severely disorganized (Van der Valk et al.

1980).

By making these arguments I do not intend to say that other techniques of cytoskeletal observation are without fault. Although glutaraldehyde fixation succeeded in preserving MTs most surely it has its limitation. For example, staminal hair cells of *Tradescantia* continue cytoplasmic streaming for 2-5 min. after immersion in the fixative. Regarding the cytoskeleton and especially motile processes associated therewith the slow fixation might induce structural modifications that do not reflect the active *in vivo* state.

The ways to solve these problems will not be easy. Ideally a method would be developed that would allow antibody localization together with structural analysis at either the light or electron microscopic level. For certain isolated small cells the use of rapid freeze fixation procedures will allow a great improvement over conventional chemical fixation in the preservation of detail. The possibility of coupling these methods to that of antibody localization should be explored. From the standpoint of membrane visualization, rapidly frozen material lends itself to freeze-fracture analysis, a method that will continue to prove important in our attempts to decipher the role of the cytoskeleton in membrane-associated processes.

A procedure that might provide impressive new insight into the dynamic aspects of the cytoskeleton is fluorescent analog cytochemistry (Taylor *et al.* 1984). In this instance one injects a fluorescently tagged protein, e.g. tubulin, actin, etc., into a living cell and examines its fate thereafter. It should be possible with an injected fluorescent tubulin analog to examine the structure and complete rearrangement of the MT cytoskeleton in a single living cell as it progresses through its cell cycle. If this technique is used together with fluorescence recovery after photobleaching (FRAP) then it may be possible to obtain important kinetic information about the incorporation of new protein subunits into the cytoskeleton.

The prospects for a bright future in plant cytoskeletal research are extremely good. There are many reasons to believe that the MT cytoskeleton participates in a variety of processes that are fundamental to development. Solving the mechanisms of cytoskeleton function will mark an important advancement in our understanding of the control of division, growth and differentiation.

10.8 SUMMARY

The cytoskeleton is intimately involved with the division, growth and differentiation of plant cells. It consists primarily of microtubules but also may contain microfilaments and membraneous components. The microtubular cytoskeleton in higher plants generally occurs in four different but interrelated locations: 1) the cell cortex, 2) the preprophase band, 3) the mitotic apparatus and 4) the phragmoplast. During the developmental cell cycle, these microtubular organelles arise at specific times and control a variety of processes including: 1) orientation of cellulose deposition, 2) the establishment of the division plane, 3) the movement of chromosomes and 4) the aggregation and fusion of cell plate vesicles.

How microtubules carry out their seemingly diverse functions is not known. One aspect of their structure that may be important in determining function is their association with membranes. Microtubules, sometimes along with microfilaments, have been observed to lie along and cross-bridge with the plasmamembrane, the nuclear envelope, the endoplasmic reticulum, and vesicular components. These associations may be important to the initial nucleation and polymerization of microtubules as well as to their subsequent function. The cross-bridges between microtubules and membranes may be mechanochemical units that are capable of generating shearing forces in the direction of the microtubule axis. The resultant motion either of the microtubules relative to the membrane or of the membrane, or component therein, relative to the microtubule may provide a common basis for understanding different processes including cellulose microfibril orientation and chromosome motion.

10.9 ACKNOWLEDGMENTS

I thank S. Lancelle and D. Callaham for permission to include some of their unpublished observations. S. Lancelle has been extremely helpful in the preparation of this manuscript. This work has been supported by grants from the National Institutes of Health (GM 25120) and The National Science Foundation (PCM 8402414).

10.10 REFERENCES

Arakawa, T. and Frieden, C. (1984). Interaction of microtubule-associated proteins with actin filaments. Studies using the fluorescence-photobleaching recovery technique. *J. Biol. Chem.* **259**, 11730-11734.

Baines, A. J. (1984). A spectrum of spectrins. *Nature* **312**, 310-311.

Busby, C. H. and Gunning, B. (1980). Observations on preprophase bands of microtubules in uniseriate hairs, stomatal complexes of sugar cane, and *Cyperus* root meristems. *Eur. J. Cell Biol.* **21**,214-223.

Bennett, V. and Davis, J. (1981). Erythrocyte ankyrin: immunoreactive analogues are associated with mitotic structures in cultured cells and with microtubules in brain. *Proc. Natl. Acad. Sci. U.S.A.* **78**, 7550-7554.

Bonsignore, C. and Hepler, P. K. (1985). Caffeine inhibition of cytokinesis: dynamics of cell plate formation-deformation in vivo. (submitted)

Clayton, L. and Lloyd, C. W. (1984). The relationship between the division plane and spindle geometry in *Allium* cells treated with CIPC and griseofulvin: an antitubulin study. *Eur. J. Cell Biol.* **34**, 248-253.

Clayton, L. and Lloyd, C. W. (1985). Actin organization during the cell cycle in meristematic plant cells. *Exp. Cell Res.* **156**, 231-238.

Davis, J. and Bennett, V. (1982). Microtubule-associated protein 2, a microtubule-associated protein from brain, is immunologically related to the subunit of erythrocyte spectrin. *J. Biol. Chem.* **257**, 5816-5820.

Emons, A. M. C. (1982). Microtubules do not control microfibril orientation in a helicoidal cell wall. *Protoplasma* **113**, 85-87.

Emons, A. M. C. and Wolters-Arts, A. M. C. (1983). Cortical microtubules and microfibril deposition in the cell wall of root hairs of *Equisetum hymenale.* *Protoplasma* **117**, 68-81.

Euteneuer, U., Jackson, W. T. and McIntosh, J. R. (1982). Polarity of spindle microtubules in *Haemanthus* endosperm. *J. Cell Biol.* **94**, 644-653.

Fineran, B. A., Wild, D. J. C. and Ingerfeld, M. (1982). Initial wall formation in the endosperm of wheat, *Triticum aestivum*: a reevaluation. *Can. J. Bot.* **60**, 1776–1795.

Galatis, B. (1980). Microtubules and guard-cell morphogenesis in *Zea mays* L. *J. Cell Sci.* **45**, 211–244.

Galatis, B., Apostolakos, P., and Katsaros, C. (1983a). Microtubules and their organizing centers in differentiating guard cells of *Adiantum capillus veneris*. *Protoplasma* **115**, 176–192.

Galatis, B., Apostolakos, P. and Katsaros, C. (1983b). Synchronous organization of two preprophase microtubule bands and final cell plate arrangement in subsidiary cell mother cells of some *Triticum* species. *Protoplasma* **117**, 24–39.

Galatis, B., Apostolakos, P., and Katsaros, C. (1984). Experimental studies on the formation of the cortical cytoplasmic zone of the preprophase microtubule band. *Protoplasma* **122**, 11–26.

Galatis, B., Apostolakos, P., Katsaros, C. and Loukare, H. (1982). Preprophase microtubule band and local wall thickening in guard cell mother cells of some Leguminosae. *Ann. Bot.* **50**, 779–791.

Goosen-de Roo, L., Bakhuizen, R., van Spronsen, R. C. and Libbenga, K. (1984). The presence of extended phragmosomes containing cytoskeletal elements in fusiform cambial cells of *Fraxinus excelsior* L. *Protoplasma* **122**, 145–152.

Goosen-de Roo, L. Burggraaf, P. O. and Libbenga, K. R. (1983). Microfilament bundles associated with tubular endoplasmic reticulum in fusiform cells in the active cambial zone of *Fraxinus excelsior* L. *Protoplasma* **116**, 204–208.

Gunning, B. E. S. (1980). Spatial and temporal regulation of nucleating sites for arrays of cortical microtubules in root tip cells of the water fern *Azolla pinnata*. *Eur. J. Cell Biol.* **23**, 53–65.

Gunning, B. E. S. (1982). The cytokinetic apparatus: its development and spatial regulation. In: *The Cytoskeleton in Plant Growth and Development*. (ed. C. W. Lloyd). pp. 229–292. Academic Press, London.

Gunning, B. E. S. and Hardham, A. R. (1982). Microtubules. *Ann. Rev. Plant Physiol.* **33**, 351–398.

Gunning, B. E. S., Hardham, A. R., and Hughes, J. E. (1978).
Evidence for initiation of microtubules in discrete regions
of the cell cortex in *Azolla* root-tip cells, and an hypothesis
on the development of cortical arrays of microtubules. *Planta*
143, 161-179.

Gunning, B. E. S. and Hepler, P. K. (1984). Investigations of
preprophase band sites. *Proc. 3rd. Int. Cong. Cell Biol.,
Tokyo, Japan.*

Hardham, A. R., Green, P. B. and Lang, J. M. (1980).
Reorganization of cortical microtubules in cellulose
deposition during leaf formation in *Graptopetalum
paraguayense*. *Planta* **149**, 181-195.

Heath, I. B. and Seagull, R. W. (1982). Oriented cellulose
fibrils and the cytoskeleton: a critical comparison of
models. In: *The Cytoskeleton in Plant Growth and
Development.* (ed. C. W. Lloyd). pp. 163-182. Academic Press,
London.

Hepler, P. K. (1980). Membranes in the mitotic apparatus of
barley cells. *J. Cell Biol.* **86**, 490-499.

Hepler, P. K. (1981a). Morphogenesis of tracheary elements
and guard cells. In: *Cytomorphogenesis in Plants.* Cell
Biol. Monographs. 0. Kiermayer. ed. Springer-Verlag,
Vienna. Vol. 3, pp. 327-347.

Hepler, P. K. (1981b). The structure of the endoplasmic
reticulum revealed osmium by tetroxide-potassium ferricyanide
staining. *Eur. J. Cell Biol.* **26**, 102-110.

Hepler, P. K. (1982). Endoplasmic reticulum in the formation
of the cell plate and plasmodesmata. *Protoplasma* **111**, 121-
133.

Hepler, P. K. and Jackson, W. T. (1968) Microtubules and
early stages of cell plate formation in the endosperm of
Haemanthus katherinae Baker. *J. Cell Biol.* **38**, 437-446.

Hepler, P. K. and Newcomb, E. H. (1964). Microtubules and
fibrils in the cytoplasm of *Coleus* cells undergoing secondary
wall deposition. *J. Cell Biol.* **20**, 529-533.

Hepler, P. K. and Palevitz, B. A. 1974. Microtubules and
microfilaments. *Ann. Rev. Plant Physiol.* **25**, 309-362.

Hepler, P. K., Wick, S. M., and Wolniak, S. M. 1981. The structure and role of membranes in the mitotic apparatus. In: *International Cell Biology 1980-1981*. (ed. H. G. Schweiger). pp 673-686. Springer-Verlag, Berlin.

Hepler, P. K. and Wolniak, S. M. 1984. Membranes in the mitotic apparatus: their structure and function. *Int. Rev. Cytol.* 90, 169-238.

Inoue, S. P. (1964). Organization and function of the mitotic spindle. In: *Primitive Motile Systems in Cell Biology*. (eds. R.D. Allen and N. Kamiya). pp. 549-598. Academic Press, New York.

Jackson, W. T. (1982). Actomyosin. In: *The Cytoskeleton in Plant Growth and Development*. (ed. C.W. Lloyd). pp. 3-29. Academic Press, London.

Lazarides, E. and Nelson, W. J. (1982). Expression of spectrin in nonerythroid cells. *Cell* 31, 505-508.

Ledbetter, M. C. and Porter, K. R. (1963). A "microtubule" in plant cell fine structure. *J. Cell Biol.* 19, 239-250.

Little, M., Paweletz, N., Petzelt, C., Ponstingl, H., Schroeter, D. and Zimmermann, H. P. (eds.) (1977). *Mitosis, Facts and Questions*. Springer-Verlag, Berlin.

Lloyd, C. W. (ed.) (1982). *The Cytoskeleton in Plant Growth and Development*. p. 457. Academic Press, New York.

Lloyd, C. W. (1983). Helical microtubular arrays in onion root hairs. *Nature* 305, 311-313.

Lloyd, C. W. (1984). Toward a dynamic helical model for the influence of microtubules on wall patterns in plants. *Int. Rev. Cytol.* 86, 1-51.

Lloyd, C. W., Slabas, A. R., Powell, A. J. and Lowe, S. B. (1980). Microtubules, protoplasts and plant cell shape. An immunofluorescent study. *Planta* 147, 500-506.

Mangeat, P. and Burridge, K. (1984). Actin-membrane interaction in fibroblasts: what proteins are involved in this association? *J. Cell Biol.* 99, 95s-103s.

McIntosh, J. R., Hepler, P. K., and Van Wie, D. G. (1969). Model for mitosis. *Nature* 224, 659-663.

O'Brien, T. P. (1983). The preprophase band of microtubules: does it block cleavage? *Cytobios* **37**, 101–105.

Ota, T. (1961). The role of cytoplasm in cytokinesis of plant cells. *Cytologia* **26**, 428–447.

Palevitz, B. A. (1980). Comparative effects of phalloidin and cytochalasin B on motility and morphogensis in *Allium*. *Can. J. Bot*. **58**, 773–785.

Palevitz, B. A. (1982). The stomatal complex as a model of cytoskeletal participation in cell differentiation. In: *The Cytoskeleton in Plant Growth and Development*. (ed. C.W. Lloyd). pp. 345–376. Academic Press, London.

Palevitz, B. A. and Hepler, P. K. (1974a). The control of the plane of division during stomatal differentiation in *Allium* I. Spindle reorientation. *Chromosoma* **46**, 297–326.

Palevitz, B. A. and Hepler, P. K. (1974b). The control of the plane of division during stomatal differentiation in *Allium* II. Drug Studies. *Chromosoma* **46**, 327–341.

Paul, D. C. and Goff, C. W. (1973). Comparative effects of caffeine, its analogues and calcium deficiency on cytokinesis. *Exp. Cell Res*. **78**, 399–413.

Pesacreta, T. C. and Parthasarathy, M. V. (1984). Microfilament bundles in the roots of a conifer, *Chamaecyparis obtusa*. *Protoplasma* **121**, 54–64.

Pickett-Heaps, J. D. and Northcote, D. H. (1966). Organization of microtubules and endoplasmic reticulum during mitosis and cytokinesis in wheat meristems. *J. Cell Sci*. **1**, 109–120.

Robinson, D. G. and Quader, H. (1982). The microtubule – microfibril syndrome. In: *The Cytoskeleton in Plant Growth and Development*. (ed. C.W. Lloyd). pp. 109–126. Academic Press, London.

Seagull, R. W. (1983). The role of the cytoskeleton during oriented microfibril deposition. I. Elucidation of the possible interaction between microtubules and cellulose synthetic complexes. *J. Ultrastruct. Res*. **83**, 168–175.

Seagull, R. W. and Heath, I. B. (1979). The effects of tannic acid on the *in vivo* preservation of microfilaments. *Eur. J. Cell Biol*. **20**, 184–188.

Schnepf, E. (1984). Pre- and postmitotic reorientation of microtubule arrays in young *Sphagnum* leaflets: transitional stages and initiation sites. *Protoplasma* **120**, 100-112.

Stephens, R. E. (1985). Evidence for a tubulin-containing lipid-protein structural complex in ciliary membranes. *J. Cell Biol.* **100**, 1082-1090.

Taylor, D. L., Amato, P. A. Luby-Phelps, K. and McNeil, P. (1984). Fluorescent analog cytochemistry. *TIBS* (March) 88-91.

Traas, J. A. (1984). Visualization of the membrane bound cytoskeleton and coated pits of plant cells by means of dry cleaving. *Protoplasma* **119**, 212-218.

Traas, J. A., Braat, P. and Derksen, J. W. (1984). Changes in microtubule arrays during the differentiation of cortical root cells of *Raphanus sativus*. *Eur. J. Cell Biol.* **34**, 229-238.

Tiwari, S. C., Wick, S. M., Williamson, R. E. and Gunning, B. E. C. (1984). Cytoskeleton and integration of cellular function in cells of higher plants. *J. Cell Biol.* **99**, 63s-69s.

Vale, R. D., Schnapp, B. J., Reese, T. S. and Sheetz, M. P. (1985). Organelle, bead, and microtubule translocations promoted by soluble factors from the squid giant axon. *Cell* **40**, 559-569.

Van der Valk, P., Rennie, P. J., Connolly, J. A., and Fowke, L. C. (1980). Distribution of cortical microtubules in tobacco protoplasts. An immunofluorescence microscopic and ultrastructural study. *Protoplasma* **105**, 27-43.

Venverloo, C. J., Hovenkamp, P. H., Wieda, A. J. and Libbenga, K. R. (1980). Cell division in *Nautilocalyx* explants. I. Phragmosome, preprophase band and plane of division. *Z. Pflanzenphysiol.* **100**, 161-174.

Wick, S. M. and Duniec, J. (1983). Immunofluorescence microscopy of tubulin and microtubule arrays in plant cells. I. Preprophase band development and concomitant appearance of nuclear envelope-associated tubulin. *J. Cell Biol.* **97**, 235-243.

Wick, S. M. and Duniec, J. (1984). Immunofluorescence
microscopy of tubulin and microtubule arrays in plant cells.
II. Transition between the preprophase band and the mitotic
spindle. *Protoplasma* <u>122</u>, 45-55.

Wick, S. M., Muto, S. and Duniec, J. (1985). Double
immunofluorescence labelling of calmodulin and tubulin in
dividing plant cells. (submitted).

Wick, S. M., Seagull, R. W., Osborn, M., Weber, K. and
Gunning, B. E. S. (1981). Immunofluorescence microscopy of
organized microtubule arrays in structurally stabilized
meristematic plant cells. *J. Cell Biol.* <u>89</u>, 685-690.

Zimmerman, A. M. and Forer, A. (eds.) (1981).
Mitosis/Cytokinesis. p 479. Academic Press, New York.

11

Microtubule rearrangement during plant cell growth and development: an immunofluorescence study

K. Roberts, J. Burgess, I. Roberts, and P. Linstead

11.1 INTRODUCTION

The sequence of structural events in plant cell growth and
development is generally pieced together from conventional
static electron micrographs, and usually from cells in intact
plants or tissues. One of the best examples is the development
of primary xylem elements or tracheids, reviewed comprehensively
by O'Brien (1981). The advantages of being able to study such
events in a more defined system are great, and the development
of the *Zinnia* mesophyll system (Kohlenbach and Schmidt 1975),
in which single cells in sterile culture can be induced to redifferentiate into tracheary elements, provides a general model
in which to combine biochemical studies with structural data
(Fukuda and Komamine 1980a, 1980b; Burgess and Linstead 1984a,
1984b). However the role of MTs in predicting the orientation
of cellulose microfibril deposition and in anticipating the
pattern of secondary wall formation during xylogenesis remains
controversial (Emons 1982; Robinson and Quadar 1982; Schnepf
et al. 1978). Although the localization of the cytoskeletal
elements in animal cells has been possible for some while the
presence of a cell wall has made such studies in plants rather
difficult until recently. In this study the technique of whole
cell indirect immunofluorescence is used to examine the MT
arrays that form in *Zinnia* cells, and to correlate their rearrangements with the timing of polysaccharide deposition in
cell wall thickenings and their subsequent lignification. Other
microtubule rearrangements, in particular those induced by
ethylene, are also presented and the data as a whole are
discussed.

11.2 MATERIALS AND METHODS

11.2.1 Plant material

Sterile mesophyll cells of *Zinnia elegans* var. Envy (Sutton
Seeds Ltd., Torquay, U.K.) were prepared as described by .
Burgess and Linstead (1984a). Cells in shaken liquid culture
were incubated in the tracheary element induction medium of
Fukuda and Komamine (1980a) under conditions as described by
Burgess and Linstead (1984a).

11.2.2 Immunofluorescence

To reveal microtubule arrays in fixed cells the following
method was used. Cells in suspension were fixed by adding to
them an equal volume of freshly prepared paraformaldehyde (3.7%)
in a microtubule stabilizing buffer (50mM PIPES pH 6.9, 5mM
$MgSO_4$, 5mM EGTA). After 10 minutes at room temperature the
cells were resuspended in complete fixative for a further 2
hours. Cells were washed 3 times (10 minutes each) in TRIS
buffered saline (TBS; 10mM TRIS pH 7.4, 0.9% sodium chloride).
Cells were then incubated in β-glucuronidase H2S (Sigma, St.
Louis, Mo.) at a concentration of 750 Fishman units/ml in 50mM
citrate buffer, pH 5.0, at room temperature. Experiments with
different incubation periods suggested that the optimum time
for permeabilizing the cell wall was 2 hours for freshly
isolated cells and 30 minutes for cultures older than four days.
Cells were then washed 3 times in TBS and air dried onto freshly
prepared acetone cleaned coverslips. As soon as they were dry,
they were placed in 0.5% Nonidet P40 in TBS for 10 minutes and
then washed in TBS. They were then incubated for 1 hour at
37°C in a rat monoclonal antibody to yeast α-tubulin (generous
gift of J. Kilmartin, MRC Laboratory of Molecular Biology,
Cambridge, now available commercially from Sera Laboratories
Ltd., Crawley Down, Sussex, U.K.). The antibody (Kilmartin
et al. 1982) was diluted 1:200 in TBS. Cells were given 3
washes in TBS of 30 minutes each before being incubated at 37°C
for 40 minutes in fluorescein isothiocyanate (FITC)-conjugated
goat anti-rat IgG diluted 1:100 in TBS (Miles Scientific, Stoke
Poges, Slough, U.K.). Cells were washed in TBS, then in 0.5%
Nonidet P40 in TBS for 10 minutes and then in TBS again for 2
more washes before mounting in a glycerol mountant containing a
non-fade additive (Chemistry Department, City University,
London, U.K.).

For simultaneous fluorescence detection of cell wall
thickenings, cells were processed as above, but before the final
washing steps they were placed in 0.01% Calcofluor White ST
(Cyanamid Corporation, U.S.) in TBS for 1 minute, washed briefly

and mounted in 50% glycerine in 10mM TRIS pH 9.0 (the non-fade
additive of the usual mountant interferes with the Calcofluor
fluorescence). For the demonstration of chromosomes and nuclei,
cells were processed as above but before mounting were placed
in 1 μg/ml DAPI (4',6-Diamidino-2-phenyl indole dihydrochloride,
Sigma, St. Louis, Mo, U.S.A.) in TBS for 1 minute, washed in
TBS and mounted in the non fade mountant. Cells were stained
for lignin in 1% phloroglucinol in the presence of 6N HCl.

11.2.3 Microscopy

A Zeiss Universal II microscope was used for epifluorescence
observations. Selective FITC fluorescence was observed using
Zeiss filter set 10 giving selective green emission at 520-560nm.
For Calcofluor White and DAPI fluorescence a new Zeiss UV filter
combination was used (exciter filter 365-376nm, beam splitter
425nm and barrier filter 450-490nm) which gives specific intense
blue fluorescence for either cell wall thickenings or for
nuclei. For FITC fluorescence the following objectives were
used: Zeiss Plan Neofluor 25/0.8 Ph, Planapochromat 40/1.0 Ph
and Planapochromat 63/1.4 Ph. The latter two do not transmit
UV light and so for UV induced fluorescence the Plan Neofluor
25/0.8 and the Plan Neofluor 63/1.25 were used. Photographs
were taken on Ilford FP4 film with typical exposures of 30
seconds for microtubule images and 4-8 seconds for the UV induced fluorescent images. Scanning electron micrographs were
taken of unfixed material prepared using a Hexland Sputter-Cryo
System on a CamScan Series 4 SEM.

11.2.4 Ethylene induced microtubule rearrangements

The methods for staining MT arrays in outer cortical and epidermal cells of pea epicotyl and mung bean hypocotyl, together
with the protocol for ethylene treatments are described in
Roberts *et al.* (1985).

11.3 RESULTS

11.3.1 Microtubule rearrangements during the mesophyll cell to tracheary element transition

Most of the spongy mesophyll cells in the leaf are destroyed in
the grinding process and palisade mesophyll cells constitute the
vast majority of those which eventually divide and differentiate
(Figs. 11.1, 11.2).

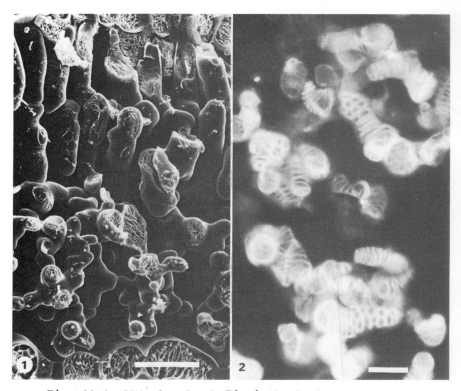

Fig. 11.1 SEM of a fresh *Zinnia* leaf, frozen and
sputter coated with gold. The upper and lower epidermis are just included and the palisade and spongy
mesophyll cells can be seen. The palisade cells,
when cultured *in vitro*, redifferentiate into tracheary
elements whose secondary cell wall patterns (Fig. 11.2)
can be seen by Calcofluor fluorescence. (bar = 40 µm).

The technique of whole cell indirect immunofluorescence renders
visible the entire MT arrays of these cells, from the time of
cell isolation to the final stages of cell death during
tracheary element formation. Freshly isolated cells can show a
variety of patterns of MTs. Some have a cortical meshwork of
MTs related to the positions of underlying chloroplasts, which
appear to act as exclusion zones, while others show a more
ordered, often helical array. As the cells are cultured several
events happen prior to xylogenesis. The majority of the cells
divide, and a period of cell enlargement follows (from 3-4 days
of culture). MT rearrangements also take place, and show two
main identifiable configurations. One, in which random mesh-
works of MT are seen (now unrelated to underlying chloroplasts)
is found mainly in isodiametric and enlarging cells (Figs.11.3-
11.5). MTs are long (ends are hard to detect) and they cross

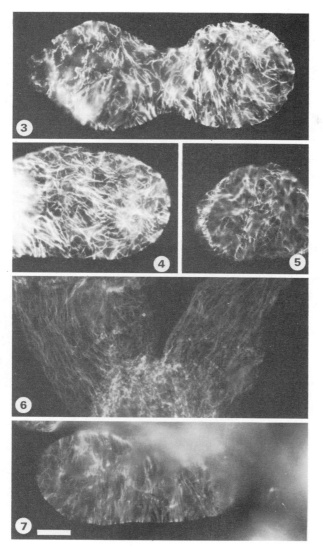

Figs. 11.3–11.7 MT arrays found in 7 day cultures
of *Zinnia* mesophyll cells are revealed by immunofluorescence. Meshwork-like arrays are seen in
Figs. 11.3–11.5, net longitudinal arrays in Fig.
11.6 and a net transverse array in Fig. 11.7.
(bar = 10 µm).

each other in a meshwork-like array. The other configuration,
found mostly, but not exclusively, in older and elongated cells,
is a polarized array which in general is parallel to one or

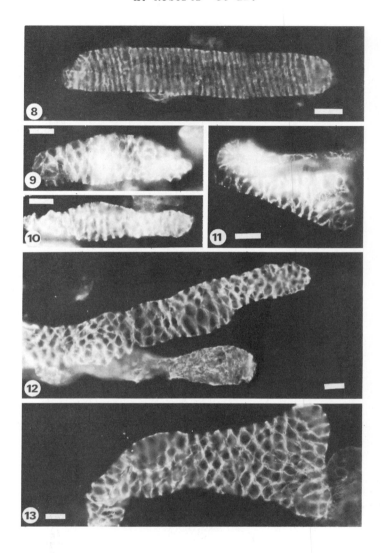

Figs. 11.8-11.13 MT arrangements in early tracheid
development in *Zinnia* are seen by immunofluorescence.
Patterns can be relatively simple (11.9-11.11), or
more reticulate (11.12, 11.13). The simple helical
arrangement of MTs found in normal tracheid development (in this case from pea epicotyl) is shown in
Fig. 11.8. By focussing only on the top of the cell,
one side of the helical array can be seen in this
early tracheid from a pea hypocotyl. Such regular
arrays are not found in the *Zinnia* system.
(bar = 10 μm).

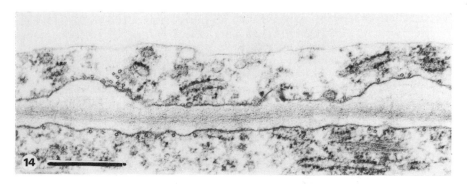

Fig. 11.14   The wall between two cells in the xylem
of a young *Zinnia* leaf.   Cell wall thickenings
associated with localized groups of MTs are seen in
this electron micrograph, which also shows the early
grouping of MTs in the lower cell prior to any wall
thickening.   (bar = 0.5 μm).

other of the cells axes.   This corresponds to the net transverse and net longitudinal helical arrays described in cortical
cells (Roberts *et al.* 1985), with the majority being net transverse and probably reflecting an underlying helical construc-
tion (see Figs. 11.6-11.7).   All those cells which eventually
become tracheids start with a net transverse helical MT array.

Those cells which are committed to become tracheary elements
undergo a final rearrangement of this transverse array.   Groups
of adjacent MTs in the array appear to aggregate laterally into
bundles or bands, leaving clear gaps between them (Figs. 11.8-
11.13).   This pattern of grouped microtubules anticipates the
subsequent pattern of cell wall thickening and correspond to
the groups of 5-15 MTs found in thin sectioning studies of xylogenesis in *Zinnia* (Fig. 11.14).   Several patterns of wall
thickening can be distinguished, each matched by corresponding
MT arrangements.   These range from a simple pattern roughly
corresponding to the underlying helical MTs (Figs. 11.8 and
11.10), to a more interconnected or reticulate pattern (Figs.
11.12 and 11.13) both of which have been described in primary
xylem as spiral and reticulate respectively (Bierhorst and
Zamora, 1965).   The proportion of each type in any one experiment is variable but in general terms smaller cells tend to
have a simple pattern of thickening while the large and irregular cells tend to show highly reticulate patterning.   Annular,
scalariform, or pitted thickenings are never encountered in our
material.   This may be related to the demonstration *in vivo*
that the proportion of spiral and pitted elements, for example,

is very dependent on exposure to visible light (Goodwin, 1942).
Examination of living cells in the final stages of MT rearrangement reveals extensive cytoplasmic streaming. Both in phase
contrast and in Nomarski optics, organelles move in streams
largely corresponding to the phase dense bands in the cell
cortex, which appear to correspond to the bands of MTs which
precede the cell wall thickenings. Occasionally organelles may
cross over from one band to another.

11.3.2   Microtubule rearrangements during cell division

By whole cell indirect immunofluorescence for MTs, coupled with
DAPI staining of nuclear DNA, the arrangement of microtubules
can be correlated with the morphological state of the chromosomes throughout the cell division cycle. By carefully corre-
lating the MT arrays with the onset of chromosome condensation
(DAPI double staining) we have demonstrated that a 'preprophase'
band of MTs is present. It is broader and more diffuse than
those found in root tip cells and also cortical MTs are not
entirely absent, several being found radiating from the nuclear
periphery to the cell cortex (Figs. 11.15-11.17). Of the
several hundred cells in which we have seen a 'preprophase' band
of MTs all also show clear signs of chromosome condensation
(Figs. 11.18-11.21). Conversely, all cells which show clear
chromosome condensation, also show evidence of the presence of
a 'preprophase' band of MTs. In other words, the beginning of
chromosome condensation and the appearance of the 'preprophase'
band appear to be synchronous events. Some cells eventually
divide, not across their shorter diameter but along their longer
axis, producing the side-by-side pairs (Fig. 11.11) described
by Fukuda and Komamine (1980b).

11.3.3   The sequence of events in tracheary element formation

In order to investigate the relative timing of the cytoskeletal
rearrangements and of the deposition of the tracheid wall
thickenings, cells were examined with indirect immunofluorescence for MTs, Calcofluor White (Roberts, 1985) to locate new
cell wall deposition, and a combination of autofluorescence,
safranin and phloroglucinol/HCl staining to detect lignin deposition. An unambiguous sequence of events emerged from these
observations. Cells may contain MTs in the tracheid banded
array, but present no evidence of wall thickening (Figs. 11.22-
11.23). Such cells after plasmolysis show no detectable wall
thickenings either by Calcofluor White staining, polarized light
microscopy, phase contrast or Nomarski optics. By contrast, all
cells which show early stages of ordered wall deposition, also
have well developed banded MT arrays (Figs. 11.26-11.27). In

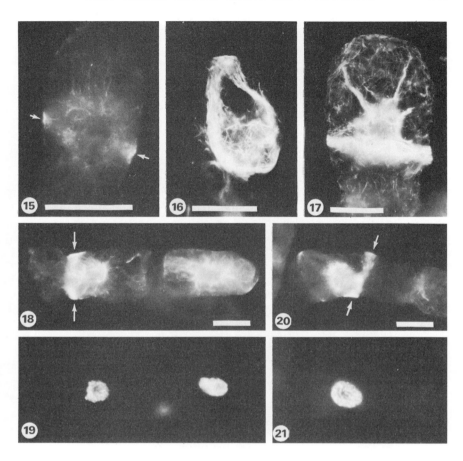

Figs. 11.15-11.21   The 'preprophase' bands of MTs
in dividing *Zinnia* mesophyll cells.   Other cortical
microtubules are still present (11.17) and MTs
radiating from the nuclear surface are seen (11.15,
11.17, 11.18).   Sometimes during squashing the entire
band of MTs becomes detached from the cell (11.16).
The same cells are shown in Figs. 11.18 and 11.19,
stained for MTs and with DAPI to show chromosome
behaviour.   Another cell is shown in Figs. 11.20
and 11.21.   (bar = 10 μm).

cells in which considerable thickening has taken place the MT
bundles eventually disappear (Figs. 11.24-11.25).   Lignin deposition, as judged by the appearance of autofluorescence in the
wall thickenings, does not occur until some time after maximum
Calcofluor fluorescence (Figs. 11.28-11.29), at a stage when
protoplast breakdown is well advanced and MTs have disappeared

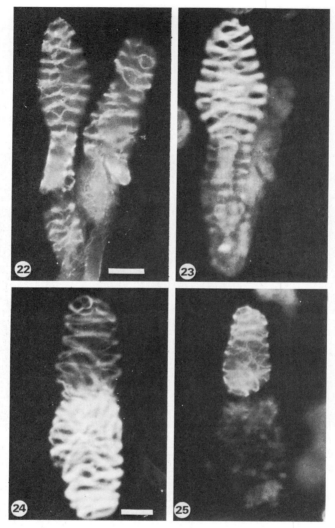

Figs. 11.22-11.25 Correlation of MT arrangements seen
by immunofluorescence, and cell wall deposition seen
by Calcofluor fluorescence. In 11.22 two cells show
the bundled MT arrangement typical of early tracheid
formation. However in 11.23, only one shows evidence
of Calcofluor staining indicating that MT rearrangement must precede new cell wall deposition. Once wall
thickening has reached a certain stage, lignin autofluorescence can be found (data not shown) and all MTs
disappear in the course of cell death (Fig. 11.25,
lower cell). The upper cell is at an earlier stage
of wall thickening. (bar = 10 μm).

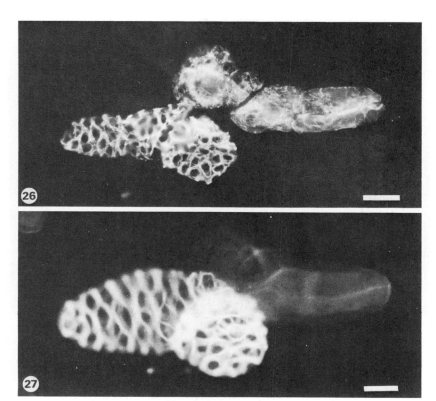

Figs. 11.26,11.27 Cell wall thickenings seen by
Calcofluor fluorescence (11.27) correlate well with
the bundled bands of MTs seen by immunofluorescence
in the same cell (11.26). Compare this with the
amorphous wall texture in the adjacent cells.
(bar = 10 μm).

(Figs. 11.24-11.25). As lignin deposition progresses in the
thickenings, and the intervening wall is degraded, the Calcofluor fluorescence weakens. This result from the partial removal of cell wall polysaccharides or by interference caused by
lignification. These results correspond well with the ultrastructural interpretation of events described by Burgess and
Linstead (1984a, 1984b).

11.3.4 Ethylene-induced microtubule reorientations

Microtubule reorientations also occur in non differentiating
cells and in that context it has been instructive to compare the

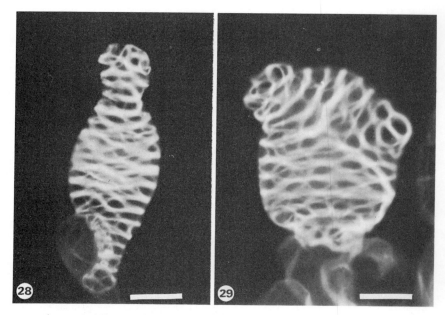

Figs.11.28, 11.29   Two *Zinnia* cells in a late
stage of xylogenesis showing the elaborate patterns
of cell wall thickening revealed by Calcofluor
fluorescence. (bar = 10 μm).

results we have obtained on *Zinnia* with those on ethylene induced reorientations of MTs in pea epicotyl cells. The complete
results have been presented elsewhere (Roberts *et al.* 1985) but
briefly it has been found that the cytoskeleton in cortical and
epidermal cells can be seen as a multistart helix of MTs of
variable pitch. In untreated cells the majority of cells have
a tightly compressed helix and the MTs are net transverse with
respect to the cell axis (Fig. 11.30). In ethylene treated
cells, however, the whole helical MT array in most cells changes
pitch rapidly, passing through an oblique array (Fig. 11.31)
and in some cells ending up in a net longitudinal array or
steeply pitched helix (Fig. 11.32). A similar effect can be induced by high osmolality.

11.4 DISCUSSION

11.4.1 Microtubules and the sequence of events in xylogenesis

The *Zinnia* system of Kohlenbach and Schmidt (1975) has recently
been elaborated into a well characterized model system in which
to study plant cytodifferentiation (Fukuda and Komamine, 1980a,

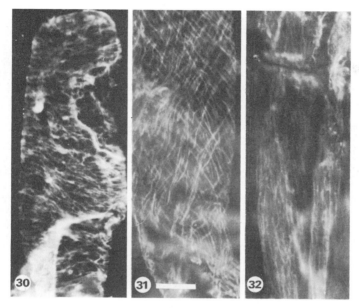

Figs.11.30-11.32   The effect of ethylene on the
MT array of epidermal cells of pea epicotyl
(Roberts *et al*. 1985). The array in controls is
net transverse (30), but following ethylene treatment oblique arrays are seen (31) followed by net
longitudinal arrays (32).   (bar = 10 μm).

1980b, 1982; Burgess and Linstead, 1984a, 1984b; Falconer and
Seagull, 1985a, 1985b) and the application of whole cell immunofluorescence to it has enlarged our appreciation of the complex
cycle of events involved as the technique allows the MT array
to be seen in its entirety and, by through-focussing, the 3-D
arrangement can be deduced directly. It had long been known
that cytoplasmic events precede wall thickening (Barkley, 1927;
Sinnot and Bloch, 1945) and the present results confirm the
conclusion, drawn from earlier EM studies of *in vivo* tracheary
element formation, that cortical MTs rearrange themselves to
form a series of bands which anticipate the characteristic cell
wall thickenings (Wooding and Northcote, 1964; Pickett-Heaps,
1966; Cronshaw and Bouck, 1965; Hepler and Newcomb, 1964).
The presence of many clear cases where MT rearrangements had
taken place but in which there was no detectable Calcofluor
fluorescence, (Figs. 11.22,11.23) and the absence of examples
to the contrary, show that the reorganization of MTs must occur
before the cell wall thickening begins. In other words the new
MT patterns predict where the cell wall thickenings will take
place. By double staining clear congruity of the two patterns

is found. This is in agreement with the recent results of
Falconer and Seagull (1985a), also using the *Zinnia* system. It
is certain that complex factors must operate in deciding the
final position of bundled MT bands, for in cases where two
daughter cells differentiate the wall thickenings in the two
cells are aligned in much the same manner as in Sinnott's
classic description (1945).

The timing of cellulose and lignin deposition has been less well
defined up till now. Several studies have suggested that lignin
formation is independent of differentiation and that old cell
cultures deposit lignin in the absence of tracheid formation
(Haddon and Northcote, 1976; Minocha and Halperin, 1976).
There is general agreement from biochemical studies that lignin
deposition occurs after the cellulose wall thickenings have been
completed (O'Brien, 1981) and indeed, in the *Zinnia* system
lignin synthesis is not detectable biochemically until 5 hours
after wall thickenings are observed in the culture (Fukuda and
Komamine, 1982). We have confirmed these conclusions by
observations on individual cells. Lignin can be detected by
safranin staining, phloroglucinol/HCl, and by its autofluorescence (in the red part of the spectrum). All three methods
identify the same population of cells. Phloroglucinol/HCl
cannot be used in the double fluorescence studies but confirmed
that autofluorescence is a sound measure of lignin deposition.
MT rearrangements precede cellulose deposition and lignin deposition follows some time later, usually after Calcofluor
fluorescence has reached its maximum and well after the MTs
have disappeared. This latter observation supports the idea
that the localization of lignin is independent of cytoskeletal
control. Hydroxyproline-rich protein is preferentially laid
down in wall thickenings in tissue cultures (Roberts and
Northcote, 1972) which lignify with age, and as lignin is
covalently linked to such proteins (Whitmore, 1978, 1982) by
peroxidase which is a wall bound marker of xylogenesis
(Fukuda and Komamine, 1982), it is probable that such wall
proteins may be involved in the localization of lignin.

Calcofluor White is not specific for cellulose, binding to
xyloglucan backbones, chitin, and mixed β1-3, β1-4 glucans in
addition (Stone, 1984), but it seems safe to assume that at
least some of the fluorescence seen in new cell wall thickenings
is due to nascent cellulose formation. Within the framework of
this assumption an unambiguous sequence of events has been
demonstrated: a) MT rearrangements, b) new cell wall polysaccharide deposition in thickenings, c) MT removal, d) lignin
deposition in thickenings.

Our demonstration of the presence of a 'preprophase' band of MTs
in isolated *Zinnia* mesophyll cells, which in the normal course

of events would not have divided again, is in agreement with
similar observations of Falconer and Seagull (1985a). These
are the only reports of the band in single dividing isolated
cells and may represent a far more universal occurrence than
has yet been appreciated (Gunning, 1982; Hepler and Palevitz,
1974). It is of note that if prophase is considered (Mazia
1961) to commence with the first visible signs of chromosome
condensation, and to end with the start of prometaphase i.e.
at nuclear envelope breakdown, then the appearance of the
'preprophase' band in *Zinnia* corresponds exactly with this
period, and suggests that the name 'prophase' band, in this
case, is far more appropriate (see also Burgess, 1969).

11.4.2 Reorientation of MT arrays

The microtubule system in the isolated mesophyll cell progresses through several distinct arrays in the course of
tracheary element formation in cortical cells and following
ethylene treatment a rapid transition between different helical
MT arrays is seen. In principal there are two ways that one
MT array can succeed another. The first, seen for example in
the transition between cortical arrays and the kinetochore
fibers in the mitotic spindle, is thought to involve a cycle of
depolymerization and repolymerization. The second is that the
entire MT array may bodily rearrange itself into the new configuration. As plant cortical MTs are cross-linked to plasma
membrane components this implies that such movements must be
accompanied either by the lateral movement in the plane of the
membrane of integral membrane proteins involved in MT attachment, or by the uncoupling of intact MTs from the membrane.
The MTs in the usual cortical arrays are probably too far away
from each other for direct MT/MT coupling, and it is far from
clear what factors are responsible for stabilizing the MTs in
any particular array. The forces responsible for changes in
orientation are equally obscure, either in xylogenesis or in
mediating the effect of ethylene. But in both cases our
evidence suggests that entire MT arrays may reorient without
intervening phases of depolymerization and repolymerization.
Recent evidence on the location of microtubule organising
centres (MTOC) in plants (Dickinson and Sheldon, 1984; Clayton,
1985; Clayton *et al*. 1985) has suggested that the outer face of
the nuclear envelope is the primary MTOC (see Figs. 11.15 to
11.18). If this is so, and the cortical MT array is a structure
initiated at the nuclear surface during cytokinesis, but
stabilized by a variety of cortical interactions, then depoly-
merization/repolymerization cycles in the absence of new
cortical MTOCs are hard to envisage. We prefer to view the
entire cortical MT array as a metastable structure, stabilized
in different arrays by a combination of free energy

considerations, interactions with membrane components and MT
associated proteins. The strongest possible evidence in favour
of bulk reorientation is found in the observation that taxol,
which both prevents any MT disassembly, and also stabilizes and
enlarges existing MT arrays, affects neither the MT 'binding'
seen during xylogenesis nor the subsequent cell wall thickening
(Falconer and Seagull, 1985b). This further suggests that the
changes in helical pitch of the whole array, and lateral aggregation of individual groups of MTs are distinct and separable
events. For further extensive discussion of the nature and
significance of helical arrays of MTs the reader is referred to
the excellent work by Lloyd (Lloyd, 1984; Lloyd and Seagull,
1985; Lloyd *et al.* 1985). The molecular events involved in
the 'bundling' of MTs during xylogenesis remain obscure, but
they might involve a positive coupling via MT associated
proteins.

11.4.3   Microtubules and cell wall deposition

The vexed question of the relationship between MTs and the
orientation of cellulose microfibril deposition is very complex
and will not be covered in detail here (for recent discussions
see Robinson and Quadar, 1982; Heath and Seagull, 1982; Lloyd,
1984; Brown, 1985). Our results with the two plant systems
described here do, however, prompt certain speculations. Our
new view of the MT cytoskeleton as being a dynamic integral
array, often based on an underlying helical structure, allows
the explanation of a wide range of observed wall textures, in
particular helicoidal walls, if some degree of parallelism is
allowed between MTs and cellulose microfibrils (Lloyd, 1984;
Lloyd *et al.* 1985). It is unlikely that MTs are required for
moving cellulose synthase complexes, as cell wall synthesis
proceeds quite readily in the presence of colchicine, with the
proviso that the usual directional constraints are removed. The
simplest model, and one with which we are in agreement, is that
of Herth (1980). In his model MTs provide passive channels,
between which cellulose synthase complexes can move. His recent
freeze fracture data on 'rosettes' (which he assumes are something to do with the cellulose synthetic machinery) show a
marked clustering in the region of the cell wall thickenings,
but the 'rosettes' do not show any evidence of accurate alignment with underlying MTs (Herth, 1985).

We therefore suggest that perhaps the simplest way of viewing
our data on *Zinnia* is that the cross linking of the MTs to the
membrane, presumably involving integral membrane proteins, im-
poses a stabilizing influence on that region of the membrane
below the MT. These stabilized linear domains of membrane would
prevent lateral movement across them of large membrane complexes

such as rosettes. These would instead be constrained to the
membrane area between adjacent MTs. During xylogenesis, the
MTs are laterally aggregated (by some unknown mechanism) and
in the process they concentrate the cellulose synthase complexes into a narrow band between them. This provides a
mechanism for localized cell wall thickening in this region
without having to invoke special controls on rates of cellulose
synthesis. The model is illustrated in Figure 11.33. No
assumptions are made as to the direction and motility of the

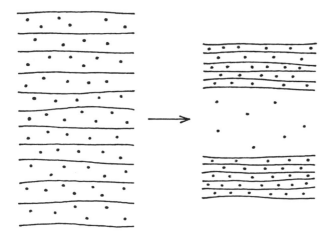

Fig. 11.33   A schematic model to show how lateral
MT association during xylogenesis might sweep the
cellulose synthase complexes (dots) along with them
between stabilized membrane domains associated with
the overlying MTs. A high concentration of complexes
over the sites of secondary cell thickening is the
natural result.

synthase complexes. A prediction from our model is that during
prophase of cell division, cellulose synthase is likely to be
switched off. This is because a similar process of MT lateral
aggregation occurs during the formation of the preprophase
band, and yet no cell wall thickening occurs. (An alternative
explanation would be that, in the case of the forming preprophase band, membrane domain stabilizing linkages are broken and
integral membrane complexes are free to laterally diffuse).
For further progress in this field we shall require probes for

both cellulose synthase and the proteins that are involved in
MT-membrane interaction.

11.5 SUMMARY

Whole cell immunofluorescence methods have been used to study
global rearrangements of the cortical microtubule cytoskeleton
in plant cells. During xylogenesis the MTs rearrange as an
integral system into a variety of arrays, the final one of
which predicts the sites of secondary cell wall thickening.
Later, independent of the MTs, the wall thickenings are lignified before the cell finally dies. The underlying pattern of
MTs in the cell is helical and this is demonstrated vividly by
the response to ethylene in which the helical array rapidly
changes pitch. The implications of these results for MT/micro-
filament orientations is discussed.

11.6 ACKNOWLEDGEMENTS

We would like to thank Dr. C. Lloyd, Dr. L. Clayton and Dr. R.
Seagull for helpful discussions and Patricia Phillips for preparing the manuscript.

11.7 REFERENCES

Barkley, G. (1927). Differentiation of vascular bundle of
Trichosanthes anguina. Bot. Gaz. 83, 173-185.

Bierhorst, D.W. and Zamora, P.M. (1965). Primary xylem elements and element associations of Angiosperms. *Am. J. Bot.* 52,
657-710.

Brown, R.M. (1985). Cellulose microfibril assembly and orientation: Recent developments. *J. Cell Sci. Suppl. 1.* 13-32.

Burgess, J. (1969). Two cytoplasmic features of prophase in
wheat root cells. *Planta* 87, 259-270.

Burgess, J. and Linstead, P. (1984a). *In vitro* tracheary element formation: Structural studies and the effect of tri-
iodobenzoic acid. *Planta* 160, 481-489.

Burgess, J. and Linstead, P. (1984b). Comparison of tracheary
element differentiation in intact leaves and isolated mesophyll
cells of *Zinnia elegans*. *Micron & Microscopia Acta* 15, 163-160.

Clayton, L. (1985). The cytoskeleton and the plant cell cycle. In: *The Cell Division Cycle in Plants*. *Soc. for Exp. Biol. Seminar Series.* 26, (eds. J.A. Bryant, D. Francis). Cambridge University Press, pp. 113-131.

Clayton, L., Black, C.M. and Lloyd, C.W. (1985). Microtubule nucleating sites in higher plant cells identified by an auto-antibody against pericentriolar material. *J. Cell Biol.* 101, 319-324.

Dickinson, H.G. and Sheldon, J.M. (1984). A radial system of microtubules extending between the nuclear envelope and the plasma membrane during early male haplophase in flowering plants. *Planta* 161, 86-89.

Falconer, M.M. and Seagull, R.W. (1985a). Immunofluorescent and calcofluor white staining of developing tracheary elements in *Zinnia elegans* L. suspension cultures. *Protoplasma* 125, 190-198.

Falconer, M.M. and Seagull, R.W. (1985b). Xylogenesis in tissue culture: Taxol effects on microtubule reorientation and lateral association in differentiating cells. *Protoplasma* (in press).

Fukuda, H. and Komamine, A. (1980a). Establishment of an experimental system for the study of tracheary element differentiation from single cells isolated from the mesophyll of *Zinnia elegans*. *Plant Physiol.* 65, 57-60.

Fukuda, H. and Komamine, A. (1980b). Direct evidence for cyto-differentiation to tracheary elements without intervening mitosis in a culture of single cells isolated from the mesophyll of *Zinnia elegans*. *Plant Physiol.* 65, 61-64.

Fukuda, H. and Komamine, A. (1982). Lignin synthesis and its related enzymes as markers of tracheary element differentiation in single cells isolated from the mesophyll of *Zinnia elegans*. *Planta* 155, 423-430.

Cronshaw, J. and Bouck, G.B. (1965). The fine structure of differentiating xylem elements. *J. Cell Biol.* 24, 415-431.

Goodwin, R.H. (1942). On the development of xylary elements in the first internode of Avena in dark and light. *Am. J. Bot.* 29, 818-828.

Gunning, B.E.S. (1982). The cytokinetic apparatus: Its development and spatial regulation. In: *The Cytoskeleton in Plant Growth and Development*. (ed. C.W. Lloyd). Academic Press, London, pp. 229-292.

Haddon, L. and Northcote, D.H. (1976). Correlation of the induction of various enzymes concerned with phenylpropanoid and
lignin synthesis during differentiation of bean callus
(*Phaseolus vulgaris* L.). *Planta* 128, 255-262.

Heath, I.B. and Seagull, R.W. (1982). Oriented cellulose
fibrils and the cytoskeleton: A critical comparison of models.
In: *The Cytoskeleton in Plant Growth and Development*. (ed.
C.W. Lloyd). Academic Press, London, pp. 163-182.

Hepler, P.K. and Newcomb, E.H. (1964). Microtubules and
fibrils in the cytoplasm of Coleus cells undergoing secondary
cell wall deposition. *J. Cell Biol.* 20, 529-533.

Hepler, P.K. and Palevitz, B.A. (1974). Microtubules and microfilaments. *Ann. Rev. Plant Physiol.* 25, 309-362.

Herth, W. (1980). Calcofluor White and Congo Red inhibit chitin
microfibril assembly of Poterioochromonas: evidence for a gap
between polymerization and microfibril formation. *J. Cell Biol.*
87, 442-450.

Herth, W. (1985). Plasma membrane rosettes involved in
localized wall thickening during xylem vessel formation of
Lepidium sativum L. *Planta* 164, 12-21.

Kilmartin, J.V., Wright, B. and Milstein, C. (1982). Rat monoclonal antitubulin antibodies derived by using a new non-
secreting rat cell line. *J. Cell Biol.* 93, 576-582.

Kohlenbach, H.W. and Schmidt, B. (1975). Cytodifferentiation
in the mode of a direct transformation of isolated mesophyll
cells to tracheids. *Z. Pflanzenphysiol.* 75, 369-374.

Lloyd, C.W. (1984). Toward a dynamic helical model for the influence of microtubules on wall patterns in plants. *Int. Rev.
Cytol.* 86, 1-51.

Lloyd, C.W., Clayton, L., Dawson, P.J., Doonan, J.H., Hulme, J.
S., Roberts, I. and Wells, B. (1985). The cytoskeleton underlying side walls and cross walls in plants: molecules and
macromolecular assemblies. *J. Cell Sci. Suppl. 1.* 143-155.

Lloyd, C.W. and Seagull, R.W. (1985). A new spring for plant
cell biology: microtubules as dynamic helices. *Trends in
Biochem. Sci.* (in press).

Mazia, D. (1961). Mitosis and the physiology of cell division.
In: *The Cell.* Vol 3. (ed. J. Brachet and A.E. Mirsky).
pp. 77-412.

Minocha, S.C. and Halperin, W. (1976). Enzymatic changes and lignification in relation to tracheid differentiation in cultured tuber tissue of Jerusalem artichoke (*Helianthus tuberosis*). *Can. J. Bot.* 54, 79-89.

O'Brien, T.P. (1981). The Primary Xylem. In: *Xylem Cell Development*. (ed. J.R. Barnett). Castle House Publications, London, pp. 47-95.

Pickett-Heaps, J.D. (1966). Incorporation of radioactivity into wheat xylem walls. *Planta* 71, 1-14.

Roberts, I.N., Lloyd, C.W. and Roberts, K. (1985). Ethylene-induced microtubule reorientations: mediation by helical arrays. *Planta* 164, 439-447.

Roberts, K. (1986). Antibodies and the plant cell surface: practical approaches. In: *Immunology in Plant Science. Soc. for Exp. Biol. Seminar Series*. (ed. T.L. Wang). (in press).

Roberts, K. and Northcote, D.H. (1972). Hydroxyproline: observations on its chemical and autoradiographical localization in plant cell wall protein. *Planta* 107, 43-51.

Robinson, D.G. and Quader, H. (1982). The microtubule-microfibril syndrome. In: *The Cytoskeleton in Plant Growth and Development*. (ed. C.W. Lloyd). Academic Press, London, pp. 109-126.

Sinnott, E.W. and Bloch, R. (1945). The cytoplasmic basis of intercellular patterns in vascular differentiation. *Am. J. Bot.* 32, 151-157.

Stone, B.A. (1984). Noncellulosic β-glucans in cells walls. In: *Structure, Function and Biosynthesis of Plant Cell Walls*. (ed. W.M. Dugger and S. Bartnicki-Garcia). Riverside, USA. pp. 52-74.

Whitmore, F.W. (1978). Lignin-protein complex catalysed by peroxidase. *Plant Sci Letters* 13, 241-245.

Whitmore, F.W. (1982). Lignin-protein complex in cell walls of *Pinus elliottii*: amino acid constituents. *Phytochemistry* 21, 315-318.

Wooding, F.B.P. and Northcote, D.H. (1964). The development of the secondary wall of the xylem in *Acer pseudoplatanus*. *J. Cell Biol.* 23, 327-337.

12

Plant cell wall formation

Werner Herth

12.1 INTRODUCTION

Cell walls of plants have a skeletal function. In principle they
are composed of a structural backbone and matrix materials.

The chemical composition of cell walls has been studied in
some detail (for reviews see Frey-Wyssling 1959; Kreger 1962;
and Northcote 1972). There is a variety of structural
polysaccharides, with cellulose being most abundant and of wide
industrial application. In some cases chitin, β-1,4 mannan, β-
1,3-xylan or β-1,3-glucan is the structural component (Preston
1968; Herth and Meyer 1977; Herth and Hausser 1984). These
structural polysaccharides have a high degree of polymerization
(Marx-Figini 1982), and form microfibrils of species-specific
dimensions. Even for one component the dimensions of the fibrils
show considerable variability. The dimensions of cellulose
microfibrils may be in the often described range of "elementary
fibrils" (Frey-Wyssling and Mühlethaler 1963; Mühlethaler 1967;
Frey-Wyssling 1969) from about 3 x 4 nm, up to ribbons of 10 x
25 nm (Harada and Goto 1982), and down to "sub-elementary"
fibrils about 1.5 nm wide (Franke and Ermen 1969; Hanna and Cote
1974; Herth and Meyer 1977). The crystallinity may be low in the
thin fibrils, with heterosaccharides included in the fibril, or
it may be high in the crystalline ribbons like *Valonia*
cellulose, which is almost a homopolymer (Preston 1964;
Zugenmaier 1981; Blackwell 1982). This diversity has led to the
distinction of "eucellulose" clearly showing the typical
physical characteristics and purity, from other cellulose
modifications (Preston 1974).

The matrix components are rather complex. They consist of a
variety of sugar monomers forming an interlinked meshwork

between the microfibrils. This has been especially well
eatablished for primary walls of higher plants (Albersheim
model: Albersheim 1975; McNeil et al. 1984). Acidic components
like pectin are frequent in the matrix. In addition to the
polysaccharide moiety, there is a certain amount of protein in
the cell wall, in part being cross-linked to the other matrix
polysaccharides via arabinosides ("extensin", Lamport and Miller
1971), in part having enzymatic functions (e.g. peroxidase:
McNeil 1984). In addition, secondary walls may be impregnated
with lignin, a complicated intermeshed polymer of coniferyl
alcohol, and the outer surface of the wall may be protected by
derivatives of fatty acids like cutin, suberin and waxes
(Kolattukudy 1981).

This short account already demonstrates that plant cell
wall formation must be a rather complex interaction of enzymatic
activities, regulated in sequence and timing with respect to
the cell cycle. Furthermore, the structural components are
laid down in specific orientations, which are controlled by the
cell, and which are necessary to establish tensile strength, the
form of the cell, etc. (Frey-Wyssling 1959; Preston 1974).

From earlier experiments such as the uptake of radioactive
precursors in vivo, and with membrane fractions in vitro, and
with autoradiography etc. (Northcote 1971, 1974) it seems well
eatablished that the matrix polysaccharides are formed in the
Golgi apparatus, and that export proteins are glycosylated also
in the Golgi system and then secreted via vesicles fusing with
the plasma membrane (Northcote 1979, 1984; Morré and VanDerWoude
1974; Leblond and Bennet 1977).

Attempts at biochemical localization of the enzymes
involved in structural polysaccharide formation have been less
successful for plant cells. In contrast to fungal and insect
chitin synthase, which is still active in vitro and forms
microfibrils with the crystallographic properties of α-chitin
(Ruiz-Herrera 1982, Cohen et al. 1982), cellulose synthesis in
vitro has not been achieved in spite of numerous attempts to
establish optimal conditions for membrane isolation and
incubation (e.g. VanDerWoude et al. 1984; Carpita 1982; Delmer
et al. 1984; Callaghan and Benziman 1984, 1985). The product of
such assays is mostly a mixture of β-1,3- and β-1,4-linked
glucan with a low degree of polymerization and no formation of
cellulose I crystals. This β-glucan synthase activity is
associated with the plasma membrane; another β-glucan synthase
activity is found in Golgi membrane fractions and is assumed to
be involved in xyloglucan formation (Ray et al. 1976). Although
claimed repeatedly (e.g. Elbein and Forsee 1973; Hopp et al.
1978) it is not yet clear whether lipid intermediates are
involved in cellulose biogenesis, or whether primers are needed

(McLachlan 1982). This relatively obscure biochemical situation seems to indicate that cellulose synthesis is a somewhat complex and easily destroyed plasma membrane bound function.

Preston (for review see Preston, 1974) had already earlier postulated that cellulose should be made by plasma membrane-bound "ordered granules", and had proposed a model of such a granule complex, with granules cooperating in simultaneous polymerization and crystallization to produce a cellulose microfibril (for discussion see also Frey-Wyssling 1969). Cytological observations on the unicellular algae *Pleurochrysis* (Brown *et al.* 1973) and *Micrasterias* (Kiermayer and Dobberstein 1973) led to the hypothesis that, according to the membrane flow concept (Morré 1975; Morré *et al.* 1971, 1979), synthases should be made in the endoplasmic reticulum or corresponding nuclear envelope regions and then be transferred to Golgi cisternae. These synthases may then either get activated in the Golgi cisternae, as in scale formation of *Pleurochrysis*, or may be transported to the plasma membrane in specific vesicles (F-vesicles of *Micrasterias*) bearing hexagonally arranged globular particles on their extraplasmic face, which start to produce microfibrillar ribbons after incorporation into the plasma membrane.

Improvements of the freeze-fracture technique have shown for *Micrasterias* (Kiermayer and Sleytr 1979; Giddings *et al.* 1980) that the F-vesicles bear hexagonal arrangements of particle rosettes on the plasmatic fracture face of the membrane (PF). These authors have suggested that rows of rosettes cooperate in the formation of microfibrils and that, in primary wall formation of *Micrasterias*, individual rosettes form thin microfibrils. These results led the authors to postulate that these rosettes might represent the cellulose synthase complex.

In *Spirogyra* where similar, but smaller, rosette aggregates are found on the ends of microfibrillar ribbon imprints (Herth, 1983), I have calculated that one rosette should synthesize a 3 x 4 nm fibril. This is the value given earlier by Frey-Wyssling and Mühlethaler (1963) for elementary fibrils. If one rosette formed one elementary fibril with 36 glucan chains, each of the 6 rosette subparticles of 8 nm width would still represent 6 glucan synthases cooperating in close proximity (Herth 1983), which may be necessary to produce cellulose I.

For higher plants, Mueller and Brown (1980) also demonstrated rosettes in the PF, but presented a model with terminal globules (tg) in the external leaflet of the membrane (EF) bearing the synthases, which should fit with their projections into the central hole of the rosettes, the terminal globule and rosette being part of a transmembrane complex.

The role of rosettes in cellulose synthesis was questioned by Willison (1983), based on frequency calculations of terminal globules and rosettes. In his view, only the terminal globules are putative candidates for cellulose synthases.

To clarify this somewhat controversial situation, we have now investigated several higher plant systems with freeze-fracture, and have analyzed the replicas for the occurrence and distribution of rosettes and terminal globules. The main results are presented in this paper (see also Herth 1984; Herth and Weber 1984; Herth and Hausser 1984; Herth 1985).

12.2 MATERIALS AND METHODS

Soy bean *(Glycine max)* suspension culture cells were cultivated by Dr. G. Weber, Max-Planck-Institute of Cell Biology, Ladenberg, as described in Herth and Weber (1984).

Cotton *(Gossypium hirsutum)* seed hairs were obtained from ovule cultures *in vitro*. Cotton plants were cultivated in the green house of Institut de Biologie Vegetale in Fribourg, Switzerland. The ovules were prepared by Dr. U. Ryser 2 – 4 days post anthesis and were here cultivated in liquid medium as described in Beasley (1984).

Cress *(Lepidium sativum)* seeds were germinated on moist filter paper in a moist chamber and the roots were used for freeze-fracture when they had reached ca. 3 cm length.

Mung bean *(Vigna radiata)* was germinated in the same way as the cress, and hypocotyl pieces were used from 5 – 6 day old germlings.

Greenhouse plants of *Limnobium stoloniferum* (syn. *Trianea bogotensis*) were obtained from the Botanical Institute in Heidelberg, and were cultivated in 20 cm ∅ Petri dishes in tap water. Roots from well growing plants were used for freeze-fracturing.

For freeze-fracturing, the specimens were transferred to a yeast-paste-coated specimen holder, sandwiched between the first and the second specimen holder (double replica holders of Balzers, Liechtenstein), immediately frozen in melting nitrogen slush, and fractured in a Balzers BAF-400-T apparatus. Further details are as described in Herth (1983) and Reiss *et al*. (1984). Observation was carried out using a Philips 400 electron microscope.

12.3 RESULTS

12.3.1 Suspension culture cells

Soy bean cells in suspension culture are spherical objects with
a high rate of cell division and cell wall formation. Cells from
the exponential growth phase were freeze-fractured. The EF of
these cells shows no obvious terminal globules and it is not so
rich in particles (Fig. 12.1). Individual rosettes were found on
the particle-rich PF (Fig. 12.2), but numerous cells did not
show identifiable rosettes at all. The rosettes consisted of six
subunit particles (Fig. 12.3).

12.3.2 Seed hairs

Cotton seed hairs cultivated *in vitro* were prepared at various
stages post anthesis. Again many fractures did not reveal any
rosettes on the PF. In some cases, seed hairs during primary
wall formation (Fig. 12.4) and during secondary wall formation
(Fig. 12.5) revealed irregularly distributed rosettes on the PF,
with up to 10 rosettes/μm^2 during secondary wall formation. The
EF showed occasionally globular elevations which might be
interpreted as terminal globules (Fig. 12.6), but which may also be
a preparation artifact. Later stages of secondary wall formation
have not yet been successfully fractured due to problems with the
thick wall and the tendency to break in the cuticle layer.

12.3.3 Xylem elements

The localized regions of helical secondary wall thickening in
developing roots of cress were hit by a few lucky fractures.
They showed a rather rough surface of the EF, with numerous
globular projections on the surface occurring in the regions
undergoing wall thickening (Fig. 12.7). These globular
elevations were ill-defined in diameter, but were mostly around
30 nm in diameter. The PF was characterized by the frequent
occurrence of rosettes, restricted to band-like regions of cell
wall thickening, and excluded from the adjacent regions of the
plasma membrane (Fig. 12.8). The distribution of the rosettes was
random (Fig. 12.9). More than 200 rosettes/μm^2 could be counted
in these regions of secondary wall thickening. This number
corresponds roughly with the elevated globular regions in the
EF, but these could not be counted so exactly.

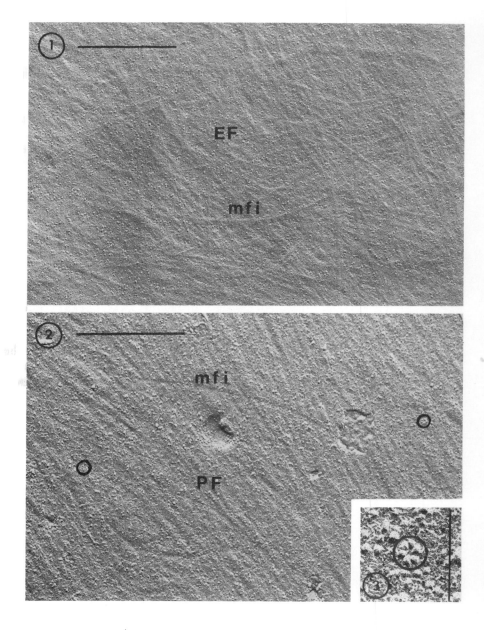

Figs. 12.1 to 12.3. Soy bean suspension culture cells,
exponential growth phase, plasma membrane fractures. mfi
– microfibril imprints.   (1) Survey EF.   (2) Survey PF –
individual rosettes (encircled).   (3) Detail from PF
showing an individual rosette with six subunit particles
(encircled).   Bars in (1 & 2) = 0.5 µm; in (3) = 0.1 µm.

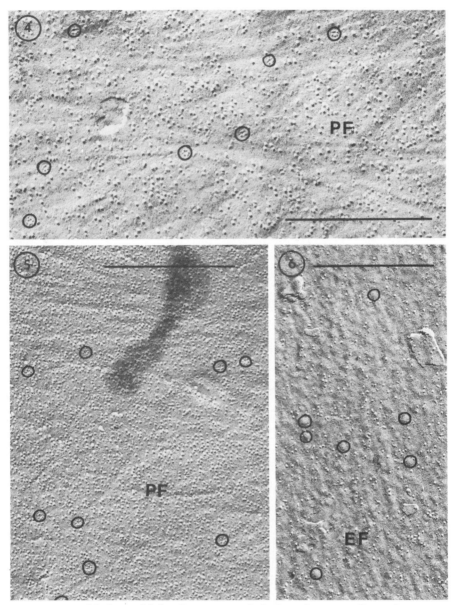

Figs. 12.4 to 12.6. Cotton seed hairs from ovule cultures
in vitro, plasma membrane fractures. (4) PF with
individual rosettes (encircled), from a stage of primary
wall formation (10 days post anthesis). (5) PF region
with higher frequency of rosettes (encircled), from a
stage of secondary wall formation (33 days post anthesis).
(6) EF stage corresponding to (4), with ill-defined
globular structures (some encircled).   Bars = 0.5 μm.

292 W. Herth

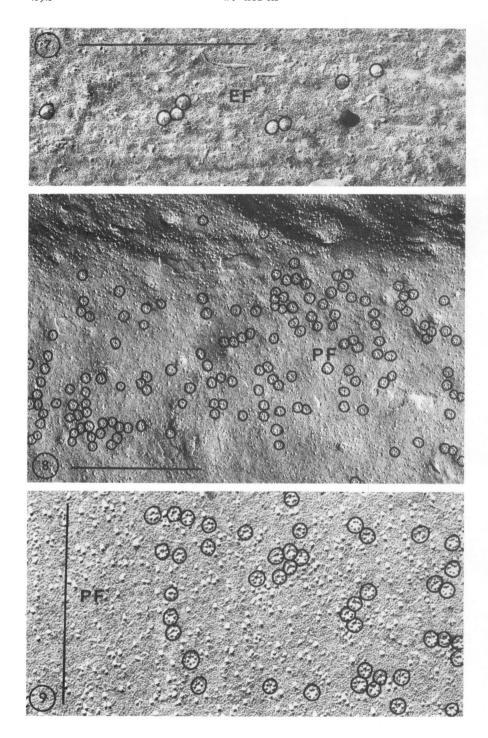

12.3.4 Hypocotyl secondary walls

During secondary wall formation of cells from the hypocotyl of
mung bean, numerous rosettes were found in the PF (Fig. 12.10).
These rosettes appear to be roughly aligned in tracks parallel
to the imprints of the recently deposited innermost microfibrils
which are seen as imprints in the plasma membrane (Fig. 12.11).
This alignment is not as exact as pearls on a string, but groups
of rosettes are clustered irregularly with a predominance of
one direction parallel to the imprints. The innermost layer of
microfibrils in such a cell (Fig. 12.12) shows groups of
microfibrils which are also not strictly parallel, but more like
bundles which show close associations in places and fringing
fibrils in other places. The EF of mung bean plasma membrane did
occasionally show some globular structures (Fig. 12.13).

12.3.5 Root cortex cells

Some, but not all cells of the cortical root parenchyma of
Limnobium stoloniferum showed individual rosettes on the PF
(Fig. 12.14). On the EF, almost no terminal-globular-like
structures were encountered (Fig. 12.15). In some cells
fractures revealed both the EF with microfibril imprints and the
cytoplasmic surface layer of the plasma membrane with cortical
microtubules (Fig. 12.16). The microtubules showed mostly
parallel orientation with those microfibril imprints, which were
found as the innermost layer. Microfibril imprints were found
either bundled in continuity with the direction of a microtubule
(Fig. 12.16, arrow), or deposited in the intermicrotubule region
of the plasma membrane (Fig. 12.16 arrowheads).

Figs.   12.7 to 12.9. (Previous page). Developing xylem
element of the cress root, plasma membrane fractures.
(7) EF of the region underlying the developing band of
secondary wall thickening, with globular projections on
its surface (some encircled). (8) PF of a plasma
membrane region underlying a developing wall thickening,
with numerous rosettes in a band-like arrangement
underlying the thickening, and absent in adjacent plasma
membrane regions. (9) Detail of PF showing the structure
of the rosettes with six subunits, and the irregular
distances between the rosettes. Bars = 0.5 μm.

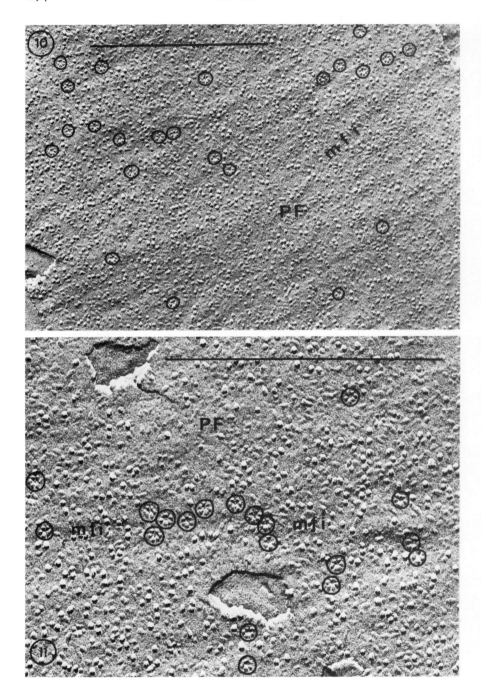

Legend on following page.

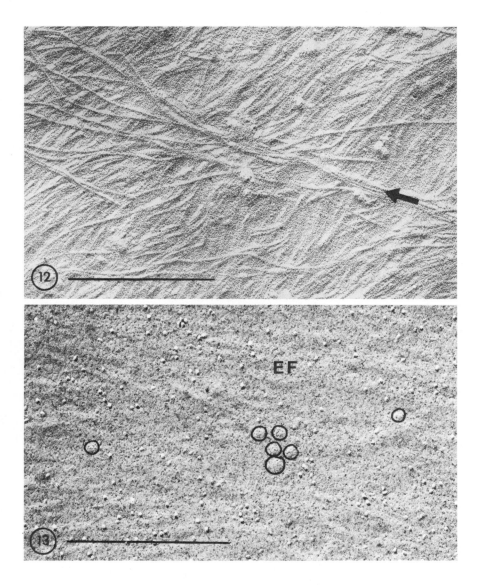

Figs. 12.10 to 12.13. Mung bean hypocotyl cell undergoing secondary wall deposition. (10) PF of the plasma membrane with numerous rosettes (encircled). (11) Detail of PF with rosette alignment parallel to microfibril imprints. (12) Inner layer of cell wall microfibrils showing small bundles (arrow). (13) EF of the plasma membrane with presence of some globules (encircled). Bars = 0.5 µm.

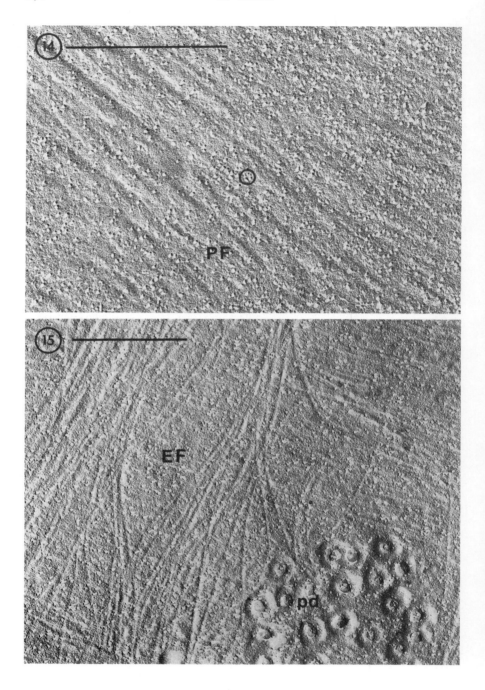

Legends for Figs. 12.14 & 12.15 on next but one page.

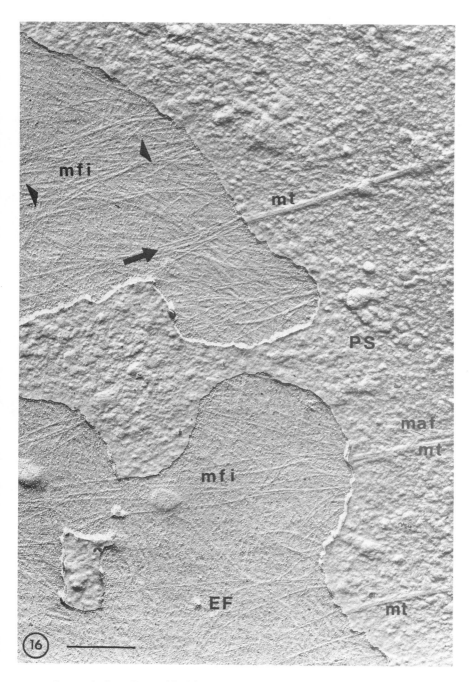

Legend for Fig. 12.16 on next page.

12.4 DISCUSSION

The results on rosette and terminal globule frequency vary
considerably, not only from object to object, but also from cell
to cell. The highest frequency of rosettes so far encountered
was in the cress xylem. In this case, the active phase of wall
thickening was obviously met. It is probable that the phase of
cellulose deposition with respect to the cell cycle is
relatively short, and that the life time of the rosettes is also
only in the range of about 10-20 min, as indicated by
calculations on *Funaria* caulonema tip cells (Reiss *et al*. 1984),
by inhibitory treatments on the same system (Schnepf *et al*.
1985) and by calculations on xylem thickening (Schneider and
Herth, in preparation). In addition, rosettes seem to be very
sensitive to preparative conditions (Mueller 1982), and seem
only to be visualized under optimal freezing conditions, which
are obtained for just some of the cells in a tissue sample.
Structures resembling terminal globules were even rarer in our
replicas. When frequent, as in the cress xylem, they were so
ill-defined that they could not be definitely counted. These tg-
structures were often absent even when rosettes were present
(see also Reiss *et al*. 1984). I therefore consider that the
rosettes are better marker structures for ongoing cellulose
synthesis, and question the supposed cellulose synthase function
of terminal globules suggested by Mueller and Brown (1980; see
also Brown *et al*. 1983; Willison 1983). The tgs might represent
either product penetrating the external leaflet, or be some kind
of tunnel protein facilitating penetration of the native glucan
chains to the exterior (Herth and Hausser, 1984). On the other
hand, the globular structures on the luminal surface of F-
vesicles of *Micrasterias* (Kiermayer and Dobberstein 1973)
possibly would not be visualized by fracturing the lipid
bilayer. As pointed out by Willison (1983), etching the outer
surface of the plasma membrane reveals more of those globular
structures. This might mean that these structures are only
occasionally seen as imprints in freeze-fracture replicas.

Figs. 12.14 to 12.16. (Previous pages). Cortical root
parenchyma cell of *Limnobium stoloniferum*. (14) PF of
the plasma membrane with individual rosette (encircled).
(15) EF of the plasma membrane, with almost no globules.
pd - plasmodesmata. (16) Fracture showing the
cytoplasmic surface (PS) and EF of the plasma membrane.
The cortical microtubules (mt) are parallel to the
innermost microfibrils, which are seen as imprints (mfi)
in the EF. Arrow - microfibril bundle in line with a
microtubule; arrowheads - microfibrils deposited in the
plasma membrane region between two microtubules; maf -
microtubule-associated filaments. Bars = 0.5 μm.

Taking together the collected data on the occurrence of rosettes from algae such as *Micrasterias, Closterium* and *Spirogyra*, mosses, pteridophytes, and higher plants (where in every carefully checked case rosettes are present) (Brown *et al.* 1983; Emons 1985; Giddings *et al.* 1980; Herth 1983, 1984, 1985; Herth and Weber 1984; Herth, this work; Kiermayer and Sleytr 1979; Mueller and Brown 1980; Mueller 1982; Reiss *et al.* 1984; Schnepf *et al.* 1985; Staehelin and Giddings 1982; Wada and Staehelin 1981), there is considerable evidence for the putative cellulose synthase function of rosettes.   This concept is further supported by the following facts: rosettes are clearly found at the ends of microfibril imprints in *Micrasterias* and *Spirogyra* (Giddings *et al.* 1980; Herth 1983); individual elementary fibrils may be too thin to leave imprints leading to the rosettes (Reiss *et al.* 1984); rosettes are mainly found in plasma membrane regions with a high rate of cellulose deposition (tip of *Funaria* caulonema, helical wall thickening region in cress); rosettes are stage-specifically incorporated into the plasma membrane in *Micrasterias* (Kiermayer and Dobberstein 1973; Giddings *et al.* 1980); rosette frequency is high in systems producing a high cellulose content such as secondary walls (this paper), and is low in systems with a low cellulose content (Reiss *et al.* 1985; Witte and Herth, in preparation). This concept is in contrast to Willison's (1983) argument that the low numbers of rosettes encountered may have been due to preparative difficulties and cell cycle specific variations in the frequency of rosettes. We hope to clarify this point further by a survey of rosette frequency with respect to the cell cycle in synchronized cell cultures.

The results indicating a short life time and extreme sensitivity of the rosettes (Reiss *et al.* 1984; Schnepf *et al.* 1985; Mueller 1982) may be interpreted as a hint to a possible short-term regulation of rosette function, allowing the plant cell to control microfibril deposition rapidly (see also Schnepf et al. 1985). On the other hand, this may be the main cause of the difficulties encountered in isolating intact cellulose synthase in membrane fractions.

Lower organisms from bacteria (*Acetobacter*) to algae such as *Fucus, Pelvetia, Oocystis, Glaucocystis* and *Valonia,* seem to have evolved another type of cellulose synthase complex. This is characterized by linear particle aggregates which in some cases may associate in parallel rows, and which are clearly found on the ends of microfibrils (Brown and Montezinos 1976; Brown and Willison 1977; Brown *et al.* 1976, 1983; Haigler and Benziman 1982, Hoh and Brown 1984; Montezinos 1982; Quader 1983; Peng and Jaffe 1976; Quatrano 1982). This type of "terminal complex" (Brown and Montezinos 1976) is found in the EF of *Oocystis* (Fig. 12.17), and spanning the membrane in *Valonia*

(Itoh and Brown 1984). Ribbon-like, extremely crystalline and
pure cellulose is made by those organisms using terminal
complexes with parallel rows of subunit particles.

Taking the size of a *Valonia* microfibril to be 10 x 20 nm
(Harada and Goto 1982), approximately 540 glucan chains would be
synthesized in a coordinated way to yield the microfibril. With
about 90 - 120 subunits in the terminal complexes (Itoh and
Brown 1984) this would mean that 5-6 glucan synthases should
reside in one subunit particle. This value is almost identical
with the assumptions about rosettes (Herth 1983). This may be
interpreted as a one subunit particle, consisting of several
synthases. being the first step of cellulose-synthase complex
formation. The trend to linear cooperating complexes then
determines the evolution of rigid microfibrillar ribbons (see
hypothetical evolution scheme in Fig. 12.18). The rosette with

Fig. 12.17. Green alga *Oocystis solitaria,* plasma
membrane EF, with "terminal complexes" (TC) on the ends
of microfibril imprints (mfi). Note the globular
subunits of the terminal complex, arranged in three
parallel rows (small arrows). Micrograph courtesy of Dr.
H. Quader. Bar = 0.5 μm.

its rather flexible "elementary fibril" product and the capacity
of rosettes to cooperate to form broader ribbons (in regular
rows as in *Micrasterias, Spirogyra* and *Closterium* (Giddings *et
al.* 1980; Herth 1983; Staehelin and Giddings 1982), or in less
regular cooperation to form fringed intermeshed microfibrils as
in the mung bean system), is a step forward in allowing flexible
adaptation of the microfibrillar dimensions (see Fig. 12.18).
This is certainly important for multicellular organisms with
cells elongating and differentiating in multiple ways. As the
rosettes are first found in those algal groups which are close

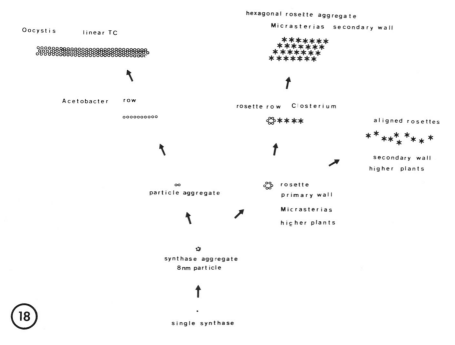

Fig.   12.18. Scheme for the hypothetical evolutionary
lines of putative cellulose synthase complexes. At the
base is the hypothetical case of a single β-1,4-glucan
synthase. The hypothetical case of up to 6 glucan
synthases grouped together in an 8 nm particle follows.
From there one trend has led to linear particle
aggregates with multiples of the 8 nm particles, in rows
or parallel rows. Another line leads to the rosette
structure consisting of six 8 nm particles. These
rosettes may again cooperate in linear rows, or parallel
rows. In higher plants they seem to function
individually during primary wall formation, and in
loosely aligned rows during secondary wall formation.

to the supposed origin of higher land plants (Mattox and Stewart 1984; Melkonian, 1984), this invention of rosettes may have been of fundamental importance for the evolution of higher plants.

The differences in the putative synthesizing complexes now demonstrated by freeze-fracture are in good correlation with the different types of cellulose found in different organisms (see Introduction). Each specific type of cellulose may be the consequence of a specific type of synthesizing complex. Primary wall cellulose seems to be made by individual rosettes working independently; secondary wall cellulose seems to be the product of many rosettes cooperating in a parallel way; bacterial cellulose is made by single rows of particles; algal cellulose ribbons are made by parallel rows of particles. These theoretical assumptions are now also backed by spectroscopic differences of different celluloses (Atallah 1984) with I α predominating in *Acetobacter* and *Valonia*, and Iβ predominating in cotton and ramie.

Orientation of microfibril deposition is under cellular control. Cytoskeletal elements have been implicated in the exertion of the control function on the synthases. Several models have been presented to explain possible interactions between microtubules, microfilaments and putative synthases in the membrane (for reviews see Heath and Seagull 1982; Robinson and Quader 1982). The experimental evidence obtained so far, mainly from uncoupling of crystallization and polymerization by dyes of high affinity for structural polysaccharides (Haigler *et al*. 1980; Herth 1980; Herth and Hausser 1984; Quader 1984), supports the view that the motive force for the synthases comes from the crystallization process acting backwards onto the enzyme complexes being mobile in the membrane. The rosette rows in *Closterium* (Staehelin and Giddings 1982) are not found exactly over cortical microtubules, and microtubule distribution is too widely spaced in *Limnobium* to explain a rail-like contractile movement of the synthases. In developing xylem, there are too few microtubules (Herth and Schneider, unpublished) to guide the observed number of rosettes directly. Recent observations of a specific microtubule dependent motility (Bray 1985; Gilbert *et al*. 1985), however might also be valid for plant cells. In this case, microtubule-associated mechanochemical structures of a still unknown nature might create some kind of oriented flow in the adjacent plasma membrane, orienting synthases as suggested for a (supposedly actin-driven) flow mechanism in the plasma membrane (Mueller and Brown 1982). It is also feasible that the rosettes are associated with such a mechanochemical component on their cytoplasmic side.

Taken together, the indirect evidence collected so far

supports the view of rosettes being cellulose synthases or associated structures. In the future we hope to use this easily identifiable marker structure in testing stabilizing conditions with the aim of their isolation in an intact functional state.

12.5 SUMMARY

Recent progress in freeze-fracture studies on putative cellulose synthase complexes located in the plasma membrane of higher plants supports the view that "rosettes" bear the synthase function or are part of the synthase complex. "Terminal globules" are not regularly observed and are rather ill-defined structures when present. Therefore the rosettes seem to be good marker structures for cellulose synthase function in higher plants. Lower plants have evolved linear synthase complexes. An evolutionary scheme for the varying cellulose synthase arrangements is presented.

12.6 ACKNOWLEDGEMENTS

The author thanks Prof. Dr. E. Schnepf and Dr. H. Quader for stimulating discussions, and Mrs B. Heck for skilful technical assistance. Part of this work was done in cooperation with Dr. U. Ryser, Institut de Biologie Vegetale, University of Fribourg, Switzerland, and with Dr. G. Weber, Max-Planck-Institute for Cell Biology, Ladenburg. This work was supported by Deutsche Forschungsgemeinschaft.

12.7 REFERENCES

Albersheim, P. (1975). The walls of growing plant cells. *Scientific Amer.* 232 (4) 80-95.

Atallah, R. H. (1984) Polymorphy in native cellulose: Recent developments. In: *Structure, function and biosynthesis of plant cell walls.* (ed. W. M. Dugger, S. Bartnicki-Garcia). Proc. VII. Ann. Symp. Bot. Riverside. Amer. Soc. Plant Physiologists, Rockeville, Maryland, pp. 381-391.

Beasly, C. A. (1984). Culture of cotton ovules. In: *Cell culture and somatic cell genetics of plants.* (ed. I. K. Vasil). Vol. 1. pp. 232-240. Academic Press, Orlando.

Blackwell, J. (1982). The macromolecular organization of cellulose and chitin. In: *Cellulose and other natural polymer systems.* (ed. R. M. Brown). Plenum Press, New York, London, pp.

403–428.

Bray, D. (1985). Fast axonal transport dissected. *Nature* 315, 178–179.

Brown, R. M., Montezinos, D. (1976). Cellulose microfibrils: Visualization of biosynthetic and orienting complexes in association with the plasma membrane. *Proc. Natl. Acad. Sci. USA* 73 143–147.

Brown, R. M., Willison, J. H. (1977). Golgi apparatus and plasma membrane involvement in secretion and cell surface deposition, with special emphasis on cellulose biogenesis. In: *International Cell Biology.* (ed. B. R. Brinkley, K. R. Porter). Rockefeller University Press, pp. 267–283.

Brown, R. M., Herth, W., Franke, W. W., Romanovicz, D. (1973). The role of the Golgi apparatus in the biosynthesis and secretion of a cellulose glycoprotein in *Pleurochrysis*: A model system for the synthesis of structural polysaccharides. In: *Biogenesis of plant cell wall polysaccharides* (ed. F. Loewus). Academic Press, New York, London, pp. 207–258.

Brown, R. M., Willison, J. H. M., Richardson, C. L. (1976). Cellulose biosynthesis in *Acetobacter xylinum* I. Visualization of the site of synthesis and direct measurement of the *in vivo* process. *Proc. Natl. Acad. Sci. USA* 73 4565–4569.

Brown, R. M., Haigler, C. H., Suttie, J., White, A. R., Roberts, E., Smith, C., Itoh, T,. Cooper, K. (1983). The biosynthesis and degradation of cellulose. *J. Appl. Polym. Sci.* 37, 33–78.

Callaghan, T., Benziman, M. (1984). High rates of *in vitro* synthesis of 1,4β-D-glucan in cell free preparations from *Phaseolus aureus*. *Nature* 311, 165–167.

Callaghan, T., Benziman, M. (1985). High rates of *in vitro* synthesis of 1,4-β-D-glucan in cell free preparations from *Phaseolus aureus*. Corrigendum. *Nature* 314, 383.

Carpita, N. C. (1982). Cellulose synthesis in detached cotton fibers. In: *Cellulose and other natural polymer systems.* (ed. R. M. Brown). Plenum Press, New York, pp. 225–242.

Cohen, E. (1982). *In vitro* chitin synthesis in an insect: formation and structure of microfibrils. *Eur. J. Cell Biol.* 16, 289–294.

Delmer, D. P., Thelen, M., Marsden, M. P. F. (1984).

Regularatory mechanisms for the synthesis of β-glucans in plants. In: *Structure, function and biosynthesis of plant cell walls.* (ed. M. W. Dugger, S. Bartnicki-Garcia). Proc. VII. Ann. Symp. in Botany, Riverside, Amer. Soc. Plant Physiol., Rockeville, Maryland. pp. 133–149.

Elbein, A. D., Forsee, W. T. (1973). Studies on the biosynthesis of cellulose. In: *Biogenesis of plant cell wall polysaccharides* (ed. F. Loewus). Academic Press, New York, London, pp. 259–296.

Emons, A. M. (1985). Plasma membrane rosettes in root hairs of *Equisetum hiemale. Planta* 163, 350–359.

Franke, W. W., Ermen, B. (1969). Negative staining of plant slime cellulose: an examination of the elementary fibril concept. *Z. Naturforsch.* 24b, 918–922.

Frey-Wyssling, A. (1959). *Die pflanzliche Zellwand.* Springer, Berlin

Frey-Wyssling, A. (1969). The ultrastructure and biogenesis of native cellulose. *Progress Chem. Organic Natural Products* 27 1–30.

Frey-Wyssling, A., Mühlethaler, K. (1963). Die Elementarfibrillen der Cellulose. *Makromol. Chemie* 62, 25–30.

Gilbert, S. P., Allen, R. D., Sloboda, R. D. (1985). Translocation of vesicles from aquid axoplasm on flagellar microtubules. *Nature* 315, 245–247.

Giddings, T. H., Brower, D. L., Staehelin, L. A. (1980). Visualization of particle complexes in the plasma membrane of *Micrasterias denticulata* associated with the formation of cellulose fibrils in primary and secondary walls. *J. Cell Biol.* 84 327–339.

Haigler, C. H., Brown, R. M., Benziman, M. (1980). Calcofluor white alters the *in vivo* assembly of cellulose microfibrils. *Science* 210, 903–906.

Haigler, C. A., Benziman, M. (1982). Biogenesis of cellulose I microfibrils occurs by cell-directed self-assembly in *Acetobacter xylinum.* In: *Cellulose and other natural polymer systems* (ed. R. M. Brown). Plenum Press, New York, pp. 273–298.

Hanna, R. B., Cote, W. A. (1974). The sub-elementary fibril of plant cell wall cellulose. *Cytobiol.* 10 102–116.

Harada, H., Goto,   T. (1982). The structure of cellulose microfibrils in *Valonia*. In: *Cellulose and other natural polymer systems*. Plenum Press, New York, pp. 383-402.

Heath, I. B., Seagull,   R. W. (1982) Oriented cellulose fibrils and the cytoskeleton: a critical comparison of models. In: *The cytoskeleton in plant growth and development* (ed. C. W. Lloyd). Academic Press, London, New York, pp. 163-182.

Herth,   W. (1980). Calcofluor white and congo red inhibit chitin microfibril assembly of *Poterioochromonas*: Evidence for a gap between polymerization and microfibril formation. *J. Cell Biol.* 87, 442-450.

Herth,   W. (1983). Arrays of plasma membrane "rosettes" involved in cellulose microfibril formation of *Spirogyra*. *Planta* 159, 347-356.

Herth,   W. (1984). Oriented "rosette" alignment during cellulose formation in mung bean hypocotyl. *Naturwiss* 71, 216-217.

Herth,   W. (1985). Plasma membrane rosettes involved in localized wall thickening during xylem vessel formation of *Lepidium sativum* L. *Planta*, in press.

Herth, W., Meyer, Y. (1977). Ultrastructural and chemical analysis of the wall fibrils synthesized by tobacco mesophyll protoplasts. *Bio. Cellulaire* 30, 33-40.

Herth, W., Hausser, I. (1984). Chitin and cellulose fibrillogenesis *in vitro* and their experiental alteration. In: *Structure, function and biosynthesis of plant cell walls* (ed. W. M. Dugger, S. Bartnicki-Garcia). Proc. VII. Ann. Symposium in Botany, University of California, Riverside, Am. Soc. Plant Physiol. Rockeville, Maryland, pp. 89-119.

Herth, W., Weber, G. (1984). Occurence of the putative cellulose-synthesizing "rosettes" in the plasma membrane of *Glycine max.* suspension culture cells. *Naturwiss* 71, 153-154.

Hopp, H. E., Romero, P. A., Daleo, G. R., Pont Leziza, R. 1978). Synthesis of cellulose precursors. The involvement of lipid-linked sugars. *Eur. J. Biochem* 84, 561-571.

Itho, T., Brown, R. M. (1984). The assembly of cellulose microfibrils in *Valonia macrophysa* Kütz. *Planta* 160, 372-381.

Kiermayer, O., Dobberstein, B. (1973). Membrankomplexe dictyosomaler Herkunft als "Matrizen" für die extraplasmatische Synthese und Orientierung von Mikrofibrillen. *Protoplasma* 77,

437– 451.

Kiermayer,   O., Sleytr,   U. B. (1979). Hexagonally ordered
"rosettes" of particles in the plasma membrane of *Micrasterias
denticulata* Breb. and their significance for microfibril
formation and orientation. *Protoplasma* 101,   133–138.

Kolattukudy,   P. E. (1981). Structure, biosynthesis, and
biodegredation of cutin and suberin. *Ann. Rev. Plant Physiol.*
32,   539–567.

Kreger,   D. R. (1962). Cell walls. In: *Physiology and
Biochemistry of algae*. (ed. R. A. Lewin). Academic Press, New
York and London, pp. 315–335.

Lamport,   D. T. A., Miller,   D. H. (1971). Hydroxyproline
arabinosides in the plant kingdom. *Plant Physiol.* 48,   454–456.

Leblond,   C. P., Bennet,   G. (1977). Role of the Golgi apparatus
in terminal glycosylation. In: *International Cell Biology 1976–
1977* (ed. B. R. Brinkley, K. R. Porter). Rockefeller University
Press, pp. 326–340.

Marx-Figini,   M. (1982). The control of molecular weight and
molecular weight distribution in the biogenesis of cellulose.
In: *Cellulose and other natural polymer systems* (ed. R. M.
Brown). Plenum Press, New York, pp. 243–272.

Mattox,   K. R., Stewart,   K. D. (1984). Classification of the
green algae: a concept based on comparative cytology. In:
*Systematics Association special vol 27 Systematics of the green
algae* (ed. D. E. G. Irvine, D. M. John). Academic Press, London
and Orlando, pp. 29–72.

McLachlan,   G. A. (1982). Does β-glucan synthesis need a primer?
In: *Cellulose and other natural polymer systems* (ed. R. M.
Brown). Plenum Press, New York, pp. 327–340.

McNeil,   M., Darwill,   A. G., Fry,   S. C., Albersheim,   P.
(1984). Structure and function of the primary cell walls of
plants. *Ann. Rev. Biochem.* 53, 625–663.

Melkonian,   M. (1984). Flagellar apparatus ultrastructure in
relation to green algal classification. In: *Systematics
Association special vol. 27, Systematics of the green algae* (ed.
D. E. G. Irvine, D. M. John) Academic Press, London and Orlando,
pp. 73–120.

Montezinos,   D. (1982). The role of the plasma membrane in
cellulose microfibril assembly. In: *The cytoskeleton in plant*

growth and development (ed. C. W. Lloyd). Academic Press, London, pp. 147–162.

Morré, D. J. (1975). Membrane biogenesis. *Ann. Rev. Plant Physiol.* 26, 441–481.

Morré, D. J., VanDerWoude, W. J. (1974). Origin and growth of cell surface components. In: *Macromolecules regulating growth and development. 30th Symposium of the Soc. for Developmental Biology.* Academic Press, New York, pp. 81–111.

Morré, D. J., Mollenhauer, H. H., Bracker, C. E. (1971). Origin and continuity of Golgi apparatus. In: *Origin and continuity of cell organelles* (ed. J. Reinert, H. Ursprung). Springer, Berlin, Heidelberg, New York, pp. 82–126.

Morré, D. J., Kartenbeck, J., Franke, W. W. (1979). Membrane flow and interconversions among endomembranes. *Biochim. Biophys. Acta* 59, 71–152.

Mueller, S. C. (1982). Cellulose microfibril assembly and orientation in higher plants cells with particular reference to seedlings of Zea Mays. *In: Cellulose and other natural polymer systems. Biogenesis, structure and degradation* (ed. R. M. Brown). Plenum Press, New York, London, pp. 87–104.

Mueller, S. C., Brown, R. M. (1980). Evidence for an intramembrane component associated with a cellulose microfibril synthesising complex in higher plants. *J. Cell Biol.* 84, 315–326.

Mueller, S. C., Brown, R. M. (1982). The control of cellulose microfibril deposition in the cell wall of higher plants. I. Can directed membrane flow orient cellulose microfibrils? Indirect evidence from freeze-fractured plasma membranes of maize and pine seedlings. *Planta* 154, 489–500.

Mühlethaler, K. (1967). Utrastucture and formation of plant cell walls. *Ann. Rev. Plant Physiol.* 18, 1–24.

Northcote, D. H. (1971). Organization of structure, synthesis, and transport within the plant during cell division and growth. *Symp. Soc. exp. Biol.* 25, 51–69.

Northcote, D. H. (1972). Chemistry of the plant cell wall. *Ann. Rev. Plant Physiol.* 23, 113–132.

Northcote, D. H. (1974). Sites of synthesis of the polysaccharides of the cell wall. In: *Plant Carbohydrate Biochemistry* (ed. J. B. Pridham), *Phytochemical Soc. Symp.* 10,

Academic Press, London, New York, pp. 165-181.

Northcote, D. H. (1979). The involvement of the Golgi apparatus in the biosythesis and secretion of glycoproteins and polysaccharides. *Biomembranes* 10, 51-76.

Northcote, D. H. (1984). Control of cell wall assembly during differentiation. In: *Structure, function and biosynthesis of plant cell walls.* (ed. W. M. Dugger, S. Bartnicki-Garcia). Proc. VII Ann. Symp. Bot. Riverside. Amer. Soc. Plant Physiologists, Rockeville, Maryland. pp. 222-234.

Peng, H. B., Jaffe, L. F. (1976). Cell wall formation in *Pelvetia* embryos: a freeze-fracture study. *Planta* 133, 57-71.

Preston, R. D. (1968). Plants without cellulose. *Scientific Amer.* 218, 102-108.

Preston, R. D. (1974). *The physical biology of plant cell walls.* Chapman and Hall, London.

Quader, H. (1983). Morphology and movement of cellulose synthesizing (terminal) complexes in *Oocystis solitaria*: evidence that microfibril assembly is the motive force. *Europ. J. Cell Biol.* 32, 174-177.

Quatrano, R. S. (1982). Cell wall formation in *Fucus* zygotes: A model system to study the assembly and localization of wall polymers. In: *Cellulose and other natural polymer systems* (ed. R. M. Brown). Plenum Press, New York, pp. 45-60.

Ray, P. M., Eisinger, W. R., Robinson, D. G. (1976). Organelles involved in cell wall polysaccharide formation and transport in pea cells. *Berichte der Deutschen botanischen Ges.* 89, 121-146.

Reiss, H.-D., Schnepf, E., Herth, W. (1984). The plasma membrane of the *Funaria* caulonema tip cell: morphology and distribution of particle rosettes, and the kinetics of cellulose synthesis. *Planta* 160, 428-435.

Reiss, H.-D., Herth, W., Schnepf, E. (1985). Plasma membrane "rosettes" are present in the lily pollen tube. *Naturwiss* 72, 276.

Robinson, D. G., Quader, H. (1982). The microtubule-microfibril syndrome. In: *The cytoskeleton in plant growth and development.* (ed. C. W. Lloyd). Academic Press, London, New York, pp. 109-126.

Ruiz-Herrera, J. (1982). Synthesis of chitin microfibrils *in vitro*. In: *Cellulose and other natural polymer systems* (ed. R. M. Brown). Plenum Press, New York, pp. 207-224.

Schnepf, E., Witte, O., Rudolph, U., Deichgräber, G., Reiss, H.-D. (1985). Tip cell growth and the frequency and distribution of particle rosettes in the plasmalemma:experimental studies in *Funaria* protonema cells. *Protoplasma,* in press.

VanDerWoude, W. J., Lembi, C. A., Morré, D. J., Kindinger, J. I., Ordin, L. (1974). β'-glucan synthases of plasma membrane and Golgi apparatus from onion stem. *Plant Physiol.* 54, 333-340.

Wada, M., Staehelin, L. D. (1981). Freeze-fracture observations on the plasma membrane, the cell wall and the cuticle of growing protonemata of *Adiantum capillus veneris* L. *Planta* 151, 462-468.

Willison, J. H. M. (1983). The morphology of supposed cellulose-synthesizing structures in higher plants. *J. Appl. Polym. Sci.* 37, 91-105.

Zugenmaier, P. (1981). Present views of the conformation and packing of cellulose molecules. In: *Cell walls '81* (ed. D. G. Robinson, H. Quader). Wiss. Verlagsges. Stuttgart, pp. 57-65.

13

Structural changes in protein bodies of cotton radicles during seed maturation and germination

E. L. Vigil, R. L. Steere, M. N. Christiansen, and E. F. Erbe

13.1 INTRODUCTION

Little question remains regarding the important role(s) played by protein bodies in storage tissue as a major source of protein for consumption. Within the plant body protein storage is of vital importance for supplying needed nutrients during early germination, particularly during the heterotrophic phase of growth in the soil. There still exists, however, conjecture regarding the process and organelle interaction for formation of protein bodies and what specific changes occur during germination that involve protein bodies directly. Support has been advanced for one of several postulated processes of protein body formation: partitioning of existing vacuoles (Craig, Goodchild and Hardham, 1979; Craig, Goodchild and Miller, 1980; Dhar and Vijayaraghavan, 1979); dilation of smooth regions of RER (Kristen and Biedermann, 1981; Neumann and Weber, 1978; Oparka and Harris, 1982; Adler and Müntz 1983); combination of either vacuole, ER or dictyosomal origin (Boulter, 1981; Campbell *et al.* 1981; Bechtel, Gaines and Pomeranz, 1982).

The most detailed analysis of protein body formation is that from Chrispeels laboratory for cotyledons of legume seeds (see review, Chrispeels, 1984). These studies focused mainly on synthesis and packaging of the lectin, phytohaemagglutinin, in *Phaseolus vulgaris*. During development of bean cotyledons, lectin is synthesized and glycosylated in RER and Golgi, and then transferred to protein bodies via Golgi vesicles. This pathway was confirmed by biochemical (Chrispeels *et al.* 1982a; Chrispeels, 1983a; Chrispeels, 1983b; Higgins *et al.*, 1983; Bollini, Van der Wilden and Chrispeels, 1982; Bollini, Vitale and Chrispeels, 1983; Mader and Chrispeels, 1984) and immunocytochemical data (Baumgartner, Tokuyasu and Chrispeels,

1980; Herman and Shannon, 1984a; Herman and Shannon, 1984b;
Craig and Goodchild, 1984). Whether all proteins destined for
packaging in protein bodies are processed in this manner is
presently unknown. Chrispeels (1984) has postulated, however,
that a passage through the Golgi complex occurs for all
protein body proteins.

In the present study we have examined the process of protein
body formation and change during seed ripening and
germination. Of primary importance to our study was utilizing
recent advances in methodology for freeze-fracture (Steere,
1981). Our specific focus was to apply this methodology to
the radicle because there is limited information for protein
body formation in radicle cells. More importantly, cellular
events associated with utilization of stored reserves during
germination most likely involve changes in radicle protein
bodies far in advance of utilization of protein from
cotyledons (unpublished observations). The sensitivity of
radicles in seeds to environmental stress in semi-tropical
plants like cotton (Christiansen, 1963; Christiansen, 1968)
affects seedling growth and, ultimately, crop yield. For our
study we used a double haploid of cotton (*Gossypium hirsutum*
L. M-8) because of the uniformity in response of seeds to
standard conditions for germination or stress.

During development, protein bodies arise initially by
partitioning and the loading of vacuoles. Dilation of
cisternae of RER occurs later, generating another potential
class of vacuoles for protein body formation. Aspects of
protein body development and change in structure were
monitored through examination of freeze-fracture replicas, as
well as thin sections of radicles from developing and
germinating seeds.

13.2 MATERIALS AND METHODS

13.2.1 Plant growth

Seeds of cotton (*Gossypium hirsutum* L. M-8) were sown in soil
in clay pots. After approximately 4-6 weeks, flowers appeared
on the plants and were tagged daily. Embryos were readily
evident in ovules by 20 days after flowering (DAF). Excised
embryos or radicles therefrom were fixed for electron
microscopy, as described below, at five day intervals until
bolls opened, approximately 60 DAF. Bolls were allowed to air
dry for several days and seeds were delinted and tested for
germination. Prior to use for germination studies, seeds were
stored in sealed containers held in a refrigerator maintained
at 278 K. Germination tests were done following standard

procedures using germinator rolls placed in an incubator
maintained at 303 K. Greater than 95% germination was
obtained routinely with each seed lot.

13.2.2 Electron microscopy

13.2.2a Chemical fixation. Radicles from embryos or imbibing
seeds were placed directly into a solution of 2%
glutaraldehyde in 0.05 M sodium cacodylate buffer, pH 7.4,
containing 5 mM $CaCl_2$. Following 2 to 4 hr fixation at room
temperature and washing over several hours in 0.1 M buffer
containing 5 mM $CaCl_2$, tissue was post-osmicated in buffered
2% OsO_4, containing 5 mM $CaCl_2$ and 0.8% $K_3Fe(CN)_6$
(Forbes, Plantholt and Sperelakis, 1977; Hepler, 1981) for 2 h
at room temperature. This fixation schedule often selectively
stains ER cisternae.

For selective staining of the endomembrane system a potassium
iodide-osmium tetroxide mixture was used (Carrapico and Pais,
1981). Following buffer washing of tissue fixed in
glutaraldehyde, as described above, samples were placed in a
1:4 dilution of 4% KI to 2% OsO_4 for 24 hr in the dark at
room temperature. All samples were dehydrated in a graded
series of acetone and embedded in Spurr's epoxy resin mixture.

13.2.2b Freeze-fracturing. Recent developments in the
handling of tissue samples for freeze-fracture were used in
this study (Steere, 1981). Fixed or fresh (radicles from dry
seeds having moisture content less than 20%) radicles were
dissected into small slivers containing mainly cortical
tissue. Fracturing of samples was done at 123 K followed by
uni-directional shadowing with platinum and multi-directional
coating with carbon. Replicas were cleaned in dilute followed
by concentrated chromic-sulfuric acid solutions and mounted
onto formvar-coated grids from distilled water or collodion
grids following use of polycarbonate films used to prevent
excessive breakage of replica film (Steere and Erbe, 1983).
Thin sections and freeze-fracture replicas were examined in a
Hitachi H-500H transmission electron microscope. Stereo
tilting of freeze-fracture replicas was at 10°. All
freeze-fracture images are printed with black shadows.

13.3 RESULTS

Samples of radicle tissue prepared by the freeze-fracture
method described above yielded large replicas for examination,
making it easier to select cortical cells and vacuoles or
protein bodies therein. Although the OsFeCN reaction did not
always stain ER cisternae, the general appearance of all
organelles was far superior to other methods we have tried

(Vigil et al., 1984).

13.3.1 Seed ripening

Embryos in ovules at the base of each locule of cotton bolls
were first visible around 20 DAF. At this time ovules
contained a small embryo with a predominant swollen radicle.
As seen in Fig. 13.1, cortical cells in the radicle contain
large vacuoles which have undergone a process of partitioning
similar to that observed by Craig, Goodchild, and Miller
(1980) in cotyledons of developing pea embryos. Several Golgi
apparatus are present along the tonoplast in Fig. 13.1.
Fusion of Golgi vesicles with the tonoplast, however, was not
observed. Between 25 and 30 DAF small crystalline masses of
protein were evident attached to the inner surface of the
tonoplast opposite small clusters of ribosomes attached to the
cytoplasmic side of the tonoplast (Fig. 13.2). Lipid bodies
were more abundant in the cytoplasm at this time. Dilations
of RER were prominent in cortical cells with ribosomes present
only on the undilated portion of the cisternae (Fig. 13.2).

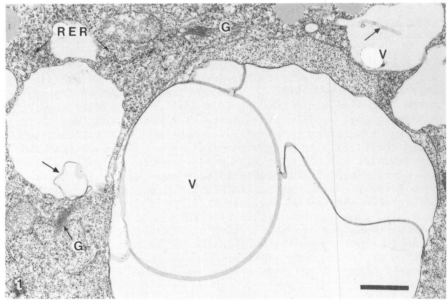

Fig. 13.1 Portion of a cortical cell in radicle of
developing cotton embryo 25 days after flowering (DAF).
Partitioning of large vacuoles (V) has occurred. Several
cisternae of rough endoplasmic reticulum (RER) and Golgi
apparatus (G) are in close proximity to the tonoplast.
Membrane sheets (arrows) present inside several smaller
vacuoles. Bar = 1.0 μm.

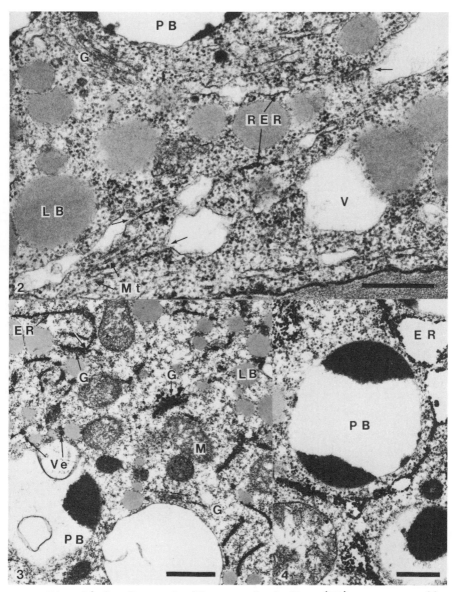

Fig. 13.2. Recognizable protein bodies (PB) appear at 30
DAF. Condensed protein is generally restricted to the
inner region of the PB membrane. Golgi apparatus (G) and
associated vesicles are situated near PB. Numerous small
vacuoles in the form of dilations of RER cisternae are
common in the cytoplasm. Arrows point to the undilated
region of RER and the dilated vacuole. Some fibrillar
material is present along the inner region of the

vacuolar membrane. Microtubules (Mt) are aligned in
parallel to the RER-vacuole complex. LB = lipid body.
Bar = 0.5 μm.
Fig. 13.3. Selective staining of RER and Golgi apparatus
(G) (28 DAF embryo) with the KI/OsO4 reaction provides
information on the connection between a cisterna of RER
and the most trans Golgi cisterna (black and white
arrow), stained Golgi vesicles (Ve) along the protein
body membrane, and absence of detectable reaction product
in protein bodies (PB). LB = lipid body. Bar = 1.0 μm.

Fig. 13.4. At higher magnification the regional dilation
of stained RER (28 DAF embryo) is evident as is the
virtual absence of staining in protein bodies (PB). Bar
= 0.5 μm.

Dilations of RER in cortical cells were observed only from 25
to roughly 35 DAF. Microtubules were common along dilated RER
(Fig. 13.2), as well as adjacent to the membrane of
recognizable protein bodies (data not shown).

Selective staining of RER, trans Golgi cisternae and associated vesicles with the KI-OsO4 reaction revealed connections
of RER with the most trans Golgi cisternae (Arrows, Fig. 13.3)
(GERL?, Marty, 1978) and free Golgi vesicles along the protein
body membrane (Fig. 13.3). The KI-OsO4 reaction also demonstrated dilations of RER which formed vacuole-like structures,
imparting a bead-like appearance to RER cisternae (Fig. 13.3
and 13.4). In view of the distinctive staining obtained with
this reaction, we anticipated obtaining evidence of staining
in developing protein bodies. As shown in Figs. 13.3 and
13.4, however, where reaction product was clearly demonstrable
in RER, Golgi cisternae and vesicles, staining in protein
bodies is not strong. Although Golgi vesicles were in close
apposition to the tonoplast/protein body membrane, there was
no evidence of fusion and only weak staining along the inside
of the tonoplast. Occasionally, fairly strong staining was
observed along the membrane of myelin-like structures within
vacuoles (data not shown).

Freeze-fracture replicas provided information for fracture
faces of the tonoplast/protein body membrane in cortical cells
of radicles at 20 (Fig. 13.5) and 25 (Fig. 13.6) DAF.
Vesicles in apposition to the tonoplast were not uncommon,
but, again, fusion was not observed. The presence of
myelin-like structures inside developing protein bodies was
quite common at 25 to 35 DAF (Figs. 13.6,7 and 9). These
membrane-like structures were generally free of membrane-
associated particles (MAPs) (Figs. 13.6,9) when compared to
the P and E faces of the tonoplast (Figs. 13.5,6 and 9).

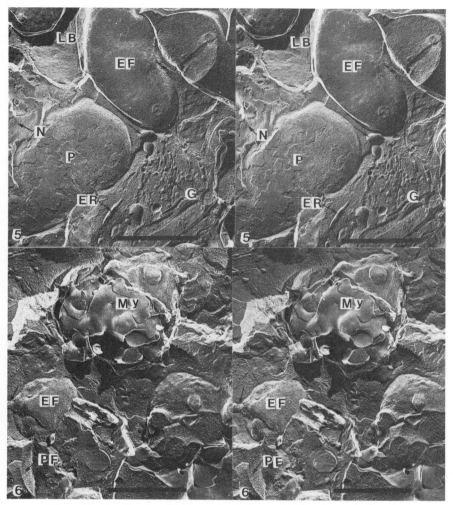

Fig. 13.5 Stereo image of E face (EF) of vacuole in
radicle of 20 DAF embryo. Membrane-associated particles
(MAPs), for the most part, are sparce and randomly
separated. There is some suggestion, however, of MAP
clustering on the EF of the two vacuoles. Vesicles
(arrows), possibly of Golgi (G) origin, are present near
the vacuoles. LB = lipid body; N = nucleus; P =
plastid. Bar = 1.0 μm.

Fig. 13.6. At 25 DAF an apparent qualitative increase in
MAP distribution in the vacuole membrane of both EF and PF
regions. One vacuole has a small tail (arrow), possibly
representing an ER connection. Myelin-like (My)
structures within vacuoles are characterized by a reduced
number of MAPs. Bar = 1.0 μm.

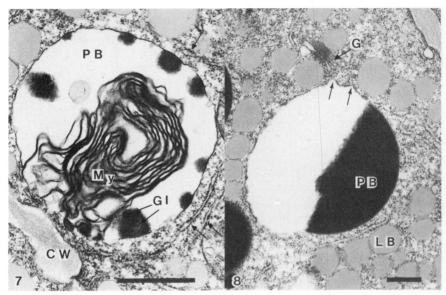

Fig. 13.7. Appearance of myelin-like (My) structure in
protein body (PB) of radicle from 30 DAF embryo. Several
small globoids (Gl) are present within compact protein
attached to the PB membrane. RER cisternae (arrows)
flank the PB except in the region near the cell wall
(CW). Bar = 0.5 µm.
Fig. 13.8. By 50 DAF filling of protein bodies (PB) is
advanced. The contour of the PB membrane is smooth only
along the region of compact protein. The remaining
regions, (arrows) are uneven and undulated, notably
opposite a Golgi apparatus (G). LB = lipid body. Bar =
0.5 µm.

Structural evidence suggesting possible fusion of Golgi
vesicles with the protein body membrane was observed (Fig.
13.8) where slight evaginations appeared along the limiting
membrane opposite a Golgi apparatus. Compaction of
crystalline protein within developing protein bodies
predominates at 45 to 50 DAF (Fig. 13.8), being complete at 55
DAF (Fig. 13.11). Smooth-surfaced vesicles (SI in Swift and
Buttrose, 1972), phytin globoids (Figs. 13.7,11 and 13) and
crystalline protein were the only predominant structures
observed in protein bodies. Globoids which were only found in
association with protein (Prattley and Stanley, 1982)
represent sites of phytin storage (Lott, 1980).
Freeze-fracture replicas showed that the limiting membrane of
vesicles were generally free of MAPs (Figs. 13.13,14).

Throughout protein body loading until crystalline protein

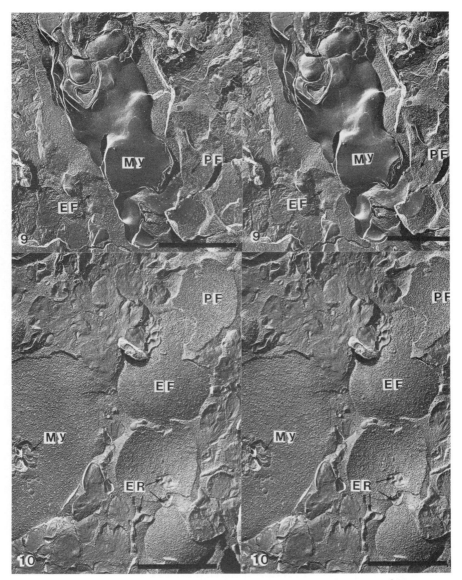

Fig. 13.9. In this stereo freeze-fracture image (33 DAF), myelin-like (My) structure seen in thin sections as a concentric whorl extends the length of the vacuole/protein body. A limited number of MAPs are present on the My in striking contrast to MAP distribution in PF and EF of the limiting membrane. Bar = 1.0 μm.

Fig. 13.10. Protein bodies in cortical cells of radicle
from 45 DAF embryo. The myelin-like (My) structure is
greatly reduced in size, consisting mainly of small
vesicles. Both PF and EF contain numerous MAPs.
Cisternae of ER are apposed to the protein body
membrane. Bar = 1.0 μm.

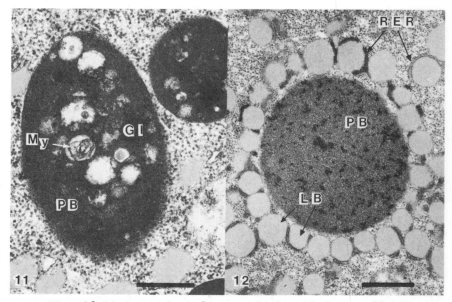

Fig. 13.11. Portion of a cortical cell from a 55 DAF
embryo, illustrating mature spherical protein bodies (PB)
which contain globoids (Gl) and remnants of myelin-like
(My) structures. Bar = 0.5 μm.
Fig. 13.12. Protein body (PB) in cortical cell following
3 hr imbibition. Selectively stained (OsFeCN reaction)
RER cisternae wrap around lipid bodies (LB). Bar = 0.5 μm.

lined the tonoplast/protein body membrane, there was constant
endocytic activity along the free regions of the membrane (see
Figs. 13.1.3). Internalized membrane undoubtedly contributed
to formation of myelin-like structures. As protein bodies
became filled, only small vesicles were found embedded within
the crystalline protein (Figs. 13.11,13 and 14).

Boll opening occured around 60 DAF at which time radicles
contained approximately 50% moisture. Protein bodies at this
time have a spherical shape and were filled with protein and
one or more globoids (Fig. 13.13). Within one day of air
drying the moisture content in the radicle drops to below 20%
and the protein bodies, presumably due to loss of water, have

a more angular and compact appearance (Fig. 13.14). Both RER
and Golgi apparatus were significantly reduced or difficult to
discern amongst the lipid and protein bodies which
collectively filled most of the cytoplasm.

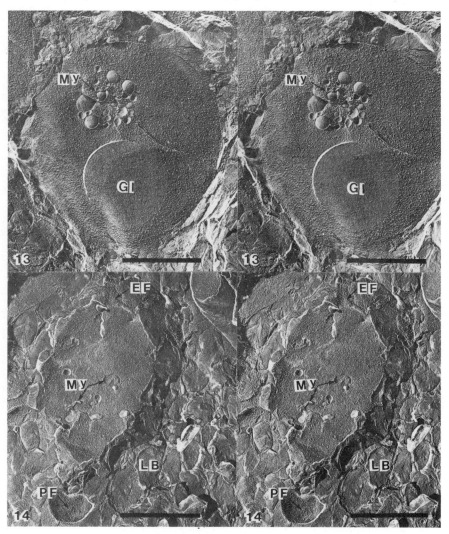

Figs. 13.13 & .14. These two cross-fractures of mature
protein bodies reveal differences between boll opening at
60 DAF (Fig. 13.13) and one day of air drying (Fig.
13.14). Considerable shrinkage has occurred in protein
bodies with the loss of water, resulting in these
organelles having an angular shape. Lipid bodies (LB) in
close apposition to the PB membrane at 60 DAF remains so

even during drying. Gl = globoid; My = myelin-like
structure. Bars = 1.0 μm.

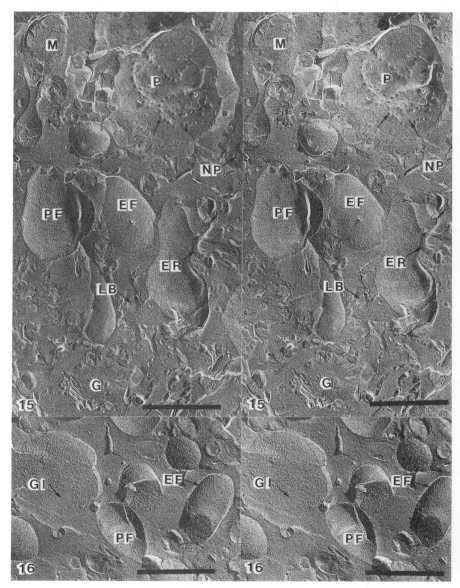

Fig. 13.15. Stereo freeze-fracture replica of cortical
cell of radicle after 4 hr imbibition. Elaboration of
membranous organelles is readily apparent. ER cisternae
are free in the cytoplasm and also in close apposition to
the PF of protein bodies. Th PF and EF of protein bodies

contain more numerous MAPs than ER, Golgi apparatus (G)
or nuclear envelope and pores (NP). Formation of cristae
in mitochondria (M) and thylakoids (arrows) in plastids
(P) complement changes in other membranous organelles.
LB = lipid body. Bar = 1.0 micrometer.
Fig. 13.16. The appearance of membranous organelles in
this stereo freeze-fracture replica from a radicle after
12 hr imbibition reveals the presence of protein bodies
in close apposition, but not fused to form a larger
vacuole-like structure. MAP distribution and frequency
in PF and EF appear similar to that observed at 4 hr
after imbibition. Globoid (Gl) still present in protein
bodies. Bar = 1.0 μm.

13.3.2 Seed germination

Imbibition under standard conditions for germination testing
effects rapid changes in cellular organization. In our
earliest time of sampling, viz. 3-4 hr, numerous short
segments of ER were observed around lipid bodies apposed to
the protein body membrane (Fig. 13.12). At this time the
protein body had regained its spherical shape and the
cytoplasm was less compact than observed in dry tissue (Vigil
et al., 1984). Freeze-fracture replicas of cortical cells at
4 (Fig. 13.15) and 12 (Fig. 13.16) hr of imbibition provided
specific information on membrane structure for various
components of the endomembrane system. With utilization of
stored protein, there was concomitant vacuolation of cortical
cells. This was especially true of several cells removed from
the meristematic region at the radicle apex (Figs. 13.17-19).
Vacuolation appears to occur by a reverse process of protein
body formation, mainly fusion, as suggested by Figs. 13.17 and
13.18.

By 24 hours cells have several recognizable vacuoles, some
containing membranous components, presumably derived from an
endocytic process similar to that observed during seed
ripening (Fig. 13.20). The first series of cell divisions
were more numerous at 24 hr at which time protein
body/vacuoles were distributed around the spindle apparatus,
primarily in association with other organelles peripheral to
the reforming nucleus and advancing cell plate (Fig. 13.21).
Partitioning of organelles during these first cell divisions
appeared equal and symmetric, effecting essentially uniform
distribution of a full organelle complement to each daughter
cell (data not shown; Birky, 1983). At 48 hr of imbibition
the cortical cells are highly vacuolate, with vacuoles often
containing membranous structures relatively free of MAPs (Fig.
13.22).

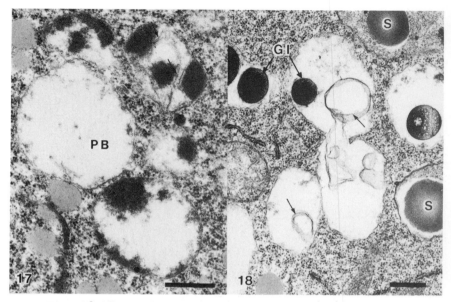

Fig. 13.17.   Cluster of protein bodies (PB) in portion of
a cortical cell from a radicle imbibed for 12 hr.   A fair
amount of condensed protein has been dissolved leaving
occasional crystalline-like strands (arrows).   Bar =
0.5 µm.

Fig. 13.18.   Several protein bodies in cortical cells 18
hr after imbibition.   Globoids (Gl) are still evident,
even though most of the stored protein is absent.
Membranous whorls (arrows) present in protein bodies are
reminiscent of myelin-like structures seen during seed
ripening.   S = starch.   Bar = 0.5 µm.

13.4 DISCUSSION

13.4.1 Protein body formation – A postulate

The cytochemical and freeze-fracture methodologies employed in
this study provide information on the participation of
vacuoles and RER in the formation of protein bodies in
cortical cells of radicles in cotton embryos.   Initially,
through a process of vacuole partitioning, the large central
vacuoles are converted into smaller structures in embryos from
20 to 25 DAF.   Loss of membrane appears to be through
endocytosis with membrane accumulating within the small
vacuole/protein body.   Loading of vacuoles with protein begins
around 25 DAF and continues to organelle maturity.

We found that, beginning around 25 DAF and extending to 35
DAF, numerous dilations of RER contributed to the formation of

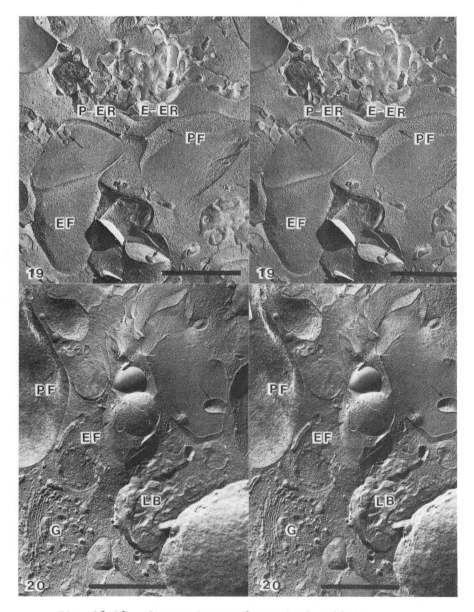

Fig. 13.19. Stereo image of cortical cell in
freeze-fracture replica of radicle imbibed for 18 hr.
Vesicles formed by invagination of the protein body
membrane (arrows) appear to be the mechanism for membrane
loss and bulk uptake of cytoplasm into the vacuole. The
different faces (P-ER and E-ER) for ER are readily
observable in this replica. Bar = 1.0 μm.

326

E.L. Vigil *et al.*

Fig. 13.20. At 24 hr of imbibition, larger vacuolar-like
structures predominate in the cytoplasm, as seen in this
stereo image. Endocytoses (arrows) contribute membrane
to the vacuole. These membranes, while initially having
a complement of MAPs in EF and PF similar to the vacuole
membrane, lose MAPs during apparent digestion within the
lumen of the vacuole. G = Golgi apparatus; LB = Lipid
Body. Bar = 1.0 μm.

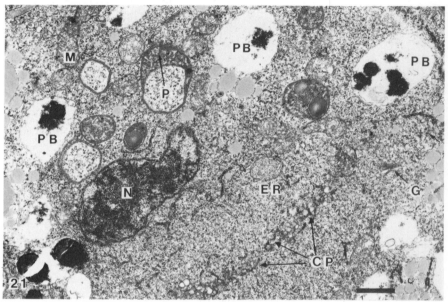

Fig. 13.21. Thin section of dividing cortical cell 24 hr
after imbibition. In Fig. 13.20 the vacuole-like
structures seen are large protein bodies (PB) apparently
derived from fusion of smaller structures. As part of
cytokinesis there is selective distribution of PB,
mitochondria (M) and plastids (P) adjacent to the nucleus
(N) and around the spindle apparatus. CP = cell plate; G
= Golgi apparatus. Bar = 1.0 μm.

an additional set of vacuoles. Beyond this 10 day window, we
did not observe any additional dilations of RER. In fact, RER
became reduced from long cisternae circulating throughout the
cytoplasm, during early stages of embryo development, to short
segments in close apposition to lipid bodies surrounding
mature protein bodies at seed maturity, 60 DAF.

The presence of myelin-like structures within vacuoles and
recognizable protein bodies coupled with active endocytosis
indicates that a significant amount of membrane and adjacent
cytoplasm is internalized. Similar structures were observed

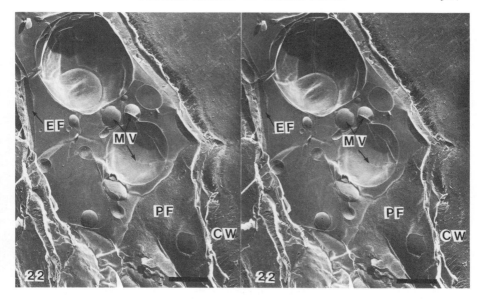

Fig. 13.22. Cortical cells become highly vacuolate by 48
hr of imbibition, as seen in this stereo freeze-fracture
image. The presence of membrane vacuoles (MV) within the
larger vacuole is common. These smaller vacuoles have
very few MAPs by comparison to the PF and EF of the
tonoplast. CW = cell wall. Bar = 1.0 μm.

by Saigo, Peterson and Holy (1983) in developing protein
bodies of oat endosperm, but an explanation for their origin
and function was not provided. We view these structures as
excess membrane, possibly derived from membrane added to the
tonoplast or protein body following vacuole partitioning or
fusion of Golgi vesicles with the limiting membrane. The fact
that protein bodies appear to develop within a constant size
range, implies that specific interplay between membrane
components within the protein body membrane proper or with
cytoplasmic organelles, such as apposed microtubules
and, possibly, other cytoskeletal structures, may be involved
in initiating endocytosis. Additional information is needed
to ascertain whether or not endocytosis is specific for
certain regions of the membrane. Our observation of a
difference in rigidity for regions of the protein body
membrane attached to crystalline protein compared to those
where protein has not been condensing (Fig. 13.8) may be
related to this phenomenon.

One of the difficult questions to answer in this model regards
the process(es) for loading protein and phytin in protein
bodies. Given that partitioning of vacuoles and dilations of
RER provide two separate populations of membrane sacs for

protein body development, it is critical to determine if both, in fact, are loaded with protein and by the same mechanism. Since small vacuoles (Adams, Norby and Rinne, 1985) or dilations of RER (Oparka and Harris, 1982; Adler and Müntz, 1983) become filled with protein, there obviously is some mechanism for effecting this change.

We have approached this problem with freeze-fracture of fixed or unfixed tissue, when possible, and selective cytochemical staining. Wherever we observed vacuoles or protein bodies in freeze-fracture replicas, the pattern of distribution of membrane-associated particles (MAPs) in P- and E-faces was fairly constant. The pattern of MAP distribution in PF and EF of ER, however, is much more dispersed. This suggests that involvement of ER in vacuole/protein body formation requires addition of a significant number of MAPs. We observed little information for fusion of dilated RER with other membranes, suggesting that addition of membrane protein may occur by direct passage from bound ribosomes or lateral migration from undilated regions of RER. Fusion with Golgi vesicles may be a source of membrane components, including protein.

Direct fusion of Golgi vesicles with the limiting membrane of forming protein bodies was not observed in thin sections of material selectively stained for such vesicles. Furthermore, support for vesicle fusion was not obtained through examination of freeze-fracture replicas at all stages of embryogenesis. If Golgi vesicles are involved in transport of glycosylated protein to protein bodies, as suggested by immunocytochemical (Herman and Shannon, 1983a,b) and biochemical (Higgins *et al.* 1983; Chrispeels, 1984) data for cotyledons, then the process is rapid and the contents readily diluted beyond level of detectability with the cytochemical stains employed in the present study.

Protein bodies of cotyledons contain hydrolytic enzymes and function, at least during germination, as autophagic vacuoles (Chappell, Van der Wilden and Chrispeels, 1980; Van der Wilden, Herman and Chrispeels, 1980). In cotton radicles the presence of myelin-like whorls in forming protein bodies and smooth membrane vesicles in mature organelles represent structural evidence in support of an autophagic function for protein bodies. Thus, a major function of vacuoles (Marty, Branton and Leigh, 1980) appears to be retained by protein bodies of radicle cells during filling and unloading, representative of critical transitions in seed ripening and germination.

13.4.2 Protein body changes during germination

The process of imbibition and subsequent germination follows a
particular course which commences with reconstitution of a
full complement of cellular organelles. Elaboration of ER and
Golgi apparatus occurs in parallel with loss of protein from
protein bodies and diminution of lipid bodies. Analysis of
changes in neutral and polar (phospho-) lipids in radicles of
imbibing seeds confirmed that there is a shift from neutral to
polar lipids with corresponding increases in phospholipids
(data not shown). This would be expected from the fine
structural data presented here for a close association between
ER and lipid bodies (3-4 hr of imbibition). The fact that RER
cisternae and lipid bodies are also in close apposition to the
protein body membrane suggests the possibility that amino
acids needed for synthesis of membrane protein may be derived
from hydrolysis of stored protein followed by short transport
to site(s) of utilization.

As organelles become more prominent in cortical cells, there
is a concomitant loss in protein in the protein bodies. These
organelles begin to take on the appearance of vacuoles and,
through a process of fusions (Weber and Neuman, 1980),
contribute to reconstitution of the vacuolar apparatus
observed within two days of imbibition. The importance of
formation of a large vacuolar system during germination may
relate directly to economy of cytoplasm, as suggested by
Matile (1983). The internalization of membrane observed in
vacuoles supports the role of vacuoles and protein bodies in
autophagy (Baumgartner et al., 1980). We consider it
important that the number of MAPs on membranes present within
protein bodies or vacuoles was always much less than present
in the limiting membrane. Additional experimental data are
needed to better explain this phenomenon.

13.5 SUMMARY

We have obtained fine structural evidence indicating that the
formation of protein bodies in radicles during seed ripening
and disappearance of stored protein during germination
involves a time-dependent series of changes implicating
vacuoles and RER along with an autophagic capability for
protein bodies in plant cells. Clearly, protein bodies do not
represent, at least in radicles, organelles only for
compartmentalization of storage protein along with a
complement of hydrolytic enzymes. Instead we view protein
bodies as dynamic organelles which, during seed ripening and
germination, undergo modifications in morphology and membrane
structure and are heterogeneous in function.

13.6 ACKNOWLEDGEMENTS

Special thanks to L. C. Frazier for preparing and photographing thin sections of all materials for CTEM and to Christopher Pooley for printing all figures. Scientific Article No. A-4169, Contribution No. 7154 of the Maryland Agricultural Experiment Station.

13.7 REFERENCES

Adams, C.A., Norby, S.W., and Rinne, R.W. (1985). Production of multiple vacuoles as an early event in the ontogeny of protein bodies in developing soybean seeds. *Crop Sci.* <u>25</u>, 255-262.

Adler, K. and Müntz (1983). Origin and development of protein bodies in cotyledons of *Vicia faba* *Planta* <u>157</u>, 401-410.

Baumgartner, B., Tokuyasu, K. T., and Chrispeels, M.J. (1980). Immunocytochemical localization of reserve protein in the endoplasmic reticulum of developing bean (*Phaseolus vulgaris*) cotyledons. *Planta* <u>150</u>, 419-125.

Bechtel, D.B., Gaines, R. L. and Pomeranz, Y. (1982). Early stages in wheat endosperm formation and protein body initiation. *Ann. Bot.* <u>50</u>, 507-518.

Birky, C.W., Jr. (1983). The partitioning of cytoplasmic organelles at cell division. *Int. Rev. Cytol.* <u>Supp. 15</u>, 49-89.

Bollini, R., Van der Wilden, W. and Chrispeels, M.J. (1982). A precursor of the reserve-protein, phaseolin, is transiently associated with the endoplasmic reticulum of developing *Phaseolus vulgaris* cotyledons. *Physiol. Plant.* <u>55</u>, 82-92.

Bollini, R., Vitale, A. and Chrispeels, M.J. (1983). *In vivo* and *in vitro* processing of seed reserve protein in the endoplasmic reticulum: evidence for two glycosylation steps. *J. Cell Biol.* <u>96</u>, 999-1007.

Boulter, D. (1981). Biochemistry of storage protein synthesis and deposition in the developing legume seed. In: *Advances in botanical research. Vol. 9.* (ed. H. W. Woolhouse) pp. 1-31. Academic Press, London.

Campbell, W.P., Lee, J.W., O'Brien, T.P. and Smart, M.G. (1981). Endosperm morphology and protein body formation in developing wheat grain. *Aust. J. Plant Physiol.* <u>8</u>, 5-19.

Carrapico, F. and Pais, M.S. (1981). Iodure de potassium, tetroxide d'osmium. Un melange d'impregnation pour la microscopie electronique. *C.R. Acad. Sc. Paris.* 292, 131-135.

Chappell, J., Van der Wilden, W. and Chrispeels, M.J. (1980). The biosynthesis of ribonuclease and its accumulation in protein bodies in the cotyledons of mung bean seedlings. *Devel. Biol.* 76, 115-125.

Chrispeels, M. J. (1984). Biosynthesis, processing and transport of storage proteins and lectins in cotyledons of developing legume seeds. *Phil. Trans. R. Soc. Lond.* B 304, 309-322.

Chrispeels, M.J. (1983a). Incorporation of fucose into the carbohydrate moiety of phytohemagglutinin in developing *Phaseolus vulgaris* cotyledons. *Planta* 157, 454-461.

Chrispeels, M.J. (1983b). The Golgi apparatus mediates the transport of phytohemagglutinin to the protein bodies in bean cotyledons. *Planta* 158, 140-151.

Chrispeels, M.J., Higgins, T.J. and Spencer, D. (1982a). Assembly of storage protein oligomers in the endoplasmic reticulum and processing of the polypeptides in the protein bodies of developing pea cotyledons. *J. Cell Biol.* 93, 306-313.

Chrispeels, M.J., Higgins, T.J.V., Craig, S. and Spencer, D. (1982b). Role of the endoplasmic reticulum in the synthesis of reserve protein bodies in developing pea cotyledons. *J. Cell Biol.* 93, 5-14.

Christiansen, M.N. (1963). Influence of chilling upon seedling development of cotton. *Plant Physiol.* 38, 520-522.

Christiansen, M.N. (1968). Induction and prevention of chilling injury to radicle tips of imbibing cottonseed. *Plant Physiol.* 43, 743-746.

Craig, S. and Goodchild, D.J. (1984). Periodate-acid treatment of sections permits on-grid immunogold localization of pea seed vicilin in ER and Golgi. *Protoplasma* 122, 35-44.

Craig, S., Goodchild, D.J. and Hardham, A.R. (1979). Structural aspects of protein accumulation in developing pea cotyledons. I. Qualitative and quantitative changes in parenchyma cell vacuoles. *Aust. J. Plant Physiol.* 6, 81-98.

Craig, S., Goodchild, D.J. and Miller, C. (1980). Structural aspects of protein accumulation in developing pea cotyledons. II. Three-dimensional reconstructions of vacuoles and protein bodies from serial sections. *Aust. J. Plant Physiol.* **7**, 329-337.

Dhar, U. and Vijayaraghavan, M.R. (1979). Ontogeny, structure and breakdown of protein bodies in *Linum usitatissimum* Linn. *Ann. Bot.* **43**, 107-111.

Forbes, M.S., Plantholt, B.A. and Sperelakis, N. (1977). Cytochemical staining procedures selective for sarcotubular systems of muscle:modifications and applications. *J. Ultrastruct. Res.* **60**, 306-327.

Hepler, P.K. (1981). The structure of the endoplasmic reticulum revealed by osmium tetroxide-potassium ferricyanide staining. *Eur. J. Cell Biol.* **26**, 102-110.

Herman, E.M. and Shannon, L.M. (1984a). Immunocytochemical evidence for the involvement of Golgi apparatus in the deposition of seed lectin of *Bauhinia purpurea* (Leguminosae). *Protoplasma* **121**, 163-170.

Herman, E.M. and Shannon, L.M. (1984b). Immunocytochemical localization of concanavalin A in developing jack-bean cotyledons. *Planta* **161**, 97-104.

Herman, E.M., Baumgartner, B. and Chrispeels, M.J. (1981). Uptake and apparent digestion of cytoplasmic organelles by protein bodies (protein storage vacuoles) in mung bean cotyledons. *Eur. J. Cell Biol.* **24**, 226-235.

Higgins, T.J.V., Chrispeels, M.J., Chandler, P.M. and Spencer, D. (1983). Intracellular sites of synthesis and processing of lectin in developing pea cotyledons. *J. Biol. Chem.* 258, 9550-9552.

Kristen, U. and Biedermann, M. (1981). Ultrastructure, origin, and composition of protein bodies in the ligule of *Isoetes lacustris* L. *Ann. Bot.* **48**, 655-663.

Lott, J.N.A. (1980). Protein bodies. In: *The biochemistry of plants. Vol. I.* (ed. N.E. Tolbert). pp. 589-623. Academic Press, New York.

Mader, M. and Chrispeels, M.J. (1984). Synthesis of an integral protein of the protein-body membrane in *Phaseolus vulgaris* cotyledons. *Planta* **160**, 330-340.

Marty, F. (1978). Cytochemical studies on GERL, provacuoles, and vacuoles in root meristematic cells of Euphorbia. *Proc. Natl. Acad. Sci. U.S.A.* **75**, 852-856.

Marty, F., Branton, D. and Leigh, R.A. (1980). Plant vacuoles. In: *The biochemistry of plants. Vol. I.* (ed. N. E. Tolbert). pp. 625-658. Academic Press, New York.

Matile, P. (1982). Vacuoles come of age. *Physiol. Veg.* **20**, 303-310.

Neumann, D. and Weber, E. (1978). Formation of protein bodies in ripening seeds of *Vicia faba* L. *Biochem. Physiol. Pflanzen.* **173**, 167-180.

Oparka, K.J. and Harris, N. (1982). Rice protein-body formation: All types are initiated by dilation of the endoplasmic reticulum. *Planta* **154**, 194-188.

Prattley, C.A. and Stanley, D.W. (1982). Protein-phytate in protein bodies and globoids. J. Food Biochem. 6, 243-253. Saigo, R.H., Peterson, D.M. and Holy, J. (1983). Development of protein bodies in oat starch endosperm. *Can. J. Bot.* **61**, 1206-1215.

Steere, R.L. and Erbe, E. (1983). Supporting freeze-etch specimens with "lexan" while dissolving biological remains in acid. In: *Proc. 41st annual meeting of the electron microscopy society of America* (ed. G.W. Bailey). pp. 618-619. San Francisco Press, San Francisco.

Steere, R. L. (1981) Preparation of freeze-fracture, freeze-etch, freeze-dry, and frozen surface replica specimens for electron microscopy in the Denton DFE-2 and DFE-3 freeze etch units. In: *Current trends in morphological techniques. Vol. II.* (ed. J.E. Johonson, Jr.). pp. 131-181. CRC Press, Cleveland.

Swift, J.G. and Buttrose, M.S. (1972). Freeze-etch studies of protein bodies in wheat scutellum. *J. Ultrastruct. Res.* **40**, 378-390.

Van der Wilden, W., Herman, E.M. and Chrispeels, M.J. (1980). Protein bodies of mung bean cotyledons as autophagic organelles. *Proc. Natl. Acad. Sci. U.S.A.* **77**, 428-432.

Vigil, E.L., Steere, R.L., Wergin, W.P. and Christiansen, M.N. (1984). Tissue preparation and fine structure of the radicle apex from cotton seeds. *Amer. J. Bot.* **71**, 645-659.

E.L. Vigil *et al.*

Weber, E. and Neuman, D. (1980). Protein bodies, storage organelles in plant seeds. *Biochem. Physiol. Pflanzen.* <u>175</u>, 279–306.

14

Intercellular transport studied by micro-injection methods

P. B. Goodwin and M. G. Erwee

14.1 INTRODUCTION

A number of approaches are available for the study of intercellular transport in plants. The presumed routes of this
transport are the plasmodesmata, which provide a pathway from
cell to cell within the plasmalemma, the cells outer membrane.
Ultrastructural studies suggest that transport occurs via the
cytoplasmic annulus of the plasmodesma, rather than the axial
desmotubule, the latter consisting of a membrane bilayer
tightly constrained to a solid rod (Overall *et al.* 1982;
Thomson and Platt-Aldia 1985). Measured plasmodesmatal
densities are consistent with transport roles, for example for
assimilate movement between mesophyll and phloem in the leaf,
with a final apoplastic step close to the sieve tube/companion
cell complex (Gamalei *et al.* 1981; Fisher and Evert 1982;
Giaquinta 1983).

Ultrastructural studies have also provided the basis for
estimates of the conductivity of plasmodesma, and a number of
workers have compared these estimates to measured solute
fluxes, for example Tyree 1970; Kuo *et al.* 1974 (leaf vascular
bundles); Gunning 1976 (sucrose movement); Gunning and Hughes
1976 (*Abutilon* nectaries); Osmond and Smith 1976 (C4 photosynthesis); Robards and Clarkson 1976 (root endodermis) and
Walker 1976 (*Chara* node). The general conclusions from this
work are that plasmodesmata probably provide the pathway of
least resistance (compared to the trans-plasmalemma route) for
the intercellular movement of small electrolytes and non-
electrolytes in a number of situations, and that diffusion
should occur at rates adequate to provide the measured fluxes,
assuming gradients in solute concentration which are quite
sustainable physiologically.

A more direct approach to the study of intercellular transport has been to supply solutes to the external solution, and to follow their uptake and transport. The best known study of this type was carried out by Arisz and coworkers, summarised in Helder 1967 and Arisz 1969. The work was carried cut on submerged leaves of the water plant *Vallisneria spiralis*; either on leaf strips 7.5 to 45 cm long, with uptake in a short compartment isolated from the rest of the leaf, the latter being in one or more transport compartments; or alternatively on intact plants, with 15 cm of one leaf in the uptake compartment. Extensive movement of label from auxin, amino acids, sugars and inorganic ions was found. Since the tissue lacks xylem vessels or a cuticle, but there was little or no loss to the external solution in the transport compartment, and since movement was not altered by inhibitors of membrane uptake supplied in the transport compartment, movement was probably occurring wholly within the plasmalemma. That is, Arisz and coworkers demonstrated long distance movement in the symplast. In some experiments (for example Arisz and Wiersema 1966) the main vascular bundles were cut out over distances up to 16 mm, leaving what were presumed to be parenchyma cell bridges. The movement of tracers was similar with vascular bundle or 'parenchyma' bridges. However, the group used leaves 5-6 mm wide, and Lumley (1976) has suggested that such leaves carry minor phloem bundles in regions assumed by Arisz to lack vascular tissues.

The symplast encompasses both cell to cell movement, and long distance movement in the phloem. Arisz and coworkers found in their studies a number of features not expected in cell to cell diffusion. Movement was very fast, often 100 cm in 24 hours, and polar in that most substances moved more readily in a basipetal direction. Sucrose applied to the leaf base reversed this polarity, as did inhibitors of membrane uptake applied to the leaf apex (Arisz 1969). The probable explanation of these results is that they are a consequence of long distance movement in the phloem, polarity being due to the basipetal flux of assimilate, and being reversed when the sucrose gradient is reversed, or uptake at the apex prevented.

More rigorous studies have been carried out by Littlefield and Forsby (1967) on $P^{32}O_4$ movement in *Chara*. Label was shown to move intercellularly in a submerged tissue without phloem or a cuticle. Bostrom and Walker (1975) carried out a detailed study of the intercellular transport of chloride in *Chara*.

A number of important studies have been carried out on the intercellular movement of fluorescein. Schumacher (1936) observed its intercellular movement in *Cucurbita* leaf hairs, Bauer (1953) in *Cucurbita* hairs and in *Nitella*. Tyree and Tammes

(1975) in *Tradescantia* staminal hairs, and Barclay *et al.* (1982) in trichomes of *Lycopersicon esculentum*. Movement showed the properties of a process limited by diffusion through the cell walls (the plasmodesmata?). Transport was non-polar in terms of cell number, and the number of cells reached was proportional to the square root of time. The rate of diffusion was slower than physical diffusion in water, and was not limited by cytoplasmic streaming. Barclay *et al.* showed that plasmolysis reduced the rate of transport by 70%. The incomplete inhibition of transport by plasmolysis is the consequence of an unfortunate property of this dye, that it has a reflection coefficient below 1; that is to say, it can pass the plasmalemma at an appreciable rate (Socolar and Loewenstein 1979; Goodwin 1983). Thus its intercellular transport, particularly in tissues protected by a cuticle, is due to a combination of symplastic and apoplastic transport.

Animal physiologists have long been interested in intercellular transport as a mechanism for the physiological integration of tissues. Animal cells are linked by gap junctions, which are formed by the close apposition of particles in the junctional plasma membranes. The gap junctions provide a low resistance pathway for the intercellular movement of both electrical and chemical signals. Chemicals shown to move intercellularly include ions such as Na^+, K^+, Cl^-, SO_4^{2-}, Co^{2+} and C^{14} labelled molecules such as tetraethylammonium (130 dalton), adenosine monophosphate (328 dalton) and sucrose (342 dalton) (see De Mellio 1982; Findlay 1984 for reviews). Recently a number of fluorescent probes have been developed. The first of these was procion yellow (697 dalton), introduced by Stretton and Kravitz (1968) to determine neuronal geometry. A far more fluorescent probe is lucifer yellow (473 dalton), developed by Stewart (1978).

Of special significance was the production of a series of probes consisting of conjugates of fluorescent dyes with amino acids or peptides. These were tested on *Chironomus* salivary gland cells (Simpson *et al.* 1977) and on cultured insect and mammalian cells (Flagg-Newton *et al.* 1979). More recently the series has been extended to include dye conjugates with glycopeptides and oligosaccharides (Schwarzmann *et al.* 1981). The largest of these molecules able to move intercellularly has a width of 1.6 to 2.0 nm for mammalian cells, and of 2.0 to 3.0 nm for insect cells (Schwarzmann *et al.* 1981). The extent of movement of the probes falls as they approach the size limit. The mammalian cell junctions appear to select against large negatively charged molecules. In molecular weight terms, the molecule of highest weight able to pass insect cell junctions is 3002 dalton, and to pass mammalian cell junctions 819 dalton.

Table 14.1

Fluorescent probes commonly used for studying
intercellular transport in plants

Probe	Molecular weight (dalton)
6-carboxyfluorescein (6COOHF)	376
lucifer yellow	457
lissamine rhodamine B (LRB)	559
F* conjugated to	
glutamic acid (FGLU)	536
glutamylglutamic acid (F(GLU)$_2$)	665
hexaglycine (F(GLY)$_6$)	749
leucyldiglutamylleucine (FLGGL)	874
pentemer of prolylprolylglycine (F(PPG)$_5$)	1678

* : Fluorescein isothiocyanate isomer I.

Note in excess of thirty other FITC conjugates have been used
     in plants, 24 by Tucker (1982).

14.2   PROPERTIES OF THE FLUORESCENT PROBES

The critical features of the fluorescent probes (Table 14.1) for
the study of membrane-bound intercellular transport in plants
are as follows:

14.2.1   Reflection coefficient

The probes pass the plasmalemma only at very slow rates.   In
soak trials lasting 2 hours no obvious loading of *Egeria* or
Elodea cells is seen (Goodwin 1983; Erwee and Goodwin 1983) and
injected dye is still visible 2 days after injection.   However,
Kanchanopoom *et al.* (1985) found 6-carboxyfluorescein to enter
protoplasts and leaf mesophyll cells after overnight incubation
in the dye.   Steinbiss and Stahl (1983) found loss of lucifer

yellow from tobacco protoplasts over 3-4 days and Santiago
(pers. comm.) found 6-carboxyfluorescein to persist in *Petunia*
leaf and stem epidermal cells for 7 days, but to be lost by
14 days after injection. Nevertheless, given that intercellular movement of injected dye can be seen within seconds,
and the average injection is assessed within 15 minutes, the
flux of dye through the plasmalemma is sufficiently slow to be
ignored in these studies.

14.2.2 Binding

Lack of binding to extraplasmalemmal structures can simply be
demonstrated in soak trials. If the tissue is left in a low
concentration of the fluorescein conjugates for up to 2 hours,
and examined directly under the epifluorescence microscope,
bound dye would be expected to be seen as a higher than
background fluorescence. This is not found, except with dyes
of the type of calcofluor M2R new, which is known to bind to
β-linked polysaccharides (Maeda and Ishida 1967). Two types
of experiment would argue the absence of binding to structures
within the plasmalemma. Firstly, if *Egeria* leaves are squeezed
by forceps, then some cells are damaged sufficiently that the
plasmalemma becomes permeable to the fluorescent dyes.
However, apart from an absence of streaming, they appear
normal. Soak trials give the same type of result as with the
intact tissue, but dye is within the cells. That is to say,
there is no suggestion of binding to any structure. The
second type of experiment is to prick cells to which dye has
moved following injection. All dye appears to rapidly escape,
even though the cytoplasm and organelles remain in the cell.

14.2.3 Toxicity

The probes do not rapidly damage the cells - and in fact it is
probable that normal development can occur. Cytoplasmic
streaming often continues in the injected cell - sometimes
after an interruption, an accumulation of particles at the site
of injection and then recovery. In nearby cells to which dye
moves intercellularly, streaming is not interrupted. We have
found that dyes can remain in cells in excess of 2 days, without
causing apparent damage - for example a change in the rate of
streaming. Steinbiss and Stabel (1983) and Palevitz and Hepler
(1985) report that cells microinjected with lucifer yellow
survive and can go on to divide.

14.2.4 Stability

Two types of test have been used to assess stability of the
fluorescein conjugates. In the first the probes have been
incubated with tissue homogenates, and then chromatographed or
electrophoresed. No breakdown products were found (Tucker 1982;
Goodwin 1983). In a more critical study (Goodwin, unpublished)
the probes 6-carboxyfluorescein, F-glutamic acid,
F-glutamylglutamic acid and F-hexaglycine were injected into
Egeria densa leaves. After 20 minutes the leaf was rinsed well
and the probe extracted by crushing the cells onto HPTLC plates.
No breakdown products were seen after chromatography. Thus any
intracellular metabolism proceeds at a relatively slow pace,
compared to intercellular movement.

14.2.5 Types available

As mentioned in the introduction, a wide range of probes are
available - the FITC conjugates are readily prepared and
purified following Simpson (1978). The LRB probes can be
prepared following Nairn (1976), but purification requires a
combination of paper chromatography, paper electrophoresis
and gel permeation chromatography. There are no reported
limitations on the peptides which can be used.

Two plant dyes of special significance are procion yellow
and lucifer yellow. Procion yellow survives normal histological procedures, enabling dye filled internal cells to be
identified in sections (see Samejima and Sibaoka 1983).
Lucifer yellow gives a higher quantum yield than
procion yellow, and is more mobile. It can be fixed by
glutaraldehyde, but in addition can be reacted with
diaminobenzidine to give an osmophilic polymer which can be
visualised in the electron microscope (Maranto 1982).

14.2.6 Injection

Since the probes are selected as being unable to pass the
plasmalemma, its is necessary to bring them into the cytoplasm
in some fashion. The preferred method is direct injection using
glass microelectrodes. For reviews on injection into animal
cells see Kater *et al.* 1973; Socolar and Loewenstein 1979;
Celis 1984. In plant cells no particular problems have been
encountered. Injection by both iontophoresis and pressure have
been used. Tucker (1982) found neither to be necessary to
obtain movement from the electrode tip into the cell. Pressure
appears to be superior for large cells or compounds of very
high molecular weight.

14.2.7 Monitoring injections

The fluorescent probes have the great advantage that the
injection and subsequent movement can be monitored as it occurs.
Fluorescein concentrations of the order of 0.06 ppm are visible
in a layer of dye 165 μm thick (Barclay *et al*. 1982), so that
in a cell 33 μm thick, a concentration of 0.3 ppm would be
visible. The most sensitive measure of fluorescence is with a
photomultiplier - for example the Leitz MPV 3 system, but in
most cases spatial resolution is required, and the best detector
becomes an image intensifier TV tube (with a threshold of
20×10^{-9} ft. candles, compared to 0.03×10^{-9} ft. candles for a
photomultiplier, Socolar and Loewenstein 1979). Photographic
film is 3-4 orders of magnitude less sensitive again. If
films are used, small format films are more sensitive. We
currently use Ektachrome P800/1600 film, rated at 3200 ASA.

It should be noted that all fluorescent dyes are prone to
fading with extended exposure to strong light. However, in our
experience fading is only a problem with exposures to the
exciting wavelength of several minutes.

14.3 CHARACTERISTICS OF INTERCELLULAR TRANSPORT
AS REVEALED BY MICROINJECTION

The probes have been used to study a number of the long
standing problems relating to cell-cell communication in plants.

14.3.1   What size molecules can move freely from cell to cell?

No system has so far been found where probes of molecular
weight larger than 874 dalton can move intercellularly in
normal circumstances. Results of the few studies carried out
to date are summarised in Table 14.2. To this might be added
that Samejima and Sibaoka (1983) found no intercellular
movement of procion yellow (697 dalton) in the vascular
parenchyma of petioles of *Mimosa pudica*. The data from the
Elodea leaf were used to estimate that the equivalent pore
diameter of the leaf epidermal cell plasmodesmata lies in the
range 3.0-5.0 nm. This is half the radius of the cytoplasmic
annulus at the neck constriction, as revealed by electron
microscopy of the very similar *Egeria* leaf (Erwee and Goodwin
1985). A similar disagreement was found for the nuclear pore
size, as determined by electron microscopy and molecular
sieving (Paine *et al*. 1975). These results are very
satisfactory for systems where different cytoplasmic
biochemistry lies on each side of the plasmodesmata, for

example in C_4 plants (Edwards and Huber 1981). However, they
leave unresolved the question of what special adaptations
viruses use to move intercellularly (see Atabekov and Dorokhov
1984 for a recent review). Furthermore, if viruses can alter
the intercellular channels to move, then surely the plant
itself can make the same alterations on   occasion, and so
allow exceptionally large molecules to move intercellularly.

Table 14.2

Largest fluorescein-peptide conjugates
able to move intercellularly in plants

Tissue	Largest mobile probe	MW	Reference
Egeria densa			
leaf epidermis	$F(GLY)_5$	674	Erwee & Goodwin, 1985
shoot apex	$F(GLU)_2$	665	" " "
stem epidermis	6COOHF	376	" " "
stem cortex	6COOHF	376	" " "
root epidermis	6COOHF	376	" " "
root cortex	6COOHF	376	" " "
Elodea canadensis			
leaf epidermis	FLGGL	874	Goodwin, 1983
Setcreasea purpurea			
staminal hairs	$F(ALA)_6$ *	834	Tucker, 1982
Silene coeli-rosa			
vegetative shoot apex	$F(GLU)_2$	665	Goodwin & Lyndon, 1983

*
No upper limit established, as larger immobile molecules
contain aromatic amino acids, shown to prevent intercellular
transport (Erwee and Goodwin 1983b) or a preponderance of
neutral amino acids, which also leads to immobility.

14.3.2 Are all plant cells linked into a single symplast?

It has long been apparent that two tissues, the guard cells and
the phloem/companion cell complex, could not be well linked to
other tissues. Both function by developing high turgor
pressures relative to the adjoining tissues. The turgor
difference would be dissapitated if there was efficient cell-
to-cell communication. On the other hand, plasmodesmata have
been reported, albeit in reduced numbers, linking these
tissues to adjoining cells. The symplastic isolation of guard
cells has now been demonstrated in four species by the use of
fluorescent probes (Erwee *et al.* 1985; Palevitz and Hepler
1985). No studies on the phloem have been reported, but Erwee
et al. (1985) found intercellular dye movement
(6-carboxyfluorescein and lucifer yellow) in the leaf of
Commelina cyanea between epidermal, spongy and palisade
mesophyll and vascular cells.

    The most studied plant in regard to symplastic movement is
Egeria densa, where Erwee and Goodwin (1985) have surveyed the
extrastellar tissues. As judged by the intercellular transport
of 6-carboxyfluorescein, all cells are not linked. Rather the
plant appears to be split into symplast domains (Fig. 14.1).
The domains differ in the size of the largest probe able to

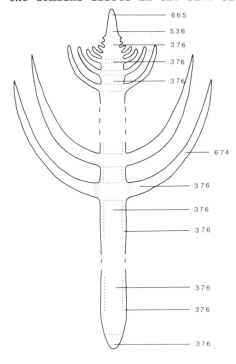

move intercellularly
(Table 14.2). The limited
results to date give a
fairly consistent picture.
No tissue has as yet been
reported with a molecular
exclusion limit in excess
of 900 dalton.

Fig. 14.1   A diagram of
Egeria densa showing the
molecular exclusion limits
of different extrastelar
tissues (in dalton) and
the apparent barriers to
movement of
6-carboxyfluorescein.

In *Egeria densa* the leaf epidermal and shoot apex symplasts are the most permeable. The leaf symplast would be expected to be permeable enough to cope with the flux of sucrose to the phloem. The shoot apex in *Egeria* is an actively growing tissue, in an aqueous environment, which lacks sieve tubes within several hundred microns (Erwee and Goodwin 1985).

Barriers to symplastic dye movement are found between the epidermis and cortex in both the stem and root; at the nodes and leaf bases in the shoot, and between the root cap and the remainder of the root. The existence of these barriers, most notably the root epidermis/cortex barrier (Fig. 14.2) flies in the face of evidence for the symplastic movement of ions into the root. Furthermore, there appears to be a matching apoplastic barrier to calcofluor (Fig. 14.3). If these barriers were fully effective, and applied to all molecules, nothing could enter the root. However, both barriers have been established using molecules far larger than hydrated ions. Carboxyfluorescein has a diameter of 1.3 nm, but the hydrated forms of K^+, Cl^- and NO_3^- have a diameter of 0.3 nm. Calcofluor M2R new has a molecular weight of 900 dalton. Thus barriers to the dyes are not necessarily barriers to water or small ions.

Tenuous evidence for the existence of a symplastic restriction in the outer cortex of roots exists. Mertz and Higinbotham (1976) found a sharp fall in membrane potential (of 5 to 49 mV) between the epidermal and first cortical cell layer in barley, and a further fall to the second cortical layer. They interpreted the difference as reflecting the characteristic ion transport of each cell, but possibly the cells are also symplastically isolated.

Evidence for apoplastic barriers is more commonly found. Hypodermal barriers have been reported in *Zea mays* (Ferguson and Clarkson 1975), *Eucalyptus* (Tippett and O'Brien 1976), *Allium cepa* (Peterson *et al.* 1978), *Hoya* (Olesen 1978), *Carex arenaria* (Robards *et al.* 1979) and citrus (Walker *et al.* 1984). Peterson *et al.* (1982) and Peterson and Perumalla (1984) have identified the hypodermal barrier as a casparian band in onion and corn, and shown that it is a barrier to apoplastic tracers. However, Clarkson *et al.* (1978) showed that the onion hypodermis was permeable to water, ions and small organic molecules. Colloidal lanthanum will move to the endodermis in barley (Clarkson and Robards 1975), as will U^{2+} (see Haynes 1980 for a review). Peterson *et al.* (1981) found the non-binding apoplastic dye PTS (508 dalton) to pass the corn hypodermis, but not the cellulose-binding apoplastic dye Tinopal CBS (563 dalton). The latter dye has a very extended configuration, and would be expected to be blocked more readily

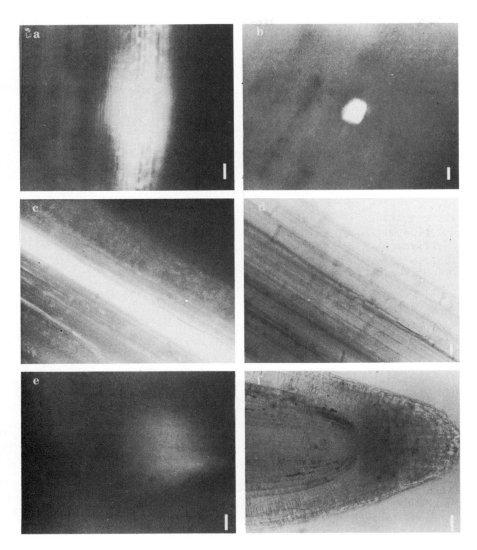

Fig. 14.2*a* Extensive movement of 6-carboxyfluorescein in root
epidermal cells of *Egeria densa*. *b*. Non-movement of F(GLU) in
the epidermis. *c*. 6-Carboxyfluorescein moves freely in the root
cortex but the dye does not move into the hypodermis.
d. Transmitted light photograph of the same area as (c).
e. 6-Carboxyfluorescein moves freely in the root cap, but not
into the root itself. *f*. Transmitted light photograph of the
same area as (e). In (a): bar = 50 µm. In (b) to (f): bar =
25 µm.

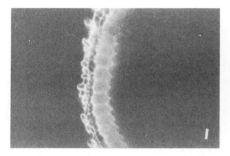

Fig. 14.3   The distal 50 mm of *Egeria densa* roots were cut
off, and the cut end blotted dry and sealed with sticky wax.
Roots were submerged in a solution of 0.01% calcofluor M2R New
in distilled water overnight, rinsed 1 hour in cold flowing
tap water, hand sectioned and then mounted in glycerine.
a. Calcofluor is restricted to the epidermis and hypodermis.
b. Transmitted light photograph of the same area. Note: the
full cortex stained when deep cuts were made through the
epidermis/hypodermis.   Bar = 25 µm.

by small pores. Thus only the *Carex* hypodermis has been shown
to be impermeable to small molecules.

The functional importance of the symplast barrier is also
uncertain. A number of studies offer good evidence for ion and
water uptake at the epidermis, with further flux via the
symplast, or at least within the plasmalemma; for example
phosphate uptake in corn roots (Malone *et al*. 1977), K^+ uptake
in corn roots (Kochian and Lucas 1983) and water transport in
barley roots (Steudle and Jeschke 1983). The induction of
epidermal transfer cells by salt treatment or iron deficiency
(Kramer *et al*. 1978, 1980) is consistent with this concept, as
is the high plasmodesmatal density at the base of root hairs
(Vakhmistrov and Kurkova 1978). Peterson and Perumalla (1984)
found that fluorescein could pass the hypodermis from an
external solution, suggesting a symplastic route. However, as
previously mentioned, fluorescein can move in both the symplast
and apoplast.

Little information is available with which to assess the
other symplast domains. It has been tentatively suggested that
a reduced plasmodesmatal pore size at the hypodermis in the
stem and root, and at the leaf base, may serve to restrict the
movement, not of small ions and water but of larger potentially
toxic organic and inorganic molecules, including Mn^{2+} and Al^{3+}
(Erwee and Goodwin 1985). Other roles may also be served by
the restriction – for example rapid isolation of the epidermal

cells in the event of damage to them, or separation of
different plasmalemma constituents in a fluid membrane. Thus,
if the epidermal plasmalemma is especially adapted for cation
uptake (see for example Van Iren and Boers-van der Sluij 1980),
for Na^+/K^+ exchange, and for Cl^- extrusion, then it is likely
to have constituents different to those in the adjacent
cortical cells. Were these cells in free communication, then
the essential membrane molecules could conceivably be lost via
the plasmodesmatal membrane to the cortex. Restricted cell-
to-cell communication might limit this loss. Techniques are
now available to test this proposition.

14.3.3 Can the permeability of the symplast be regulated?

Three factors have been shown so far to be able to regulate the
permeability of the symplast: group II ions, plasmolysis, and
aromatic amino acids.

14.3.3a Group II ions. The permeability of the *Egeria* leaf
symplast is reduced by the injection of group II ions
(Fig. 14.4), Erwee and Goodwin 1983. It is possible to inject
cations before the anionic probe by loading the microelectrode
with a mixture of both chemicals. Iontophoresis with a
positive current flow injects the cation into the cell. After

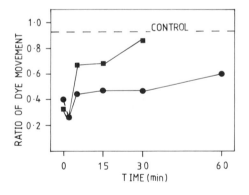

Fig. 14.4 Ratio of the
number of injections where
$F(GLU)_2$ movement occurred,
to the total number of
injections as influenced by
time after cation injection.
Squares, Ca^{2+}; circles, Mg^{2+}.

an appropriate delay, a negative current injects the probe into the same position in the cell. Group II ions (Mg^{2+}, Ca^{2+}, Sr^{2+}), but not K^+ or Na^+ inhibit intercellular movement of the probe, having maximum effect when there is a delay of one minute between injection of the ions and dye. With a 30 minute delay, Ca^{2+} ions have little effect on dye movement. The inhibition applies even to the dye of lowest molecular weight used (6-carboxyfluorescein, 376 dalton). Inhibitors expected to increase the cytoplasmic level of group II ions have a similar effect. Thus both the group II ionophore A23187 and the herbicide trifluralin (which inhibits mitochondrial uptake of Ca^{2+}: Hertel et al. 1980), when given as a ten minute soak prior to probe injection, block intercellular movement. Calcium applied extracellularly has no effect.

Calcium ions are known to play a major role in regulating cellular metabolism in both plants and animals (Clarkson 1984; Rasmussen and Goodman 1977), and in particular, to regulate cell-cell communication between animal cells (Rose et al. 1977). Calcium is found to induce the appearance of electron dense granules around the plasmodesmatal neck regions, which may contain a Ca-ATPase (Belitser et al. 1982). Thus calcium ions may act directly on the plasmodesmata to regulate intercellular communication.

Since cytoplasmic levels of free group II ions are generally very low, with higher concentrations in the mitochondria and chloroplasts, any damage to the plasmalemma, or to the biochemical integrity of the organelles is likely to increase the cytoplasmic level of group II ions, and isolate the damaged cell from adjoining healthy cells. Healthy cells too can use this simple mechanism to isolate themselves from their neighbours.

14.3.3b Aromatic amino acids. Tucker (1982) showed that probes containing an aromatic amino acid did not move intercellularly in Setcreasea purpurea staminal hairs. It has been shown that, in Egeria at least, such probes also block the transport of coinjected probes which would normally be mobile, and that the free aromatic amino acids have the same effect (Erwee and Goodwin 1984). The barrier is not at the tonoplast, as the aromatic conjugates fill the entire cell. The site of action is within the cell, in that bathing the tissue in an aromatic amino acid has no effect on intercellular transport.

14.3.3c Plasmolysis has long been accepted as a method for interrupting cell-cell communication, and Erwee and Goodwin (1984) have shown by direct injection of probes into plasmolysed Egeria leaf cells that intercellular transport is blocked. More unexpected results were obtained on

deplasmolysis. Within ten minutes full intercellular transport
is restored, and within 2 hours there is a major increase in
symplast permeability, compared to controls neither plasmolysed
or deplasmolysed (Table 14.3). The increase involves the more
extensive movement of probes which usually show limited
movement; the movement of probes up to 1678 dalton, which
normally never move; movement of probes becoming relatively
insensitive to Ca^{2+} ions; and the symplast becoming permeable
to probes containing aromatic amino acids. The symplast does
not return to its normal permeability within 20 hours of
deplasmolysis. The results show that deplasmolysis disrupts

Table 14.3

Movement of dyes in leaves of *Egeria densa* which have
to be subjected to plasmolysis (120 minutes in 0.7 M
mannitol) and deplasmolysis (120 minutes). R is the
fraction of injections where dye movement occurred.
The numbers in ( ) are the total number of injections
per treatment. The cell number is the average
    number of adaxial epidermal cells containing dye,
excluding the injected cell, and counting only those
injections where movement occurred.

Dye	Molecular weight (dalton)	Control R	Control cell number	Deplasmolysed R	Deplasmolysed cell number
$F(GLU)_2$	665	0.93(15)	9.1	1.00(15)	15.2
$F(GLY)_6$	749	0.00(15)	0.0	1.00(15)	10.9
FLGGL	874	0.00(15)	0.0	0.50(20)	5.0
$F(PPG)_5$	1,678	0.00(14)	0.0	0.75(16)	3.7
LRB-insulin A chain	4,158	0.00(10)	0.0	0.00(14)	0.0
$F(Tyr)$	571	0.13(15)	5.0	1.00(8)	16.3
$F(Phe)$	555	0.07(15)	2.5	1.00(8)	22.8
$F(Trp)$	594	0.13(16)	3.5	1.00(8)	20.0

the normal regulation of symplast permeability, and makes the *Egeria* symplast more permeable. If this is the normal result of deplasmolysis (in those cases where deplasmolysis does lead to the resumption of intercellular transport), then studies on the importance of cell isolation, where isolation was achieved by temporary plasmolysis, need to be reassessed. Thus Wetherell (1984) found increased numbers of adventitious embryos after brief plasmolysis of cultured wild carrot cells. The result, attributed to cell isolation, could be due to increased intercellular communication.

14.3.4 Are there changes in symplast permeability with development?

Changes in symplast permeability with development have been shown in three cases. In vegetative apices of *Egeria densa* there is a gradient in permeability, with the molecular exclusion limit falling from 665 dalton at the tip, to 536 in the region with young leaf primordia, and to 376 once internode expansion begins. At the same time, nodal and hypocotyl barriers appear (Fig. 14.5; Erwee and Goodwin 1985).

Vegetative apices of *Silene coeli-rosa* are permeable to probes of size up to 665 dalton. Induced apices, examined on the eight inductive long day, are permeable only to probes of 536 dalton (Goodwin and Lyndon 1983). There is a strong synchronisation of division on the eighth day, but this is not clearly linked to the change in permeability, which is first seen after four long days. Cell cycle changes can be detected on the first long day (Ormond and Francis 1985).

In the third case, Palevitz and Hepler (1985) showed that the cells in the stomatal mother cell complex are symplastic-ally linked, but that mature guard cells are isolated. It will be noted that in all the cases cited, the young meristematic cells are well linked, the more developed or mature cells are more isolated. A similar situation is found in the development of animal zygotes, where the cells in the young embryo are well linked while undergoing basic differentiation, but in the more developed embryo the tissues become isolated (Lo 1982). This suggests that in each case the signals controlling the integrated differentiation of the organism move intercellularly within the plasmalemma.

We are but at the beginning of our attempts to understand the nature of the plant symplast, its regulation, its role in transport, and its influence on plant development. However, it is already evident that the injected fluorescent probes will provide very useful tools in this study. It will be possible,

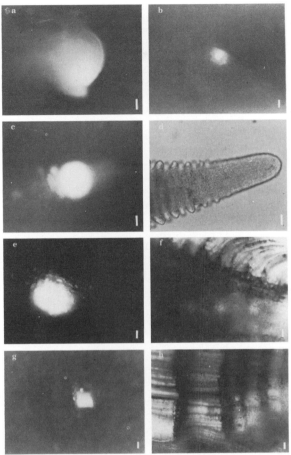

Fig. 14.5*a*   Extensive movement of 6-carboxyfluorescein in the vegetative apex of *Egeria densa*.   *b*. Limited movement of $F(GLY)_6$ in the shoot apex.   *c*. Extensive movement of F(GLU) in the young leaf primordia and unexpanded stem. *d*. A transmitted light photograph of the same area as (c). *e*. Movement of F(GLU) across the nodes and internodes of the unexpanded region of the stem.   *f*. Transmitted light photograph of the same area as (e).   *g*. F(GLU) moves in the internode, but does not cross the node when injected into an expanding internode.   *h*. Transmitted light photograph of the same area as (g).   a, c, d, g, h, bar = 50 µm; b, e, f, bar = 25 µm.

by extending the chemical range of the probes tested, to learn more about the chemical and physical requirements for intra- and inter-cellular transport. Moderately simple experiments will teach us much about the regulation of this transport. The knowledge acquired will help provide techniques to more effectively probe the developmental problems.

14.4 ACKNOWLEDGEMENTS

M.G.E. acknowledges support by the Thomas Lawrance Pawlett Postgraduate Scholarship. This work was supported by the Australian Research Grants Committee.

14.5 REFERENCES

Arisz, W.H. (1969). Intercellular polar transport and the role of the plasmodesmata in coleoptiles and *Vallisneria* leaves. *Acta Bot. Neerl.* <u>18</u>, 14–38.

Arisz, W.H. and Wiersema, E.P. (1966). Symplastic long distance transport in *Vallisneria* leaves investigated by means of autoradiograms. *Proc. Kon. Akad. V. Wetensch. Amsterdam, Series C.* <u>69</u>, 223–41.

Atabekov, J.G. and Dorokhov, Yu. L. (1984). Plant virus-specific transport function and resistance of plants to viruses. *Adv. Virus Res.* <u>29</u>, 313–64.

Barclay, G.F., Peterson, C.A. and Tyree, M.T. (1982). Transport of fluorescein in trichomes of *Lycopersicon esculentum*. *Can. J. Bot.* <u>60</u>, 397–402.

Bauer, L. (1953). Die Frage der Stoffbewegungen in der Pflanzen mit besonderer Berücksichtingung der Wanderung von Fluokochrome. *Planta* <u>42</u>, 367–451.

Belitser, N.V., Zaalishvili, G.V. and Sytnianskaya, N.P. (1982). Ca^{2-} binding sites and Ca^{2+}-ATPase activity in barley root tip cells. *Protoplasma* <u>111</u>, 63–78.

Bostrom, T.E. and Walker, N.A. (1975). Intercellular transport in plants. 1. The rate of transport of chloride and the electric resistance. *J. Exp. Bot.* <u>26</u>, 767–82.

Celis, J.E. (1984). Microinjection of somatic cells with micropipettes: comparison with other transfer techniques. *Biochem. J.* 223, 281-91.

Clarkson, D.T. (1984). Calcium transport between tissues and its distribution in the plant. *Plant, Cell and Environment* 7, 449-56.

Clarkson, D.T. and Robards, A.W. (1975). The endodermis, its structural development and physiological role. In: *The development and function of roots* (ed. J.G. Torrey and D.T. Clarkson). pp. 415-36. Academic Press, London.

Clarkson, D.T., Robards, A.W., Sanderson, J. and Peterson, C.A. (1978). Permeability studies on epidermal-hypodermal sleves isolated from roots of *Allium cepa* (onion). *Can. J. Bot.* 56, 1526-32.

De Mello, W.C. (1982). Cell-to-cell communication in heart and other tissues. *Prog. Biophys. molec. Biol.* 39, 147-82.

Edwards, G.E. and Huber, S.C. (1981). The C4 pathway. In: *The biochemistry of plants a comprehensive treatise.* (ed. P.K. Stumpf and E.E. Conn) Vol 8, pp. 228-81. Academic Press, New York.

Erwee, M.G. and Goodwin, P.B. (1983). Characterisation of the *Egeria densa* Planch. leaf symplast. Inhibition of the intercellular movement of fluorescent probes by group II ions. *Planta* 158, 320-8.

Erwee, M.G. and Goodwin, P.B. (1984). Characterisation of the *Egeria densa* leaf symplast: response to plasmolysis, deplasmolysis and to aromatic amino acids. *Protoplasma* 122, 162-8.

Erwee, M.G. and Goodwin, P.B. (1985). Symplast domains in extrastelar tissues of *Egeria densa* Planch. *Planta* 163, 9-19.

Erwee, M.G., Goodwin, P.B. and Van Bel, A.J.E. (1985). Cell-cell communication in the leaves of *Commelina cyanea* and other plants. *Plant, Cell and Environment* 8, 173-8.

Ferguson, I.B. and Clarkson, D.T. (1975). Ion transport and endodermal suberization of the roots of *Zea mays*. *New Phytol.* 75, 69-79.

Findlay, I. (1984). Microtechniques for investigating direct cell-to-cell communication in isolated tissue fragments. In: *Investigative microtechniques in medicine and biology.* (ed. J. Chayen and L. Bitensky). pp. 169-194. Marcell Dekker Inc, New York.

Fisher, D.G. and Evert, R.F. (1982). Studies on the leaf of *Amaranthus retroflexus* (Amaranthaceae): ultrastructure, plasmodesmatal frequency, and solute concentration in relation to phloem loading. *Planta* 155, 377–87.

Flagg-Newton, J., Simpson, I. and Loewenstein, W.R. (1979). Permeability of the cell-to-cell membrane channels in mammalian cell junction. *Science* 205, 404–7.

Gamalei, Y.V. and Pakhomova, M.V. (1981). Distribution of plasmodesmata and parenchyma transport of assimilates in the leaves of several dicots. *Sov. Plant Physiol.* 28, 649–61.

Giaquinta, R.T. (1983). Phloem loading of sucrose. *Ann. Rev. Plant Physiol.* 34, 347–87.

Goodwin, P.B. (1983). Molecular size limit for movement in the symplast of the *Elodea* leaf. *Planta* 157, 124–30.

Gunning, B.E.S. (1976). The role of plasmodesmata in short distance transport to and from the phloem. In: *Intercellular communication in plants: studies on plasmodesmata.* (ed. B.E.S. Gunning and A.W. Robards). pp. 203–7. Springer-Verlag, Berlin-Heidelberg-New York.

Gunning, B.E.S. and Hughes, J.E. (1976). Quantitative assessment of symplastic transport of pre-nectar into the trichomes of *Abutilon* nectaries. *Aust. J. Plant Physiol.* 3, 619–37.

Haynes, R.J. (1980). Ion exchange properties of roots and ionic interactions within the root apoplasm: their role in ion accumulation by plants. *Bot. Rev.* 46, 75–99.

Helder, R.J. (1967). Translocation in *Vallisneria spiralis.* In: *Handbook of plant physiology.* (ed. W. Ruhland), 13, 30–43. Springer, Berlin.

Kanchanapoom, K., Brightman, A.O., Grimes, H.D. and Boss, W.F. (1985). A novel method of monitoring protoplast fusion. *Protoplasma* 124, 65–70.

Kater, S.B., Nicholson, C. and Davis, W.J. (1973). A guide to intracellular staining techniques. In: *Intracellular staining in neurobiology.* (ed. S.B. Kater and C. Nicholson). pp. 307–25. Springer-Verlag, Berlin-Heidelberg-New York.

Kochian, L.V. and Lucas, W.J. (1983). Potassium transport in corn roots. II. The significance of the root periphery. *Plant Physiol.* 73, 208–15.

Kramer, D., Anderson, W.P. and Preston, J. (1978). Transfer cells in the root epidermis of *Atriplex hastata* L. as a response to salinity: a comparative cytological and x-ray microprobe investigation. *Aust. J. Plant Physiol.* 5, 739–47.

Kramer, D., Romheld, V., Lansberg, E. and Marschner, H. (1980). Induction of transfer cell formation by iron deficiency in the root epidermis of *Helianthus annuus* . *Planta* 147, 335–49.

Kuo, J., O'Brien, T.P. and Canny, M.J. (1974). Pit-field distribution, plasmodesmatal frequency, and assimilate flux in the mestome sheath cells of wheat leaves. *Planta* 121, 97–118.

Littlefield, L. and Forsberg, C. (1965). Absorption and translocation of phosphorous-32 in *Chara globuralis* Thuill. *Physiologia Pl.* 18, 291–6.

Lo, C.W. (1982). Gap junctional communication compartments and development. In: *The functional integration of cells in animal tissue.* (eds. J.D. Pitts and M.E. Finbow). pp. 167–79. Cambridge University Press, New York.

Lumley, P. (1976). Open discussion. In: *Intercellular communication in plants: studies on plasmodesmata.* (ed. B.E.S. Gunning and A.W. Robards). pp. 224–5. Springer-Verlag, Berlin-Heidelberg-New York.

Maeda, H. and Ishida, N. (1967). Specificity of binding to hexopyranosyl polysaccharides with fluorescent brightener. *J. Biochem.* 62, 276–8.

Malone, C.P., Barke, J.J. and Hanson, J.B. (1977). Histo-chemical evidence for the occurrence of oligomycin-sensitive ATPase in corn roots. *Plant Physiol.* 60, 916–22.

Maranto, A.R. (1982). Neuronal mapping: a photooxidation reaction makes lucifer yellow useful for electron microscopy. *Science* 217, 953–5.

Mertz, S.M. and Higinbotham, N. (1976). Transmembrane electro-potential in barley roots as related to cell type, cell location, and cutting and aging effects. *Plant Physiol.* 57, 123–8.

Nairn, R.C. (1976). *Fluorescent protein tracing.* Churchill Livingstone, Edinburgh.

Olesen, P. (1978). Studies on the physiological sheaths in roots. I. Ultrastructure of the exodermis in *Hoya carnosa* L. *Protoplasma* 94, 325–40.

Ormrod, J.C. and Francis, D. (1985). Effects of light on the cell cycle in the shoot apex of *Silene coeli-rosa* L. on the first day of floral induction. *Protoplasma* 124, 96–105.

Osmond, C.B. and Smith, F.A. (1976). Symplastic transport of metabolites during C4-photosynthesis. In: *Intercellular communication in plants: studies on plasmodesmata.* (ed. B.E.S. Gunning and A.W. Robards). pp. 229–41. Springer-Verlag, Berlin-Heidelberg-New York.

Overall, R.L., Wolfe, J. and Gunning, B.E.S. (1982). Inter-cellular communication in *Azolla* roots. 1. Ultrastructure of plasmodesmata. *Protoplasma* 111, 134–50.

Paine, P.L., Moore, L.C. and Horowitz, S.B. (1975). Nuclear envelope permeability. *Nature* 254, 109–14.

Palevitz, B.A. and Hepler, P.K. (1985). Changes in dye coupling of stomatal cells of *Allium* and *Commelina* demonstrated by microinjection of lucifer yellow. *Planta* (in press).

Peterson, C.A., Emanuel, M.E. and Humphreys, G.B. (1981). Pathway of movement of apoplastic fluorescent dye tracers through the endodermis at the site of secondary root formation in corn (*Zea mays*) and broad bean (*Vicia faba*). *Can. J. Bot.* 59, 618–25.

Peterson, C.A., Emmanuel, M.E. and Wilson, C. (1982). Identification of a casparian strip in the hypodermis of onion and corn roots. *Can. J. Bot.* 60, 1529–35.

Peterson, C.A. and Perumalla, C.J. (1984). Development of the hypodermal casparian band in corn and onion roots. *J. Expt. Bot.* 35, 51–7.

Peterson, C.A., Peterson, R.L. and Robards, A.W. (1978). A correlated histochemical and ultrastructural study of the epidermis and hypodermis of onion roots. *Protoplasma* 96, 1–21.

Rasmussen, A. and Goodman, D.B.B. (1977). Relationship between calcium and cyclic nucleotides in cell activation. *Physiol. Rev.* 57, 421–509.

Robards, A.W. and Clarkson, D.T. (1976). The role of plasmod-esmata in the transport of water and nutrients across roots. In: *Intercellular communication in plants: studies on plasmodesmata.* (ed. B.E.S. Gunning and A.W. Robards). pp. 181–201. Springer-Verlag, Berlin-Heidelberg-New York.

Robards, A.W., Clarkson, D.T. and Sanderson, J. (1979). Structure and permeability of the epidermal/hypodermal layers of the sand sedge (*Carex arenaria* L.). *Protoplasma* 101, 331-47.

Rose, B., Simpson, I. and Loewenstein, W.R. (1977). Ca^{+2} ion produces graded changes in permeability of membrane channels in cell junction. *Nature* 267, 625-7.

Samejima, M. and Sibaoka, T. (1983). Identification of the excitable cells in the petiole of *Mimosa pudica* by intracellular injection of Procion Yellow. *Plant and Cell Physiol.* 24, 33-9.

Schumacher, W. (1936). Untersuchungen über der Wanderung des Fluoresceins in den Haaren von *Cucurbita pepo*. *Jahrb. Wiss. Bot.* 82, 507-33.

Schwarzmann, G., Wiegandt, H., Rose, B., Zimmerman, A., Ben-Haim, D. and Loewenstein, W.R. (1981). Diameter of the cell-to-cell junctional membrane channels as probed with neutral molecules. *Science* 213, 551-3.

Simpson, I. (1978). Labelling of small molecules with fluorescein. *Analytical Biochem.* 89, 304-5.

Simpson, I., Rose, B. and Loewenstein, W.R. (1977). Size limit of molecules permeating the junctional membrane channels. *Science* 195, 294-6.

Socolar, S.J. and Loewenstein, W.R. (1979). Methods for studying transmission through permeable cell-to-cell junctions. In: *Methods in membrane biology*. (ed. E.D. Korn). pp. 123-79. Plenum Press, New York.

Steinbiss, H.H. and Stabel, P. (1983). Protoplast derived tobacco cells can survive capillary microinjection of the fluorescent dye lucifer yellow. *Protoplasma* 116, 223-7.

Steudle, E. and Jeschke, W.D. (1983). Water transport in barley roots, measurements of root pressure and hydraulic conductivity of roots in parallel with turgor and hydraulic conductivity of root cells. *Planta* 158, 237-48.

Stewart, W.W. (1978). Functional connections between cell as revealed by dye-coupling with a highly fluorescent naphthalimide tracer. *Cell* 14, 741-59.

Stretton, A.O.W. and Kravitz, E.A. (1968). Neuronal geometry: determination with a technique of intracellular dye injection. *Science* 162, 132-4.

Thomson, W.W. and Platt-Aloia, K. (1985). The ultrastructure of the plasmodesmata of the salt glands of *Tamarix* as revealed by transmission and freeze-fracture electron microscopy. *Protoplasma* 125, 13–23.

Tippett, J.T. and O'Brien, T.P. (1976). The structure of eucalypt roots. *Aust. J. Bot.* 24, 619–32.

Tucker, E.B. (1982). Translocation in the staminal hairs of *Setcreasea purpurea*. I. A study of cell ultrastructure and cell-to-cell passage of molecular probes. *Protoplasma* 113, 193–201.

Tyree, M.T. (1970). The symplast concept. A general theory of symplastic transport according to the thermodynamics of irreversible processes. *J. Theor. Biol.* 26, 181–214.

Tyree, M.T. and Tammes, P.M.L. (1975). Translocation of uranin in the symplasm of staminal hairs of *Tradescantia*. *Can. J. Bot.* 53, 2038–46.

Vakhmistrov, D.B. and Kurkova, E.B. (1978). Symplastic connections in the rhizodermis of *Trianea bogotensis* Karst. *Sov. Plant Physiol.* 26, 763–71.

Van Iren, F. and Boers-Van der Sluij, P. (1980). Symplasmic and apoplasmic radial ion transport in plant roots. Cortical plasmalemmas lose absorption capacity during differentiation. *Planta* 148, 130–7.

Walker, N.A. (1976). Transport of solutes through the plasmodesmata of *Chara* nodes. In: *Intercellular communication in plants: studies on plasmodesmata.* (ed. B.E.S. Gunning and A.W. Robards). pp. 165–79. Springer-Verlag, Berlin-Heidelberg-New York.

Walker, R.R., Sedgley, M., Blesing, M.A. and Douglas, J.J. (1984). Anatomy, ultrastructure and assimilate concentrations of roots of citrus genotypes differing in ability for salt exclusion. *J. Expt. Bot.* 35, 1481–94.

Wetherell, D.F. (1984). Enhanced adventive embryogenesis resulting from plasmolysis of cultured wild carrot cells. *Plant Cell Tissue Organ Culture* 3, 221–7.

Index

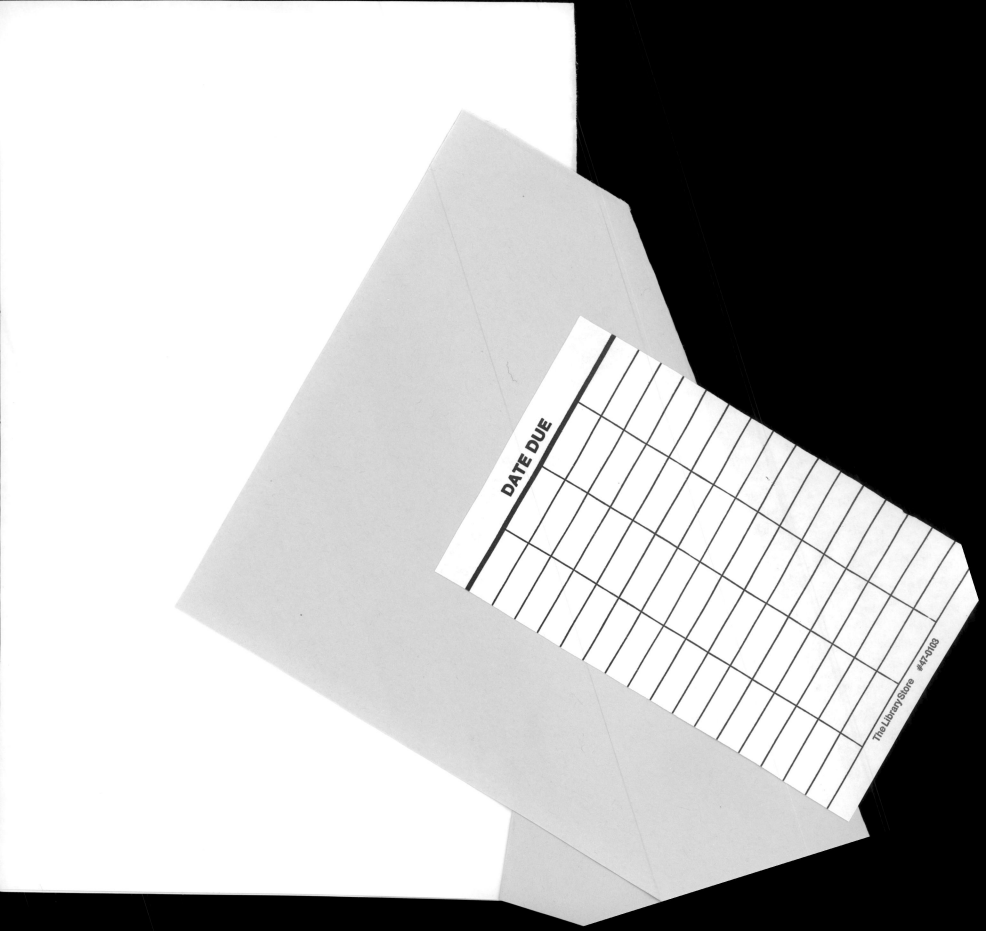

DATE DUE

The Library Store   #47-0103

DATE DUE
